D1286390

Naturalists in Paradise

John Hemming

NATURALISTS IN PARADISE

Wallace, Bates and Spruce in the Amazon

79 illustrations, 20 in color

For my family, Sukie, Bea, Henry and Helena, who have all been with me in Brazil

1 (Frontispiece) Henry Bates collected in forests throughout his eleven years in Brazil. Here he gathers a curl-crested toucan. When the wounded bird shrieked, Bates was mobbed by a flock of its fellows; but they vanished when he killed it.

On the jacket: (front, left to right) Wallace, Bates and Spruce (Bates: Photo Royal Geographical Society/with IBG; Spruce: Royal Botanic Gardens, Kew); (front flap) umbrella bird from Bates, *Naturalist*, second edition. Other images taken from inside the book.

First published in 2015 in hardcover in the United States of America by Thames & Hudson Inc., 500 Fifth Avenue, New York, New York 10110

thamesandhudsonusa.com

Library of Congress Catalog Card Number 2014944634

ISBN 978-0-500-25210-9

Printed and bound in China by Everbest Printing Co. Ltd

CONTENTS

CHAPTER 1
A naturalist's paradise
10

CHAPTER 2
Antecedents
13

CHAPTER 3
Enterprising young men
31

CHAPTER 4
The mouth of the Amazon
57

CHAPTER 5
Into Amazonia
85

CHAPTER 6
Manaus
113

CHAPTER 7
Wallace on the Rio Negro
146

CHAPTER 8
Into the unknown
185

CHAPTER 9
Calamities
212

CHAPTER 10
Bates on the Amazon
235

CHAPTER 11
Spruce climbs to the Andes
267

CHAPTER 12
Later lives: the Amazonian legacy
296

Timeline
339

References
342

Bibliography
356

Acknowledgments
363

Illustration credits
364

Index
365

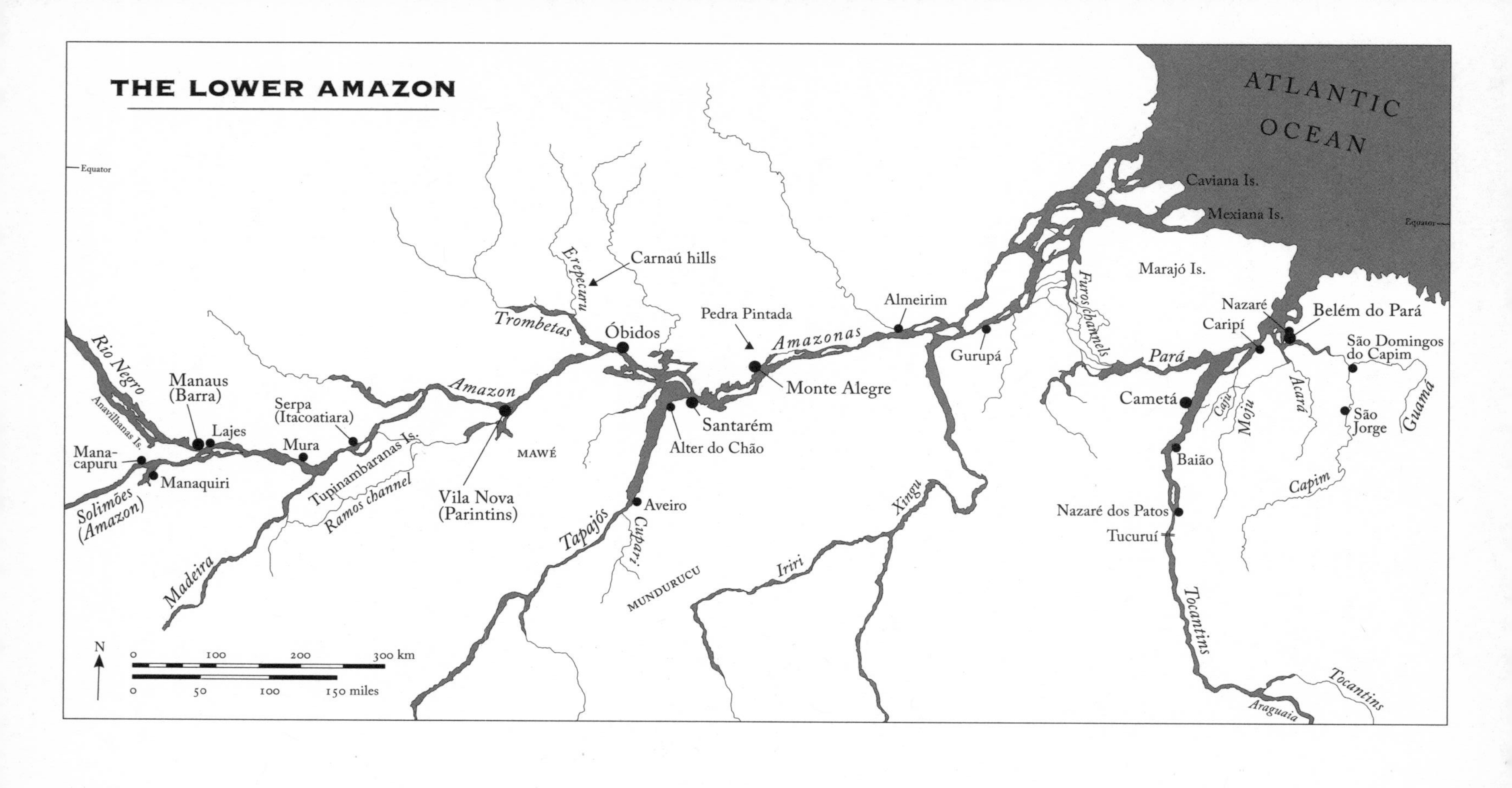
THE LOWER AMAZON
ATLANTIC OCEAN
Equator
Caviana Is.
Mexiana Is.
Marajó Is.
Furos channels
Nazaré
Caripí
Belém do Pará
São Domingos do Capim
São Jorge
Guamá
Acará
Moju
Caju
Capim
Pará
Cametá
Baião
Nazaré dos Patos
Tucuruí
Tocantins
Araguaia
Almeirim
Gurupá
Pedra Pintada
Amazonas
Monte Alegre
Santarém
Alter do Chão
Aveiro
Cupari
Tapajós
MUNDURUCU
Xingu
Iriri
Carnaú hills
Erepecuru
Trombetas
Óbidos
Amazon
MAWÉ
Vila Nova (Parintins)
Tupinambaranas Is.
Ramos channel
Serpa (Itacoatiara)
Mura
Manaus (Barra)
Lajes
Rio Negro
Anavilhanas Is.
Mana-capuru
Manaquiri
Solimões (Amazon)
Madeira
N
0 100 200 300 km
0 50 100 150 miles

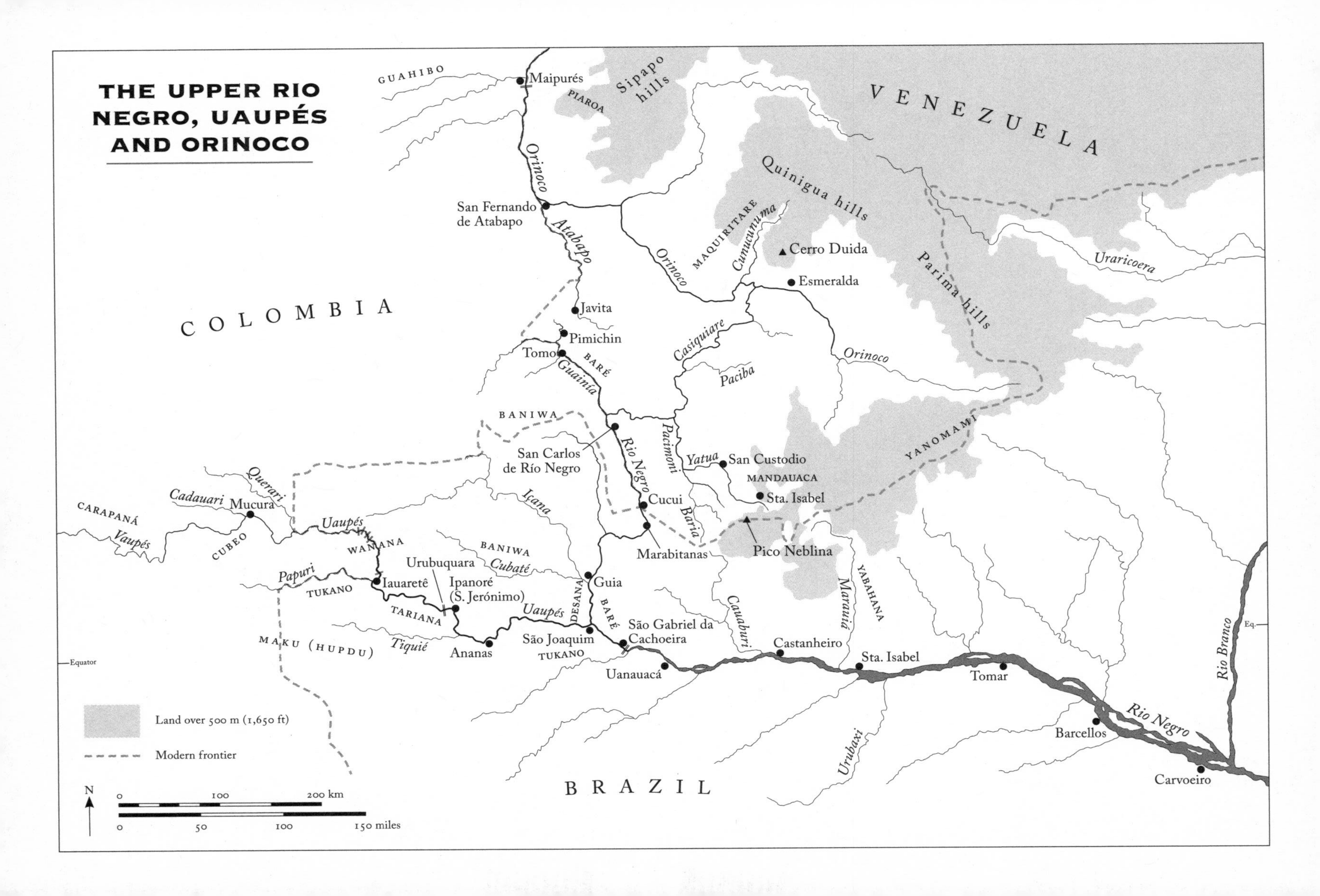
THE UPPER RIO NEGRO, UAUPÉS AND ORINOCO
VENEZUELA
COLOMBIA
BRAZIL
GUAHIBO
Maipurés
PIAROA
Sipapo hills
Orinoco
San Fernando de Atabapo
Atabapo
Quinigua hills
MAQUIRITARE
Cunucunuma
Cerro Duida
Esmeralda
Parima hills
Uraricoera
Javita
Pimichin
Tomo
BARÉ
Guainía
Casiquiare
Paciba
BANIWA
San Carlos de Río Negro
Río Negro
Pacimoni
Yatua
San Custodio
MANDAUACA
Sta. Isabel
YANOMAMI
Querari
Cadauari
Mucura
CARAPANÁ
Vaupés
CUBEO
Uaupés
WANANA
Içana
Cucui
Baria
Pico Neblina
Marabitanas
Urubuquara
Ipanoré (S. Jerónimo)
Cubaté
Papuri
Iauaretê
TUKANO
TARIANA
DESANA
Guia
São Joaquim
São Gabriel da Cachoeira
Cauaburi
Marauiá
YABAHANA
Castanheiro
Sta. Isabel
Tomar
Rio Branco
Eq.
MAKU (HUPDU)
Tiquié
Ananas
Uanauacá
Barcellos
Rio Negro
Carvoeiro
Urubaxi
Equator
Land over 500 m (1,650 ft)
Modern frontier
N
0 100 200 km
0 50 100 150 miles

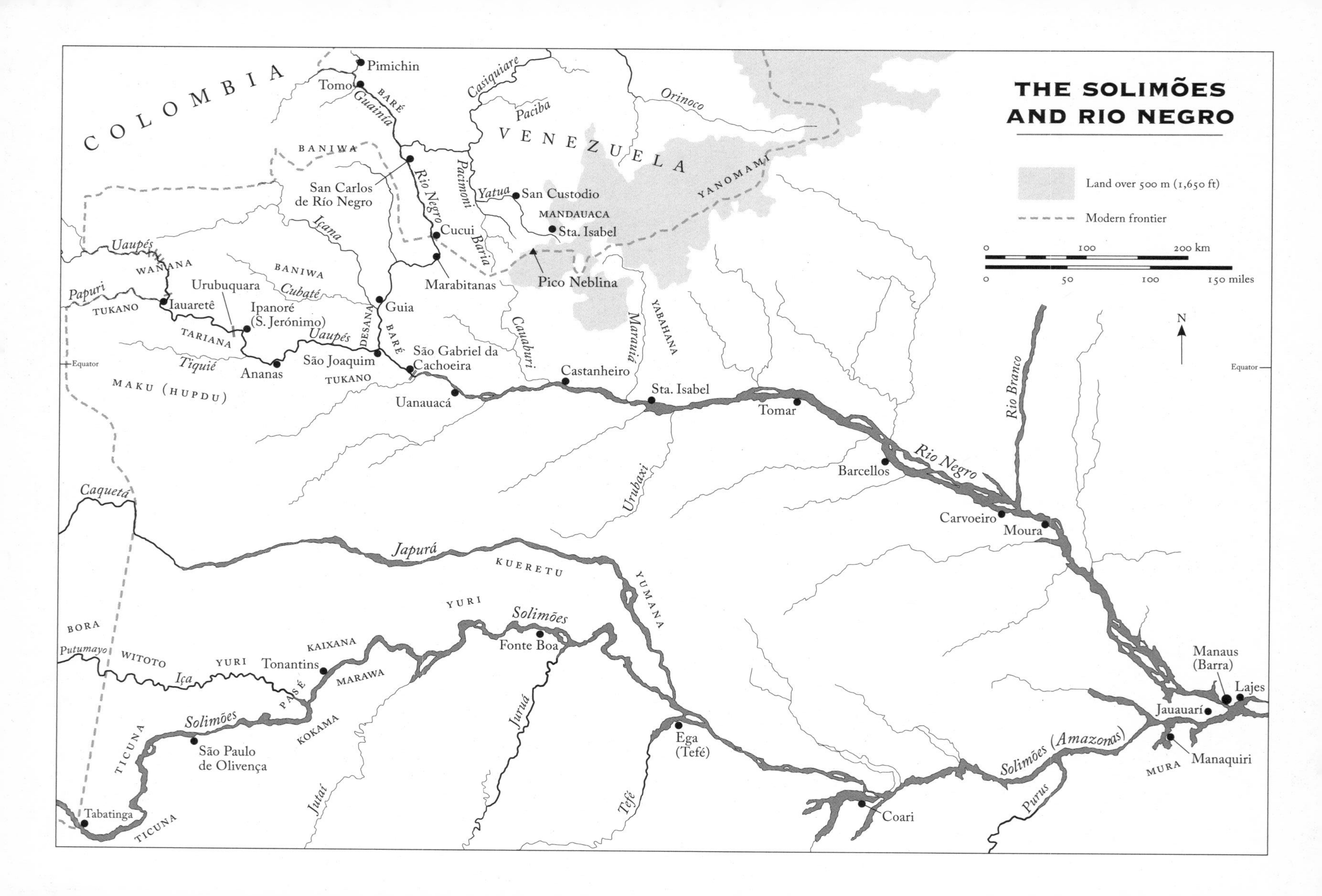
THE SOLIMÕES AND RIO NEGRO
Land over 500 m (1,650 ft)
Modern frontier
0 100 200 km
0 50 100 150 miles
N
Equator
COLOMBIA
VENEZUELA
Pimichin
Tomo
BARÉ
Guainía
BANIWA
San Carlos de Río Negro
Rio Negro
Casiquiare
Paciba
Orinoco
Pacimoni
Yatua
San Custodio
MANDAUACA
Sta. Isabel
YANOMAMI
Cucui
Baria
Marabitanas
Pico Neblina
Içana
Uaupés
WANANA
Papuri
TUKANO
Iauaretê
Urubuquara
Cubaté
Ipanoré (S. Jerónimo)
TARIANA
Tiquié
Ananas
São Joaquim
DESANA
Guia
São Gabriel da Cachoeira
MAKU (HUPDU)
Uanauacá
Cauaburi
Castanheiro
Marauiá
YABAHANA
Sta. Isabel
Tomar
Rio Branco
Barcellos
Carvoeiro
Moura
Urubaxi
Caquetá
Japurá
KUERETU
YUMANA
YURI
Solimões
Fonte Boa
BORA
Putumayo
WITOTO
Iça
Tonantins
KAIXANA
MARAWA
PASÉ
KOKAMA
TICUNA
São Paulo de Olivença
Tabatinga
Jutaí
Juruá
Ega (Tefé)
Tefé
Coari
Solimões (Amazonas)
Purus
Manaus (Barra)
Lajes
Jauauarí
Manaquiri
MURA

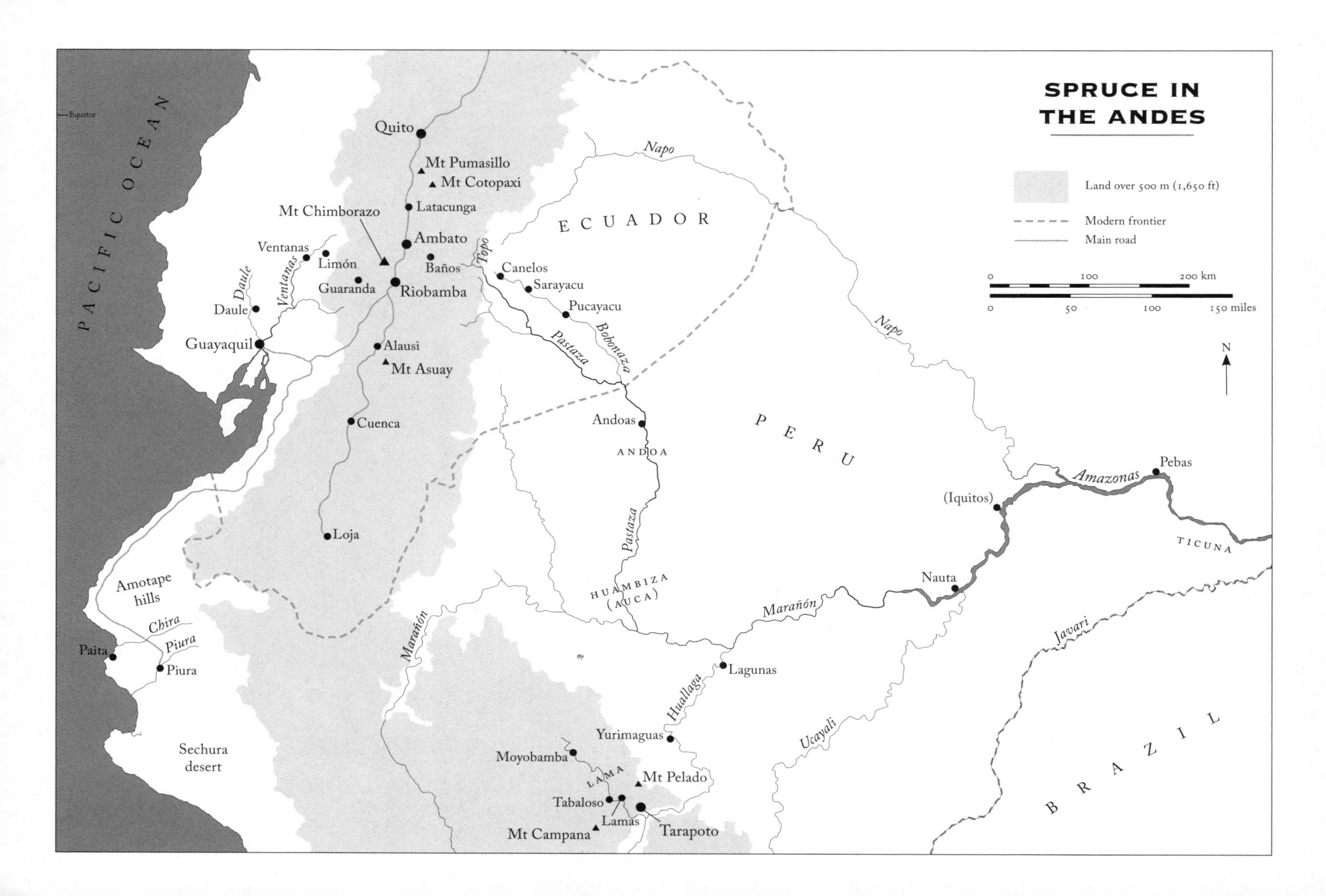
SPRUCE IN THE ANDES
Land over 500 m (1,650 ft)
Modern frontier
Main road
0 100 200 km
0 50 100 150 miles
N
Equator
PACIFIC OCEAN
ECUADOR
PERU
BRAZIL
Quito
Mt Pumasillo
Mt Cotopaxi
Latacunga
Mt Chimborazo
Ambato
Baños
Ventanas
Limón
Guaranda
Riobamba
Daule
Guayaquil
Alausi
Mt Asuay
Cuenca
Loja
Amotape hills
Chira
Piura
Paita
Sechura desert
Napo
Topo
Canelos
Sarayacu
Pucayacu
Bobonaza
Pastaza
Andoas
ANDOA
HUAMBIZA (AUCA)
Marañon
Lagunas
Huallaga
Yurimaguas
Moyobamba
LAMA
Mt Pelado
Tabaloso
Lamas
Mt Campana
Tarapoto
Ucayali
Nauta
(Iquitos)
Amazonas
Pebas
TICUNA
Javari

1

A NATURALIST'S PARADISE

On 26 May 1848 two young Englishmen sailed into the mouth of the Amazon river and landed at its largest town, Belém do Pará ('Bethlehem of the Pará river'). The older of the two, Alfred Russel Wallace, was twenty-five and his friend Henry Walter Bates twenty-three. Both travellers had had only rudimentary education, their families were without wealth or influence, they had very little experience of life outside provincial country towns and neither had any money. But they were passionate about natural history and they had the enthusiasm – almost hubris – of youth to discover nature in what they knew was its most exuberant environment. They had come to the Amazon hoping to learn more about the origin of species, and they planned to support themselves by selling specimens of its exotic flora and fauna to British museums and collectors.

The young visitors were at first disappointed by the tropical forest. Bates admitted that 'the vegetation was not so surprising as the glowing picture I had been conjuring up… during the tedium of the sea voyage'. 'The number and beauty of the birds and insects did not at first equal our expectations.' The few birds on view were small and dull-coloured, as in England. 'The whole country originally is a lofty uninterrupted and gloomy forest, without flowers and almost without sounds of life. When a bird's note disturbs the silence, the echoes startle one in one's solitary walks. Insects are found always sparingly flitting about in rays of sunshine which peer through the foliage.' Wallace had read so much about 'the surpassing beauty of tropical vegetation, and of the strange forms and brilliant colours of the animal world, that I had wrought myself up to a fever-heat of expectation'. So his first impression was also disappointment, when he found himself in quiet woods that looked superficially like those of his native country. Newcomers entering their first tropical rain forest still react as Wallace and Bates did.

The two Englishmen were, however, good naturalists. They soon appreciated where they had been deceived, realizing that they were seeing disturbed woods near a town, and at the extreme eastern edge of the endless mass of Amazon vegetation. They found that a forest that seems empty and lifeless at midday erupts into life at dawn and dusk. With more experience, they learned that much animal life is nocturnal, that many insects and birds are bafflingly camouflaged, and that the majority of natural life thrives in the canopy far above human heads. Before long, Bates was writing that 'the forest scenery is glorious beyond imagination: in some places every fifth tree is a palm, shooting up slender stems to a great height, and suspending their feathery leaves amongst the branches of still loftier trees.' 'The leafy crowns of the trees, scarcely two of which… are of the same kind, were now far away above us, in another world as it were. We could only see at times, where there was a break above, the tracery of the foliage against the clear blue sky.' Below, the tree trunks were linked by a tangle of lianas, twisted like cables, 'contorted into every variety of shape, entwining snake-like round the tree trunks, or forming gigantic loops and coils… others of zigzag shape, or indented like the steps of a staircase, sweeping from the ground to a giddy height.'

Bates was an entomologist at heart and even in the town itself he had the 'treat of seeing a great variety and number of handsome things flashing about'. On one walk, there were cassia trees with elegant pinnate foliage and conspicuous yellow blossoms, and clusters of arborescent arums, while 'over the whole fluttered a larger number of brilliantly-coloured butterflies than we have yet seen; some wholly orange or yellow (Callidryas [now *Phoebis*]), others with excessively elongated wings, sailing horizontally through the air, coloured black, and varied with blue, red, and yellow (Heliconii). One magnificent grassy-green species (Colaenis Dido [*Philaethria dido*]) especially attracted our attention.'*

Wallace became philosophical when he finally penetrated virgin forest. Like many later visitors, he compared the mighty trunks to the piers of a cathedral. Far overhead, the canopy was dense enough to block the sun. 'There is a weird gloom and a solemn silence, which combine to produce a sense of the vast – the primeval – world of the infinite. It is a world in which man seems an intruder, and where he feels overwhelmed by the contemplation of the ever-acting forces, which… build up the great mass of vegetation which overshadows, and almost seems to oppress the earth.'

In July 1849, fourteen months after the arrival of Wallace and Bates, another Englishman went to join them. The Yorkshireman Richard Spruce was, at thirty-two,

* In the nineteenth century, writers used a capital letter for every plant, insect or animal, but did not always italicize scientific names. Direct quotations are as written. But otherwise, for taxonomic names I follow the modern usage of italics with a capital letter for the genus only.

slightly older than the others and he was already an experienced botanist. But his origins, education and circumstances were similarly humble. And he was just as eager to study the Amazonian environment, and to survive as a professional collector living off the sale of his specimens. All three wanted to earn a living by doing what they loved most.

The seasoned botanist Spruce was also dazzled by his first experience of true tropical rain forest. 'There were enormous trees, crowned with magnificent foliage, decked with fantastic parasites, and hung over with lianas.' The more he penetrated the Amazon, the greater was his awe and delight. After a couple of years, he wrote excitedly to a botanist friend: 'The largest river in the world runs through the largest forest. Fancy if you can *two millions of square miles of forest* [Spruce's emphasis] uninterrupted save by the streams that traverse it.' The scale could be stupendous. 'Nearly every natural order of plants has here *trees* among its representatives. Here are grasses (bamboos) of 40, 60, or more feet in height, sometimes growing erect, sometimes tangled in thorny thickets, through which an elephant could not penetrate. Vervains forming spreading trees with digitate leaves like the horse-chestnut. Milkworts, stout woody twiners ascending to the tops of the highest trees, and ornamenting them with festoons of fragrant flowers.... We have here... violets the size of apple trees... daisies (or what might seem daisies) borne on trees like alders.'

Bates became equally thrilled. He wrote to his brother in a cascade of enthusiasm: 'The charm and glory of the country are its animal and vegetable productions. How inexhaustible is their study! ...It is one dense jungle: the lofty forest trees, of vast variety of species, all lashed and connected by climbers, their trunks covered with a museum of ferns, Tillandrias, Arums, Orchids, &c. The underwood consists mostly of younger trees – great variety of small palms, mimosas, tree-ferns, &c., and the ground is laden with fallen branches – vast trunks covered with parasites, &c.' Years later, when he knew Amazonia intimately, he could still write with undiminished enthusiasm that it was 'a region which may be fittingly called a Naturalist's Paradise.' It was in this paradise that the three young men were destined to spend years. For each of them, this would be the defining formative experience in his life.

2

ANTECEDENTS

Alfred Russel Wallace and Henry Walter Bates met in Leicester in 1844, four years before their venture to the Amazon. Neither recorded exactly where they ran into one another, but it was probably in the public library. Their friendship was clearly based on a mutual passion for natural history – Wallace more inclined to botany and Bates to entomology. They were both avid readers, largely self-taught, and they loved talking about the theories of life, nature and the origin of species.

When they met, the nineteen-year-old Bates was on home territory. Leicester was a town of 40,000 inhabitants, overcrowded and slummy, an unhealthy place with poor sanitation and an inadequate water supply. Its industry was entirely devoted to hosiery: it had for two centuries been England's centre of stocking manufacture. This was the Bates family's trade. Young Henry's father, also called Henry, had inherited and built up a prosperous little business that dyed the knitted stockings, and trimmed them so that they could be shaped on steam presses. Both Henry's mother and grandmother were also from hosiery families.

Bates was born in 1825, the oldest of four boys, and had a happy childhood. His lifelong friend Edward Clodd described his mother Sarah as intelligent and gracious, 'a woman of sweet, loving and unselfish nature, but of feeble constitution... and delicate health'. His father Henry was more robust, and 'an able, upright, modest man' known as Honest Harry Bates. He rose to an advanced level in the Unitarian church. So when his eldest son was nine he was sent to board in an Academy run by a dissenting minister, Dr H. Screaton. This highly regarded school, at Billesdon 14 kilometres (9 miles) from Leicester, aimed to teach the sons of respectable gentlemen, with an emphasis on subjects 'necessary to trade and commercial pursuits'. The three younger Bates boys went to day schools, so they missed their older brother. One of them, Frederick, later wrote how much they used to look forward to Henry's return for holidays, since he was their 'dear "guide, philosopher and friend" who... ever sought to lead us, in his kindly, genial way, to better and higher thoughts and deeds'. Frederick also noted that in those days the scholastic education of tradesmen's sons ceased at thirteen or fourteen: so Bates left the Academy in 1838 when only thirteen and a half.

The future Amazon naturalist immediately started an apprenticeship in the hosiery warehouse of his father's friend Alderman Gregory. This was hard work, from seven in the morning to eight at night, six days a week. But the boy was amazingly industrious, a true swot. As his brother said, it was at this time that he 'laid the foundation of all that he later became', and his friend Edward Clodd wryly remarked that his 'slender school education... did not suffice to dull his desire for knowledge'. Leicester was fortunate to have a Mechanics' Institute, which had 'developed into a large and magnificent educational institution' under enlightened teachers. Henry Walter entered classes of Greek, Latin, French, Drawing and Composition (essay writing), and won prizes in most of them. His enthusiasm and 'capacity for work at this time was prodigious.... It was no uncommon thing for him to work till midnight [after his thirteen-hour working day] and yet be up again at 4 a.m.' He wrote in the flyleaf of his Latin grammar: 'I am as fond of Latin, as women are of satin.' Two of his essays have survived from those classes, one on ancient Britons, the other on forms of government – both 'remarkably able productions for a boy of fifteen'. A voracious reader, he devoured Gibbon's *Decline and Fall of the Roman Empire* and much of the Institute's 'good library'. He sang in the glee club as a baritone, and learned to accompany himself on a guitar; and he later grew to love opera music – from Wagner to Sullivan.

There are three aspects of Bates's early years that are worth noting. One is his father's Unitarianism. The Unitarians were a dissenting (later called nonconformist) branch of the Church of England that believed in the supremacy of a single God rather than of the Trinity. Proscribed during the eighteenth century, Unitarianism gained some distinguished adherents and was legalized in 1813. But although Unitarians were God-fearing, hard-working and upright people, they were considered political radicals who approved of the French Revolution. They later had little difficulty in accepting Darwinian views of creation. Henry Walter Bates was never strongly religious, but his early worship in Leicester's Unitarian chapel may have helped to make him an independent thinker.

Another important influence was the Mechanics' Institute. Started in the 1820s, these admirable Institutes spread rapidly to almost every city in Britain (there were 700 by 1850) and also to Canada and Australia. They were charities, with their own buildings, to be used as libraries and for adult education (or in Bates's case secondary schooling).

Local businessmen felt that it was in their interest to fund Mechanics' Institutes because these trained engineers and skilled workers – essential in those early days of the industrial revolution. Some of the Institutes evolved into public libraries and museums, then to polytechnics, and even in the late twentieth century to universities.

The third factor was poor health. As a boy, Bates's circulation was thought to be bad, so he was massaged with coarse gloves; his face was covered in spots for which he took potions of quinine; and his digestion 'was never good'. These weaknesses may have been inherited from his mother; or 'the terrible overwork he had voluntarily undergone had injured him'; or his ailments were simply due to the rudimentary medical knowledge and lack of treatment of that period.

In 1840, when Bates was fifteen, there was an exhibition in Leicester's New Hall to raise funds for its Mechanics' Institute. Admission was sixpence, to see among other things thousands of 'preserved specimens of Foreign and English quadrupeds, birds and insects'. He was particularly excited by the insects. Thus, the most significant achievement of Bates's teen years was not so much his hosiery apprenticeship nor his remarkable self-education: it was his developing passion for natural history. Edward Clodd described him as a born naturalist. In the Mechanics' Institute he met two boys who were kindred enthusiasts. (The boys were brothers, who went on to careers in natural science and ornithology.) The young collectors were fortunate that the elderly Earl of Stamford was relaxed about rearing game birds for shooting on his Bradgate Park estate near Leicester, so he did not mind the boys searching for bugs in his Charnwood Forest. Bates's younger brother Frederick later joined in. He recalled how eagerly they awaited Good Friday, the end of the shooting season and start of their collecting. 'I can still see the happy group starting on one of those mornings, full of glee and joyous spirits', armed with home-made collecting nets, and with 'easy access to all the rich woods and places where insects most abounded.' Bates made notes about each collecting foray, with sketches and descriptions of the captured insects and wild flowers – fine training for his later remarkable talent as a collector. His father encouraged him. With business success, the older Henry Bates had in 1841 built a larger house for the family, closer to the town centre. At first young Bates stored his bugs in any cupboard, drawer or washstand he could find; but his father then gave the boys a study room with a table and drawers for their insects.

Henry Bates started, 'like most boys', with butterflies. Collecting butterflies and moths – Lepidoptera – had been the preserve of a few gentlemen in the previous century, but was now becoming a craze at all levels of society. In later decades, houses in Victorian England regularly had cabinets of pretty insects, and enthusiasts subscribed to weekly fascicules or part-works with pictures of them. Bates soon moved on to beetles, Coleoptera, and was amazed by the profusion of species so near his birthplace. Identifying beetles was difficult at that time. 'There was nothing much to enable the worker to determine his species but [James] Stephens's 'Manual [of British Coleoptera or Beetles]', and all who have puzzled over that book will know the difficulties.' Despite this, and with youthful self-confidence, Bates corresponded with leading entomologists. Aged seventeen, he submitted a short paper, 'Note on Coleopterous insects frequenting damp places,' that was published in January 1843 in the first issue of a new journal, the *Zoologist*.

Years later, his friend Grant Allen wrote that Bates was a beetle-hunter almost from his cradle. Allen was surprised that Bates, with his broad interests and knowledge, should have concentrated on necessary but arcane and unexciting taxonomy. Bates explained, in a fireside chat with his friend: 'When I was a young man I wanted to be a naturalist; but very soon I saw the days of naturalists were past, and that if I wanted to do anything, I must specialise: I must be an entomologist. A little later, I saw the days of entomologists, as such, were numbered, and that if I wanted to do anything I must be a coleopterist [beetle expert]. By-and-by, when I got to know more of my subject, I saw no man could understand *all* the coleoptera, and now I'm content to try and find out something about the longicorn beetles.' But Allen wrote that, despite such protestations, Bates could not have been further from 'the ordinary type of narrow specialist'.

Alderman Gregory died in 1843, before Bates had completed his apprenticeship. For a while the young man ran the Gregory warehouse, and he then spent a few months in two other hosiery businesses. But he was not happy. So in 1845 another entomologist friend arranged for him to escape textiles and take an office job at Allsop's Brewery in Burton-upon-Trent. This town is 40 kilometres (25 miles) northwest of Leicester, and the move may have been Bates's first journey away from home.

Before Bates left for his new job – and its slightly different location for beetle-collecting – he met Alfred Russel Wallace.

Two years older than Bates, Wallace also came from what we would now call the lower middle class. He was born in 1823 in the town of Usk in the southeastern corner of Wales.*

* Usk was in Monmouthshire. In the nineteenth century there was ambiguity about whether this was part of England or Wales. Finally in 1974 Monmouth was defined as being in Wales, but it ceased to be a county and became part of the larger county of Gwent.

When Wallace was christened, his middle name Russell was sloppily entered into the baptismal register with only one L: Wallace clung to this mistake throughout his life, proudly putting Alfred Russel Wallace on all his writings.

His father Thomas was a gentleman who claimed descent, like so many Scots, from the thirteenth-century patriot Sir William Wallace. He qualified as a lawyer, but never practised. He inherited enough to yield an income of £500 a year, but he himself never worked commercially and he dissipated his small fortune in ill-judged business ventures. So, a few years before Alfred's birth, the family had to leave a comfortable house in Hertford, north of London, where his mother's grandfather had been mayor. The Wallaces moved to a more rural and cheaper cottage on the Usk river. This was where Alfred was born. His father was happy in this rural life, growing fruit and vegetables for his family, raising poultry, and tutoring his nine children (three of whom died young). But in 1828, when Alfred was five, his mother inherited a house in Hertford and some money from her stepmother. So the family moved back from rural Wales to that town in the English home counties, living in a succession of modest terrace houses. The first of these, 11 St Andrew's Street, now bears a plaque in honour of Alfred Russel Wallace.

For six and a half years, Alfred attended Hertford Grammar School. Four masters taught eighty boys, all in one room. Lessons started at seven in the morning, and on three days a week continued until five o'clock – on dark winter afternoons each pupil had to bring his own candle. Alfred learned arithmetic, rudimentary French and 'wearisome' Latin, some geography (reciting place names) and history (memorizing names and dates). In his autobiography, he wrote that he learned more from his father and older brothers than he did from school. Thomas Wallace loved reading to his family: Shakespeare's plays and a wealth of other literature, which he got from a book club. He then found a job as librarian of Hertford's 'fairly good town library' and Alfred spent Saturday and half-holiday afternoons there, squatting on the floor in a corner and reading voraciously. He read novels like *Tom Jones* and current works by Fennimore Cooper, Byron, Scott and others, and a remarkable range of more difficult classics, from Pope's *Iliad* to Spenser's *Faerie Queene*, Dante, Milton, Walton and Cervantes. From his five-years-older brother John he grasped basic mechanics and 'the pleasure and utility' of being as self-sufficient as possible.

Financial matters worsened. Thomas Wallace lost his final savings in a property venture. His wife's uncle became bankrupt, and ruined her because he had secretly borrowed against her small inheritance of which he was executor. So in 1836 the near-destitute couple moved to a small cottage in the village of Hoddesdon. The various children had to leave school and go out into the world. The oldest son William, aged twenty-six, had learned to be a land surveyor; the surviving daughter Fanny worked as a governess and sent whatever she could to her parents; John became an apprentice carpenter in a workshop in London; and the fourteen-year-old Alfred was removed from the Grammar School and sent to London, where he shared a room (and bed) with his brother John.

Unlike his brother and Bates, Wallace was not expected to work as an apprentice. But he watched the working men and joined them in a Mechanics' Institute. The Institute that John and Alfred frequented, near the Hampstead Road in north London, was very different to the tranquil educational one in Leicester. Its members were inspired and influenced by the great radical and visionary social reformer Robert Owen. Now aged sixty-six, Owen had had a remarkable career. He started as a successful textile-mill manager in Manchester, married a girl from Glasgow, and then worked in his wife's family's mill at New Lanark outside that city. Owen gradually introduced a revolutionary style of management, to turn New Lanark into a model workplace where labourers and their families were cared for and respected. The result was splendid. The workforce was healthy, sober and contented, their children happy and well behaved, and the venture was initially a commercial success making produce of high quality. New Lanark was visited and praised by social reformers, politicians and even the Tsar of Russia. But the utopian vision did not endure. Owen himself fell out with his partners and left New Lanark in 1828, later basing himself in London. He coined the term 'socialism', launched the cooperative movement, and tried to form a Consolidated Trades Union. In the event, all his experiments failed because they were too advanced, ahead of their time, or impractical. Owen was also an agnostic. He argued that everyone was equal until shaped by upbringing and environment. This contradicted the Christian belief that individuals were responsible for their own morality and would be judged if they sinned. So Owen became aggressively secularist, arguing that religions were based on a 'ridiculous' notion that man was 'a weak, imbecile animal; [either] a furious bigot and fanatic; or a miserable hypocrite'.

By the time young Wallace came to the capital, in 1837, Robert Owen's writings and speeches had made such an impact that one commentator felt that his teachings had become the creed of the majority of working men. The teenaged Wallace spent only a few months with his brother in London. But he saw artisans at work, and in the evenings he and his brother attended Owenite meetings in a hall on Charlotte Street near Fitzroy Square. When Wallace wrote his autobiography almost seventy years later, he felt that 'it was here that I first made acquaintance with some of Owen's writings, and especially with the wonderful and beneficent work he had carried out for many years at New Lanark. I also received my first knowledge of the arguments of sceptics, and read among other books Paine's "Age of Reason".' This was a time when Owenism was expanding (in competition with the more political Chartists) so they would have welcomed the teenager as representing the future. Young Alfred was doubtless influenced by the friendly group of working men and by the daring ideas being put to them. However it was not until much later, in the 1870s, that Wallace fully espoused socialism. He then explained that as a boy he had been 'an ardent admirer of Robert Owen' but had reluctantly concluded that socialism 'was impracticable, and also to some extent repugnant to my ideas of individual liberty and home privacy.'

In mid-1837 Alfred's oldest brother William took him on as a helper in his surveying work in Bedfordshire, north of London. The fourteen-year-old Alfred was now 6 feet (1.8 metres) tall and still growing, and 'exceedingly fair'. Although the brothers were fourteen years apart in age, they got on well. Alfred admired William, a serious, somewhat humourless, but very competent young man. The pair were constantly on the move, staying in village after village while William sought land-surveying jobs. They spent seven years together, during which Alfred learned survival in the harsh conditions of rural England at that time. He tramped across miles of countryside, becoming physically tough, acquiring an interest in nature, and being shown the rudiments of geology by the self-taught William. Alfred also learned surveying, partly by practice with his brother, but also by reading everything he could on the subject – like Bates, he was an obsessive natural scholar. He adored making maps and noticing natural, particularly geological, boundaries. After a couple of years the Wallace brothers moved west to Shropshire and then to Wales. An act of Parliament had regularized the payment of tithes to the Church of England, and this meant that every parish had to be surveyed so that the amount due from its various properties could be assessed. This often led to the enclosure of common lands. Later in life Wallace became outraged by landowners denying poor farmers access to commons; but in the 1840s he was too engrossed in survey work to worry about the morality of what they were doing. William was a religious agnostic and influenced his young brother in this direction. So it was ironic that their work was for the benefit of the established church. Late in 1841 they moved to Neath, a town near Swansea in south Wales. Work was in short supply so they tried anything, including civil engineering – building warehouses and improving navigation in Neath harbour.

Those years of crossing rough country for survey work made Alfred aware of the variety, beauty and mystery of plants. It reminded him of botanical rambles with his father when he was a small boy not far away in Usk. So he started to collect wild flowers. Not content with the plants' appearance, he wanted to know more about their classifications.

A friendly bookseller in Neath sold him a small book describing the natural orders of British flora, and Wallace learned to identify all the common flowers. But in the wildernesses of Wales he found rare wildflowers, and his interest in botany outgrew his simple book. He therefore splashed over half a week's pay on John Lindley's *Elements of Botany*. To his dismay, this book by one of the country's foremost botanists dealt with taxonomy or systematics of the plant world in general, so it was no help whatever in classifying British flowers. But it did Wallace a far greater favour: it introduced him to the exciting concept of scientific methodology. He learned how the Swedish Carl Linnaeus (later ennobled as Carl von Linné) had developed the discipline of taxonomy by organizing flora and fauna into classes, orders, genera (families), species, and sometimes taxa (ranks). In his *Systema Naturae*, *Philosophia Botanica* and *Species Plantarum*, all published in the mid-eighteenth century, Linnaeus had also established 'binomial nomenclature', whereby each plant had a double name: first the genus, then a specific (species) epithet that could either be descriptive or perhaps the name of its discoverer. Wallace borrowed an encyclopaedia of British plants from the bookseller, and then spent months copying the genera and species of all native flora and inserting these into the appropriate places in his copy of *Elements of Botany*. This laborious undertaking turned the self-taught Alfred Wallace into an impressive amateur botanist.

Another book that made an important impression on Wallace was the Reverend Thomas Malthus's *Essay on the Principle of Population*. This argued that population growth, in both human beings and animals, was dependent on availability of resources. This introduction to philosophical biology was – years later – to give Wallace a 'long-sought clue to the effective agent in the evolution of organic species'. Both Lindley and Linnaeus had assumed that plants and animals were divinely created in all their diversity, without questioning their origins or evolution. However, in 1842 Wallace bought William Swainson's *Treatise on the Geography and Classification of Animals*, which aroused his interest in how animals (particularly birds) were geographically distributed.

In 1843 Alfred's father Thomas Wallace died, aged seventy-two, and the penniless family scattered. Their widowed mother Mary Anne became a housekeeper; William and Alfred continued together as jobbing surveyors for the rest of that year; John was still working in carpentry in London; Fanny sailed to America and became a teacher in Macon, Georgia; and Herbert, aged fourteen, had to leave school and became apprenticed to a luggage-maker in London.

In January 1844 William could no longer afford to pay Alfred £1 a week as his assistant, so the young man had to find a job – no easy task at that time of depression. He tried to go into teaching through an agency in London, but his own education was too rudimentary for most schools. Then, luckily, a vacancy occurred in Leicester for someone who could teach art as well as surveying and mapping, in addition to the usual primary school subjects. Alfred got the position, hastily brushed up his Latin and mathematics, and moved to Leicester with his scanty possessions – a few books, his beloved sextant, and a collection of plants.

As a schoolmaster, Wallace struggled to keep ahead of the lessons he was teaching; but the young pupils liked him. The Reverend Abraham Hill who ran the school was also fond of Alfred – perhaps unaware that he was not a believer. As Wallace wrote in *My Life*, any religious beliefs had by now 'vanished under the influence of philosophical or scientific scepticism.... By the time I came of age I was absolutely non-religious, I cared and thought nothing about it.' This agnosticism was despite his parents having been very religious in an 'orthodox C-of-E way' and having taken him to church (or occasionally dissenters' chapels) frequently and twice on Sundays. Alfred had briefly been fervent about religion, but 'as there was no sufficient basis of intelligible fact or connected reasoning to satisfy my intellect, this feeling soon left me, and never returned.' He was of course influenced by his admiration for Robert Owen and other 'religious skeptics' such as Tom Paine in his *Age of Reason*, and by his brother William. But, with his enquiring mind, Wallace reached his own conclusions. He worried that if God were able to prevent evil but was unwilling to do so, he was not *benevolent*. But if, on the other hand, he was powerless to prevent evil, pestilence, catastrophes or the devil, he was not *omnipotent*. In contrast to this intellectual agnosticism or atheism, Wallace became interested in two fashionable pseudosciences: phrenology (testing the bumps on a cranium to determine people's characters) and mesmerism (he gave demonstrations of hypnotism to small audiences). This led in later life to a passionate espousal of spiritualism.

In the small community of Leicester's young nature-lovers, Alfred Wallace heard that Henry Bates was an enthusiastic entomologist. They probably met in the library, and Wallace 'found that [Bates's] speciality was beetle collecting, though he also had a good set of British butterflies.' Bates proudly showed his collection, and Wallace was amazed to find that hundreds of species of beetle had been collected around Leicester. So Wallace also eagerly took up beetle collecting. He obtained a collecting bottle, pins, a store-box, and Stephen's *Manual of British Coleoptera*, 'which henceforth gave me almost as much pleasure as Lindley's Botany... had already done.' He gathered the insects during his Wednesday- and Saturday-afternoon walks, often accompanied by two or three of his pupils.

1846 was another hard year for the Wallaces. William, aged thirty-seven, died of 'lung congestion after a severe cold' – medical knowledge was so rudimentary at that time that we cannot tell which disease actually killed this fit man. John, twenty-eight, and Alfred, twenty-three, went to Neath to wind up their older brother's affairs. They settled there for a year, working to survey a proposed railway line, then were briefly back in London for another rail survey, as well as any other engineering jobs they could find. This included designing a modest brick house for Neath's Mechanics' Institute (which still stands, in Church Place, and is now a public library).

The Wallace children did their best to look after one another. Their sister Fanny returned from the United States and lodged for a while with her brothers in Neath. She knew Paris, and took John and Alfred for a visit there. Wallace loved his first trip abroad. He wrote enthusiastically to Bates about the elegance of Paris and how its museums and galleries were freely open to all – in contrast to London where little could be seen 'without favour or payment'. (Fanny Wallace later married Thomas Sims, who became a professional photographer and a friend of Wallace's. Brother John tried running a small dairy farm but, when this failed, sailed in 1849 to California for the gold rush, settled there and raised a family.)

It was, surprisingly, a letter of introduction from young Bates that gave Wallace entrée to the insect collections in the British Museum (now the Natural History Museum) in London.* All this inspired Alfred to do more than mere local collecting. He wanted to study one genus or family of insects in depth 'principally with a view to the theory of the origin of species', hoping to achieve 'some definite results'. He asked his friend to help him choose an insect family in which he might succeed in collecting 'the greater number of the known species'.

At this time, both Wallace and Bates read a provocative anonymous book that had just appeared: *Vestiges of the Natural History of Creation*. (This work was soon shown, by Charles Darwin, to have been by an Edinburgh editor called Robert Chambers). The book argued that organic life had evolved over millennia, by laws of nature that were not understood but were expressions of God's will. The idea that life forms were extremely ancient was, of course, blasphemous to many. But Chambers said that it was demeaning to the Deity to imagine that He had arranged the detailed creation of the myriad species on the

* In the nineteenth century, the natural history collections were a department of the British Museum – which is how the three naturalists referred to them. For many decades they were called the British Museum (Natural History). In 1866 there was a petition to the government (signed among others by Charles Darwin, Alfred Wallace and Thomas Huxley) to separate the two museums; but it was almost a century – 1963 – before this split was formalized. In the meantime, the collections moved into their own superb neo-Gothic premises on Cromwell Road, London, designed by Alfred Waterhouse. In 1992 the name was changed to the Natural History Museum. This is the name that will be used throughout this book, even though Wallace, Bates and Spruce did not know it.

planet. Chambers imagined an upward progression, with many species having common ancestors, and with man at the top of his family tree. Wallace wrote excitedly to Bates about this book's revolutionary ideas; Bates apparently replied that he thought that its theory of progressive development was 'a hasty generalisation'. Wallace insisted that this was an ingenious hypothesis, but one that needed far more research because Chambers had not attempted to explain why species differed from one another. 'It furnishes a subject for every observer of nature to attend to; every fact [such a man] observes will make either for or against it.'

Being back in the Welsh countryside rekindled Wallace's love of natural history. He botanized, and corresponded with his friend Bates (using the new penny post) 'chiefly on insect collecting' and about the differences between insect 'species' and 'varieties'. At some time in autumn 1847 Bates travelled across England to stay with Wallace in Neath. They obviously discussed their passion for natural history, and seriously considered whether they could make a living as collectors by selling specimens to museums and amateurs. Wallace had lectured on scientific subjects at his local Mechanics' Institute, and he eagerly attended the British Association (for the Advancement of Science) when it conveniently decided to hold its annual conference in nearby Swansea. So they shared in the flowering of intellectual enthusiasm in Britain at that time, with learned societies being formed in every branch of science. And they were well aware of the fashion among the rich intelligentsia to amass private collections.

Bates credited Wallace with the idea of becoming professional collectors in the Amazon. He wrote that in late 1847 'Mr. A. R. Wallace... proposed to me a joint expedition to the river Amazons, for the purpose of exploring the Natural History of its banks.' They would amass personal collections, sell duplicates in London to pay their expenses, and 'gather facts, as Mr. Wallace expressed it in one of his letters "towards solving the problem of the origin of species", a subject on which we had conversed and corresponded much together.' Wallace also recalled that, returning from his week in Paris, he had spent a day in the insect room at the Natural History Museum and was dazzled by 'the overwhelming numbers of beetles and butterflies I was able to look over'. He wrote to his friend that his favourite subject was 'the variations, arrangements, distribution, etc., of species'. So, with youthful hubris, he 'firmly believed that a full and careful study of the facts of nature would ultimately lead to a solution of the mystery... of the great problem of the origin of species.'

A few books swung them towards the Amazon. The first was the travel journal of the great German polymath, Baron Alexander von Humboldt, published in Paris in thirteen volumes between 1816 and 1831 and translated into English in those same years. Humboldt and his French companion Aimé Bonpland had in 1800–1801 travelled up the Orinoco in Venezuela and across a short and level watershed to the forests of the upper Rio Negro. But they were refused entry to the Portuguese colony of Brazil and were forced to retrace their route down the Orinoco into Spanish South America.* Humboldt's famous *Personal Narrative of Travels to the Equinoctial Regions of the New Continent* lyrically described the majestic beauty of tropical rain forests. Another inspiration for the young naturalists was Charles Darwin's *Journal and Remarks* (on the second expedition of HMS *Beagle*), published in London in 1839. Darwin did not actually enter Amazonian forests: *Beagle*'s mission was to chart the coasts of the southern cone of South America and the Galápagos islands. But Wallace wrote to Bates that he read and re-read Darwin's book. 'As the journal of a scientific traveller it is second only to Humboldt's Narrative; as a work of general interest, perhaps superior to it. He [Darwin] is an ardent admirer and most able supporter of [the geologist] Mr Lyell's views. His style of writing I very much admire, so free from all labour, affectation or egotism, and yet so full of interest and original thought.'

There were only a handful of other books about Amazonia available in England at that time. This was largely due to the paranoia of small Portugal that other powers would try to seize its great Brazilian colony. All foreigners had been rigorously excluded from Brazil. This isolation started to cease in 1808 when a Napoleonic invasion of Portugal caused the Portuguese court to flee to Rio de Janeiro, where it remained contentedly almost until Brazilian independence in 1822. During this period a few European scientists were admitted, notably in 1816 in the entourage of the first French ambassador after Napoleon's defeat, and in 1818 with an Austrian princess who went out to marry the royal heir. Most of these learned and artistic visitors travelled only in southern Brazil. The first foreign scientists to be allowed on the Amazon itself were two Bavarians, the zoologist (ichthyologist) Johann Baptist von Spix and the botanist Carl von Martius. In eight months in 1819–1820 they travelled far up the Amazon, Solimões and Japurá rivers and wrote *Reise in Brasilien*, a fine account of their journey. This was published in German in three volumes; but only the first volume was translated into English, in 1824, and this did not cover the Amazon part of their travels. So Wallace and Bates could not properly have read Spix and Martius about Amazonia, although they did know and admire Spix's

* The order for this refusal had come all the way from Lisbon and Rio de Janeiro to the tiny frontier post of Marabitanas deep in the forests of the upper Rio Negro, because it was known that the German scientist admired the French Revolution and the authorities feared that he might infect Brazilian Amazonia with his subversive ideas.

scientific work on Amazonian fishes and Martius's on flora. In those same years the delightfully eccentric Englishman Charles Waterton travelled deep into British Guiana (now Guyana) and even crossed into the northernmost tip of Brazil. Waterton was the first Englishman to write in praise of tropical rain forests, and his *Wanderings in South America* (1825) was a deserved best-seller. But, although Wallace and Bates knew Waterton's work, they did not take it seriously. They were equally unimpressed with two books by English naval lieutenants about rapid voyages down the Amazon: Henry Lister Maw (published in 1829) and William Smyth and Frederick Lowe (in 1836).

There had been a few other notable travellers in Amazonia. The German Georg Langsdorff (a diplomat for Russia in Brazil after the Portuguese court fled there) led a fine expedition down the Tapajós in 1828–1829; but he himself went mad and its reports lay in St Petersburg archives and were not published until long after his death. Between 1829 and 1835 the Austrian zoologist Johann Natterer made amazing journeys on the southern and northern tributaries of the Amazon, during which he studied seventy-two indigenous peoples; but his notes were tragically lost in a fire in Vienna and never published. The anglicized German Robert Schomburgk was sent by the Royal Geographical Society to survey the frontiers of British Guiana. In 1839 he crossed into Brazil and did a remarkable year-long circuit to the upper Orinoco, down the Rio Negro, and back up the Rio Branco. This was published in the Society's *Geographical Journal*. Wallace and Bates knew about this great explorer, but had apparently not read his reports. They also knew about an exploration up the Xingu river in 1842–1843 by Prince Adalbert of Prussia (accompanied by an aide-de-camp, the future 'Iron Chancellor' Count Otto von Bismarck); but the Prince's book was not translated into English until 1849.

The clinching inspiration for Wallace and Bates to go to the Amazon was a slim travel book, *A Voyage up the River Amazon*, by a young American law student called William Edwards. He spent a few months in 1846 in Belém (then called Pará*) and going up the river as far as Manaus (then Barra). Edwards loved what he saw of these towns and the great river, wrote enthusiastically about the flora, fauna, people and climate, and made friends with the few non-Portuguese Europeans who had settled there. He exaggerated the luxuriance of exotic animals, trees and fruits in the world's richest ecosystem. But what appealed most to the poor Englishmen was the remark that 'Almost everything grows to

* Pará means 'large river' in the Tupi-Guarani language. It was the name given to the broad but short river that flows into the Atlantic Ocean south of Marajó Island, as the southern mouth of the Amazon. When the Portuguese founded a fort and settlement there in 1616, they called it Belém do Pará. The entire region was also known as Pará (a Captaincy in colonial times, then a Province under the Brazilian Empire (1822–1889), and a State in the present Republic). In the nineteenth century the town was often called Pará, but it is now the large city of Belém, and to avoid confusion this is the name used in this book.

hand that man requires; living is cheap and the climate delightful.' Wallace confirmed how much they enjoyed Edwards's description of 'the beauty and grandeur of the vegetation' and the agreeable people, 'while showing that expenses of living and travelling were both moderate.... [So] Bates and myself at once agreed that this was the place for us to go if there was any chance of paying our expenses by the sale of our duplicate collections.'

They agreed to meet in London in March 1848. They wanted to study the collections in the Natural History Museum. Its butterfly curator Edward Doubleday was particularly encouraging, declaring that the Lepidoptera of northern Brazil (the Amazon) were very little known and full of rarities. The aspiring collectors studied the Museum's holdings of South American fauna, and noted their gaps. They also visited the Royal Botanic Gardens at Kew to see living South American plants in its hothouses and pressed specimens in its Herbarium. Kew's curators taught them more about field collecting, and indicated that it might wish to buy unique botanical specimens. Wallace and Bates had an interview with Kew's director, Sir William Hooker, who then sent them its printed 'Instructions to plant collectors'. Hooker also fulfilled their request for a letter that would show 'that we were the persons we should represent ourselves to be, and might facilitate our progress into the interior'. This letter helped them to obtain necessary passports.

The young men also wanted to buy a few reference books, collecting apparatus, and tropical outfits, as well as enquire about a passage – traffic between England and northern Brazil was rare at that time. Wallace bought new spectacles and had himself vaccinated. By good fortune, they actually met the travel writer William Edwards. The young American was in London to see his publisher, the formidable John Murray, who specialized in publishing travel and scientific books. Edwards gave the travellers useful introductions to people in Belém: he had written about 'their kindness and hospitality to strangers, and especially [that] of the English and American merchants.' He encouraged the young men, and urged them to reach Brazilian Amazonia as soon as possible in order to enjoy the whole of its dry season.

The two aspiring collectors then had a stroke of luck. As Wallace later wrote, 'We were fortunate in finding an excellent and trustworthy agent in Mr. Samuel Stevens, an enthusiastic collector of British Coleoptera and Lepidoptera, and brother of Mr. J[ohn] C. Stevens, the well-known natural history auctioneer of King Street, Covent Garden. He continued to act as my agent during my whole residence abroad, sparing no pains to dispose of my duplicates to the best advantage, taking charge of my private collections, insuring each collection as its dispatch was advised, keeping me supplied with cash and with such stores as I required, and, above all, writing me fully as to the progress of the sale of each collection, what striking novelties it contained, and giving me general information on the progress of other collectors and on matters of general scientific interest.' Stevens really proved to be a prince of agents. Among many other qualities, we shall see how he cleverly planted letters from Wallace and Bates in semi-learned journals such as the *Zoologist* and the *Phytologist*, with fairly blatant advertising of the specimens they had for sale. Wallace later said that during fifteen years of working together, he and Stevens never disagreed about anything; and Bates was equally enthusiastic about their admirable agent.

Probably during that time in London, twenty-five-year-old Alfred Wallace had his picture taken. He was by now tall (1.88 metres or 6 feet 2 inches) and his thick fair hair had turned brown, with budding sideburns. He wore round granny glasses, and looked slightly thick-lipped and swarthy. There is no portrait of Bates from this time, but he drew himself a few years later. Of average height, he too wore round glasses. His mop of fuzzy hair came down into sideburns that almost met under his chin, and he grew a moustache.

Wallace then spent a farewell week with Bates's parents in Leicester. The collectors practised shooting, skinning and stuffing birds – a foray into taxidermy that was woefully inadequate: they had to learn this skill when in the field and from books such as Swainson's. Each young man invested about £100 into their enterprise. Wallace had earned this from his surveying and teaching. Bates's father supported them with additional funding, but his mother was dubious about the remote destination and whether they could earn a living as collectors – she was won over when their family doctor opined that the change of climate might help young Henry's health.

The pair then set off towards Liverpool by stagecoach, 'a cold and rather miserable journey' because they were riding on top to save money. They broke the trip in Derbyshire, to visit the famous orchid and palm collection at the Duke of Devonshire's house Chatsworth. (Part of their disappointment on first entering Amazon forests was because orchids in the wild are far rarer and often less gaudy than the luxuriant display in Chatsworth's greenhouses.) They then pressed on to Liverpool. On 20 April 1848 they embarked as the only passengers on the small 192-ton trading barque *Mischief*. The 'berths and other accommodation [were] of the scantiest' and for a time Wallace was prostrated with sea-sickness. But after a twenty-nine-day voyage they sailed up the Pará river to the town of Belém, and their great adventure began.

The third young Englishman to go to collect in Brazil, Richard Spruce, had had a similar upbringing to that of Wallace and Bates. He was equally poor, of modest middle-class origin; like Bates of delicate health; and like Wallace mildly socialist. Spruce was also educated only to primary-school level, but he too became a meticulous and diligent student. Above all, he shared the others' passion for natural history.

Richard Spruce was born in 1817 in the village of Ganthorpe, at the edge of the great Castle Howard estate 22 kilometres (14 miles) northeast of York. His father, also called Richard Spruce, was the highly respected schoolmaster of Ganthorpe village school, and later of Welburn also bordering Castle Howard. His mother, of a York family, died when Richard was an infant, so the boy was educated wholly by his father. He learned mathematics and beautiful handwriting from his parent, and classics from another teacher. The elder Richard Spruce remarried and had eight daughters, all but two of whom died of scarlet fever and other diseases; so he was 'unable to do anything for his son but help him to follow his own profession' as a schoolmaster. The boy helped his father, taught briefly at another village school, and in 1839 when he was twenty-two became maths master at the Collegiate School in York. He did this for almost five years, until the school itself went out of business.

Spruce was offered another teaching job, but it was full-time and residential. So he turned it down, because it involved 'too much mental strain for his very delicate health.... His lungs were affected [by the damp York weather], and he believed that he should not have lived another year' had he remained there. He caught repeated illnesses, particularly in winter with 'severe colds and constantly recurring winter cough' – somewhat alleviated by wearing a poultice called 'a perpetual blister'. He suffered from ailments that would not be recognized in modern medicine (such as 'a serious attack of congestion of the brain') and more familiar gallstones that caused excruciating pain and weakness.

The true reason why Spruce abandoned teaching was his love of botany and the outdoors. Walking on the Yorkshire moors as a boy, he had constantly observed and collected. In 1834 he had made a neatly written list of 403 species of plant from around Ganthorpe; and three years later he listed 485 species of flora from the Malton district. He was no mere collector, but was becoming a careful botanist. In 1841 he found and identified a rare sedge and then a moss, both new to Britain. That was a Damascene moment. A few miles from his village, he found 'one of the uncinate Hypna [a genus of moss] in splendid fruit. His love of plants... returned with such force that he vowed on the spot that henceforth the study of plants should be the great object of his life.'

Equally importantly – and also like Wallace and Bates – the young Spruce corresponded with leading botanists. Some friends and correspondents were local, such as the tin-smith Sam Gibson, 'one of a considerable number of North-country working-men botanists of the early nineteenth century'. Three of Spruce's mentors were the authors of or contributors to *Muscologia Britannica* (British mosses): Thomas Taylor of Killarney in southwest Ireland, William Watson of Warrington in Lancashire, and particularly William Borrer of Henfield in Sussex, 'one of the most acute and enthusiastic of British botanists'. During the following years, these moss enthusiasts either visited Spruce or he went to stay with them. In the five years before his South American journey Spruce wrote sixty-six letters to Borrer, and they reveal his growing confidence as an authority on mosses. 1845 witnessed Spruce's first published papers: descriptions of twenty-three new British mosses in the *Journal of Botany*; and a list of the mosses and hepatics of Yorkshire which also recorded twenty-three species new to Britain, in a new journal called the *Phytologist* (an old word for botanist). Spruce loved hepatics as much as he did mosses. (These tiny liverworts were so called because their leaves were slightly liver-shaped and because they were thought to yield a hepatic medicine.) During his teaching years, Spruce devoted every spare moment to collecting mosses (from all over Europe) and examining them under a microscope, so that 'he could give from memory the distinctive characters of almost every species'.

Through William Borrer and his own publications, Spruce became known to Sir William Hooker, who was creating the world's greatest botanic gardens at Kew, and to his influential friend George Bentham. These three great botanists discussed the future of the bright young Yorkshireman, and decided that he would do best to make a collecting expedition to the French Pyrenees. Bentham had seen the botanical potential of those mountains during a visit. So he made Spruce a loan for an expedition, secured against the collections he would accumulate there. Spruce spent a happy year from April 1845 to 1846 botanizing in the Pyrenees, and the remainder of 1846 naming, arranging, selling and dispatching the alpine plants he had gathered. A charming account of his journey, from two letters to Hooker, was published in the *London Journal of Botany*. Spruce then spent two years back in Yorkshire compiling the scholarly 'The Musci and Hepaticae of the Pyrenees', which occupied 114 pages of the *Transactions* of the Botanical Society of Edinburgh. He was now a major authority in the recondite world of mosses and liverworts; and his health was much improved by the mountain air.

Down in London in 1848, to help the family of his late friend Thomas Taylor sell his herbarium and books, Spruce saw Sir William Hooker and George Bentham. They encouraged him to undertake nothing less than 'the botanical exploration of the Amazon valley'. All were influenced by the letters from Wallace and Bates that Stevens had persuaded the *Zoologist* to publish. These reports enthused how delightful the people, climate and above all the natural wonders of the Amazon were. So Spruce decided to go. He was back at Kew in spring 1849 and spent the next months preparing for this momentous move.

George Bentham made a fine offer. He agreed to receive all the botanical collections that Spruce sent from Amazonia, to sort them by genera, name those already known to science and describe novelties. He would then send the plants to 'subscribers' (collectors who joined a syndicate) in Britain and other parts of Europe. Bentham would find these collectors, get payments from them, keep all accounts, and send the proceeds out to Spruce in Brazil. 'In return for which invaluable services he was to receive [no commission, other than] the first complete set of the plants collected.' This gentlemen's agreement worked admirably during the coming years. At first there were only eleven subscribers; but soon, thanks to Hooker's published reports about Spruce, and Bentham's great reputation as a botanist, there were twenty; and a few years later 'when the great novelty of the collections and their admirable condition as specimens became more widely known, [these] increased to over thirty.' Spruce was most grateful.

So the thirty-two-year-old botanist sailed across the Atlantic in the small brig *Britannia* (217 tons and a twelve-man crew). He was accompanied by Herbert Wallace, who was going to join his older brother Alfred, and a young man called Robert King, 'who had agreed to brave the wilds of the Amazon as my companion and assistant'. They arrived in Belém on 12 July 1849 after a five-week voyage from Liverpool.

3

ENTERPRISING YOUNG MEN

The agent Samuel Stevens placed an advertisement on the cover of the journal *Annals and Magazine of Natural History* to market the first consignment he received from Brazil. He got the journal's editor to write: 'Messrs. Wallace and Bates, two enterprising and deserving young men, left this country last April on an expedition to South America to explore some of the vast and unexamined regions of the province of Pará, said to be so rich and varied in its productions of natural history.' Stevens then tempted his readers by announcing the arrival of 'two beautiful parcels of insects of all orders, containing about 7000 specimens in very fine condition, and a vast number of novelties, besides other very rare species [scarcely] known to the entomological world… and a few shells and bird-skins.'

The 'enterprising young men' rapidly established themselves in their new environment. Coming from the gloom of England's industrial revolution, they were thrilled by Mediterranean architecture in the tropics. Their first view of Belém at sunrise was 'pleasing in the highest degree'. Bates loved its white buildings roofed with red tiles – typical of town houses throughout Portugal – and 'the numerous towers and cupolas of churches and convents, the crowns of palm trees reared above the buildings, all sharply defined against the clear blue sky, giving an appearance of lightness and cheerfulness which is most exhilarating. The perpetual forest hems the city in on all sides landwards.' But Belém was a town of only 15,000 inhabitants. Its commercial district was a few streets of 'tall, gloomy, convent-looking buildings near the port'. The public buildings – a governor's palace, fort, cathedral and some convents – were handsome but damaged by decay and poor repairs. Between them were bits of garden and waste ground; and the few squares were lined with elegant palms but looked more like 'village greens than parts of a great city'. The main street, Rua dos Mercadores ('Merchants' Street'), had open houses, some with yellow or blue wash as a change from white. Both visitors liked the mile-long 'well-macadamized' main suburban road, the Estrada das Mongubeiras, lined with splendid monguba (*Pachira aquatica*) trees. But on closer inspection Belém was a shabby little place. 'The city consists of about a score of horribly paved streets; beyond these are most magically beautiful lanes fringed with palms, and utterly overgrown with most magnificent foliage.'

The young Englishmen knew little about Belém's history. Founded when the Portuguese first reached the mouth of the Amazon in 1616, it became the home or base of several hundred settlers. During the next two centuries, these few colonists utterly destroyed the indigenous peoples of the main Amazon with their slaving expeditions and unwittingly imported, but devastatingly lethal, diseases. The town's status changed in 1750 when the Treaty of Madrid redefined the boundaries between the two Iberian kingdoms. Under the old Line of Tordesillas of 1494, Spain controlled almost all South America, while Portugal had only the eastern bulge of the Atlantic seaboard. But the 1750 treaty changed this radically. It decreed that each kingdom should get the regions that its men had penetrated. So Portugal obtained most of the Amazon basin, the vast forests and rivers ransacked by its slavers and missionaries. This huge area was considered to be of minimal commercial value. Thus Portuguese Brazil acquired roughly its modern boundaries and covered half South America. To improve the small capital of its new acquisition, the Portuguese authorities brought in Antonio Landi, a genial and admirable architect from Bologna. He laid out Belém's tree-lined streets and designed its baroque buildings, while a German engineer improved the water supply and drainage.

During the second half of the eighteenth century, there was a determined but ultimately failed attempt to make a financial success of Brazilian Amazonia. In 1759 Jesuit missionaries were expelled, and their mission villages were disastrously entrusted to lay 'directors'. The indigenous peoples suffered terribly. This, together with political unrest following Brazilian independence in 1822, and hatred of the Portuguese from Europe and Freemasons, erupted in 1835 in the ferocious Cabanagem rebellion. This was the greatest popular uprising in Brazilian history. The oppressed 'Cabanos' – dwellers in cabin shacks on the mudflats – twice captured Belém, and for three years their rebellion spread throughout the Amazon basin. Tens of thousands were killed, on both sides of the conflict. The great river and its tributaries, already decimated by centuries of disease and slavery, were further denuded. Belém fell into weedy disrepair and, although the rebellion was crushed by 1839, the town did not fully recover during the 1840s. Its population had fallen from about 25,000 in 1820 to 15,000 in 1848.

2

3

2, 3 Riverside vegetation, as seen by Wallace, Bates and Spruce during their months of travel on Amazonian rivers. Above is forest beside the upper Rio Negro; seasonally flooded igapó in the same part of Brazil is shown below.

4

There are hundreds of species of palm trees in Amazon forests. Spruce wrote a monumental book about palms, and Wallace a shorter work about their commercial uses.

4 Majestic buriti (*Mauritia flexuosa*), which yield nutritious fruits, fibres for hammocks, fronds for baskets and thatching, and trunks for building.

5 An exhausted scientist beside a paxiúba (*Socratea exorrhiza*) raised on stilt roots.

6 Tall and slender açaí (*Euterpe oleracea*), another versatile palm whose vitamin-rich juice makes the most popular drink of Amazon dwellers.

7 A small black-water river in the Rio Negro basin, similar to the Pacimoni that Spruce explored for the first time in 1854.

8 Most Amazonian rivers were empty, but the travellers occasionally saw an Indian in his ubá dugout canoe.

5

6

7

8

9

10

11

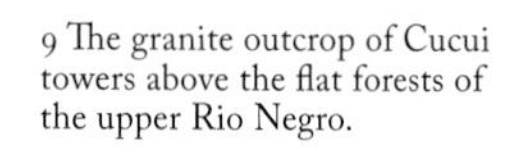

9 The granite outcrop of Cucui towers above the flat forests of the upper Rio Negro.

10 A beautiful *Monstera dubia*, now known as the Swiss-cheese plant.

11 A sapling of the quinine-yielding *Cinchona succirubra* growing out of a stump. This is exactly how Spruce found these trees in 1859, as profligate bark collectors had felled almost every large tree.

Wallace and Bates had read a little about the Cabanagem in William Edwards's book. This gave the ruling class's view of the convulsion: that the provincial president had been assassinated 'as were many private individuals of respectability.... Everywhere the towns were sacked, cities despoiled, cattle destroyed, and slaves carried away.... The disastrous effect of these disturbances is still felt [in 1846] and a feeling of present insecurity is very general.' Had they read the travels of Prince Adalbert of Prussia, who was in the Amazon in the early 1840s, they would have seen his different judgement – that 'these disturbances were the fruits of the ceaseless oppression which the white population had, from the very first, exercised on the poor natives.'

On arrival, the would-be collectors were looked after by Daniel Miller, their ship's consignee and also the British vice consul. He lodged them in his *rocinha* – a little *roça*, or forest clearing – a kilometre (half a mile) from the town centre. They bought hammocks, as used by everyone, and these were to be their beds during the coming years. They also got cooking pots, and their little roça house needed only 'a few chairs and tables and our boxes' as furniture. The house's rooms were spacious but empty: most activity took place on its large verandah. They also hired an elderly black freed-slave called Isidoro, whose 'usual deshabillé' was nothing but a pair of cotton trousers. He proved to be a good cook and servant-of-all-work, and an admirable character. Isidoro also knew much about nature, so that he was the first to tell the collectors about local trees.

Bates wrote to his friend Edwin Brown that they were at first overwhelmed by the complications and abundance of the natural history they had come to investigate. They were also 'not a little confused... by this strange and wonderful country... the language, the manners, mixture of races, and novel occupations of the people: the city is in a constant state of confusion from the rollicking propensities and religious amusements of the people; fireworks, noisy music, processions, and jingling of bells... almost every morning and evening.... On first landing, the exuberance of life everywhere burst upon us at once; in the streets, a moist, hot, earthy smell arose, and on the walls scores of lizards were scampering about.' Our two Englishmen were admirable observers. Neither had been abroad before (apart from Wallace's week in Paris), so everything was new, and they noticed details that no Brazilian would have remarked upon. (There were in fact no native travel-writers at that time: only a few histories in which this remote region was scarcely mentioned, or dry official annual reports by provincial governors.) It is a delight to glimpse mid-nineteenth-century Brazilian Amazonia through the eyes of these intelligent and enthusiastic young men.

Whites were of course at the top of the social ladder. They dressed in neat cotton clothes of spotless purity – there was no lack of washerwomen. But the more pompous 'adhere to the black cloth coat and cravat, and look most uncomfortably clad with the thermometer from 85° to 90° in the shade.' Bates wrote to his brother Frederick about their absurd dress coats, top hats and polished boots. He and Wallace were attired more sensibly, in coloured shirts, denim trousers, ordinary boots and slouch hats.

All Europeans enjoyed good health – which surprised Bates, because the town was surrounded by tidal mudflats and, like most people, he thought that malaria (*mal aria*, or 'bad air') was caused by 'miasmas' from swamps. Bates wrote to his brother that he himself had never felt better or stronger, and that Englishmen who had resided there for decades retained their 'florid complexions and "John Bull" appetites'. European merchants and the few foreign planters in residence complained about the scarcity and therefore the expense of labour – this prevented them from 'making the extraordinary riches of the country available'. It was unthinkable for white settlers to exert themselves physically. Everything depended on human labour, in a land with almost no machinery. But slavery of Indians was forbidden, and depopulation resulting from the Cabanagem upheaval made free native workers very scarce. The Amazon region had generally been too poor to afford to buy African slaves; and now these were very expensive because of efforts by the British Navy to stamp out the slave trade.

As time went by, Bates got to know some upper-class 'educated Brazilians' and found them to be courteous, lively and intelligent people. He made an interesting observation about changing attitudes to women among this elite. They were 'gradually weaning themselves of the ignorant, bigoted notions' inherited from the Portuguese, who would not let their women go out in society or receive an education. But by 1848 'Brazilian ladies were only just beginning to emerge from this inferior position' and fathers were educating their daughters. Bates felt that the previous 'degrading position held by women' had made relations between the sexes unsatisfactory – it led to 'promiscuous intercourse... among all classes, and intrigues and love-making [were] the serious business of the greater part of the population.' But that was in Belém – life in the 'pure' small towns of the interior was still very conservative.

Brazilians of both sexes wore white, and with women this contrasted attractively with 'their glossy black or brown skins'. Some women had 'dark expressive eyes, and remarkably rich heads of hair', but in the heat they 'dressed in a slovenly manner, barefoot or shod in loose slippers'. Bates was struck by the contrast of this 'squalor... with the luxuriance and beauty of these women'. All of them, even black slave girls, wore massive golden-coloured chains, earrings and necklaces of large gold beads. He particularly liked the blouses of acculturated Indian women. These had wide buttoned sleeves, but 'they were made of lacework or muslin so fine that it shades rather than covers the breast.' Wallace noticed, wistfully, that 'many dark eyes glanced at us as we passed' latticed house shutters. But the young Englishmen could admire Brazilian ladies only from afar. Families in Amazonia were protective of their women in an almost Moorish manner, as were Indian tribes; and there were no street girls in the tiny riverbank towns and villages. So life for the collectors was as celibate as it had been in England.

Wallace and Bates were scarcely class-conscious in England and even less so in Brazil. They liked the ordinary people, who were very gentle, polite and friendly. Their town seemed quiet and orderly: 'No class of people carry knives or other weapons, and there is less noise, fighting, or drunkenness in the streets both day and night than in any [comparable] town in England.' This peacefulness was due to the people's good nature – and was despite the great racial mix, and the fact that rum was sold on every street corner for a mere tuppence a pint. These inhabitants were idle, jovial and pleasure-loving. They seemed to do little farming because 'the earth and river produce all the necessaries of life spontaneously'.

There were words for *mestizos* of every shade of colour or racial mix between Indians, blacks and whites – *mulatto* (white-black), *mameluco* (white-Indian), *cafuzo* (Indian-black); detribalized Indians were known as *tapuio* and ordinary riverbank and forest people are to this day called *caboclos*. In the early morning, the visitors enjoyed the boisterous scene at public wells located in wet hollows around the town. Noisy black women washed a mass of household linen. Men filled their water-carts – 'painted hogsheads on wheels, drawn by oxen' – jabbering continually, having had a morning dram of *cachaça* rum in dirty street-corner wine-shops. But the 'motley life of the place' was charming, with idle soldiers in shabby uniforms carelessly hefting their muskets, priests, black women balancing earthenware water-jars on their heads, and solemn Indian women with naked babies on their hips. Bates generalized about black people, most of whom were slaves, that they were 'dreadfully independent and shrewd, but good-humoured'. He saw black men greet one another in the street with formal hand kisses. One evening, when he was searching for beetles, a troop of children came up to him. 'The first, a sweet little Indian girl, begged my hand, which she pressed to her lips *en passant* and all the others did the same, saying "passa bem, Senhor" ["Fare well, Sir"].... From this and a hundred other strange sights, a walk through the streets is the greatest possible pleasure for me.'

After a few nights in Mr Miller's rocinha, the Englishmen rented a house from an elderly Portuguese gentleman called Danin. This was 2 kilometres (a mile and a half) from the riverside heart of the town, in a hamlet called Nazaré (now a city square and bus terminal in the centre of the sprawling metropolis of modern Belém). The bungalow was 'a beautiful place' of four rooms surrounded by a broad verandah under a tiled roof that was 'cool and pleasant to sit and work in'. The house had a small garden of coffee, manioc and fruit trees. It was on a grassy square, around which were a few houses and palm-thatched huts, and the chapel of Our Lady of Nazareth. The Protestant Bates described the statue of this Virgin as a handsome doll about 1.2 metres (4 feet) high, wearing a silver crown and a vestment of blue silk studded with golden stars. She was a great favourite of local people, responsible for many miracles (surrounded by models of the limbs she had cured), and the focus of a great pilgrimage and celebration every October. Bates described every detail of this, the grandest of all the festivities. After all the religious ceremonial came the 'holiday portion of the programme' which was like 'a fair, without the humour and fun, but [also] without the noise and coarseness of similar holidays in England.'

This was by no means the only religious holiday in Belém: these were celebrated throughout the year, and each could last for days. Soon after their arrival there was that of the Holy Spirit in the Cathedral, then the Trinity in a smaller church of that name. Each fiesta consisted of fireworks, gaudily dressed girls selling *doces* (sweets, cakes and fruit), open-air gambling stalls with rattling dice-boxes and roulette wheels, showy processions with floats, saints and crucifixes, special masses, kissing of images and relics, church bells, and then more fireworks. Wallace (the Anglican turned atheist) watched, amused, impressed and slightly patronizingly, 'a miscellaneous crowd of negroes and Indians, all dressed in white, thoroughly enjoying the fun, and the women in all the glory of their massive gold chains and earrings. Besides these, a number of the higher classes and foreign residents grace the scene with their presence.... Music, noise, and fireworks are the three essentials to please a Brazilian populace.' Bates was more enchanted. He noted that women were out in force, 'their luxuriant black hair decorated with jasmines, white orchids and other tropical flowers. They are dressed in their usual holiday attire, gauze chemises and black silk petticoats,' and of course adorned with plenty of golden jewelry. 'The soft tropical moonlight lends a wonderful charm to the whole.... There is no boisterous conviviality, but a quiet enjoyment seems to be felt everywhere, and a gentle courtesy rules amongst all classes and colours.' After the mass there was a peal of bells, a shower of rockets, and 'the bands strike up, and parties of coloured people in the booths begin their dances. About ten o'clock the Brazilian national air is played, and all disperse quietly and soberly to their homes.' (But after a fortnight, the English visitors had almost had enough of these noisy amusements, for the people also fired off 'guns, pistols and cannon from morning to night'.)

Wallace and Bates happily settled down to three months of professional collecting in their fine rented house. It had forest on three sides, with overgrown trails. They were delighted to learn that one of these led to a house used thirty years earlier by the Bavarians Spix and Martius – the first foreign naturalists allowed on the Amazon. Another road led east towards the province of Maranhão, but was now so overgrown as to be almost impassable. Their routine was to rise at dawn, to a cup of coffee made by their servant Isidoro. He would then go to the town for the day's provisions, and the collectors spent two hours on ornithology – trapping or shooting birds. 'At that early hour… all nature was fresh, new leaf and flower-buds expanding rapidly. Some mornings a single tree would appear in flower amidst… the uniform green mass of forest – a dome of blossom suddenly created as if by magic. The birds were all active…'. At about ten o'clock, after breakfast prepared by Isidoro, they spent four or five hours on entomology, since the best time for insects was before the midday heat. They were fortunate that one of the three men who tended their house's trees and garden was celebrated as a catcher of insects, reptiles and small mammals. This Vincente, 'a fine stout handsome' black man, once brought Wallace a gigantic hairy spider, probably a tarantula, that he had dug out of its hole. By two in the afternoon, 'every voice of bird or mammal was hushed' and the only sound was the occasional whirr of a cicada in a tree. 'The leaves, which were so moist and fresh in early morning, now became lax and drooping; the flowers shed their petals'; and the Indian or mulatto farm hands slept in their hammocks or sat in the shade 'too languid even to talk'. Wallace and Bates had their main meal at four and then spent the evening preparing their specimens and taking notes; or they might stroll into Belém to watch its streets come to life or to chat with their European friends. Night invariably falls at six on the equator. So the Englishmen had tea at seven and retired to their hammocks for a sound sleep.

Bates wrote lyrically of the daily weather pattern in June and July, immediately after their arrival. In late afternoon, the sea breeze would die away. 'The heat and electric tension of the atmosphere then become almost insupportable. Languor and uneasiness would seize on every one; even the denizens of the forest betraying it by their motions.' Black clouds from the Atlantic ocean obscured the sun. 'Then a rush of mighty wind is heard through the forest, swaying the tree-tops; a vivid flash of lightning bursts forth, then a crash of thunder, and down streams the deluging rain. Such storms soon cease… and all nature is refreshed…. Towards evening life revives again, and the ringing uproar is resumed from bush and tree.'

The naturalists noted one of the great differences between tropical and temperate forests: the former were evergreen and constantly in motion throughout the year. Different species of deciduous trees do not flower or shed their leaves simultaneously, nor do birds moult, pair or breed at the same times. 'In the equatorial forests the aspect is the same or nearly so every day in the year: budding, flowering, fruiting, and leaf shedding are always going on in one species or other. The activity of birds and insects proceeds without interruption, each species having its own separate times.... It is never either spring, summer, or autumn, but each day is a combination of all three.' No one in the nineteenth century appreciated an important result of this lack of seasonality. Because of it, there is no winter pause during which humus accumulates. Instead, virtually all falling nutrients are recaptured by the growing biomass; and the soils under tropical rain forests are woefully weak and acidic.

During these happy and productive months, the young men took the essential step of learning Portuguese. Almost no Brazilian spoke anything else. But Wallace and Bates were diligent students; their knowledge of basic Latin was a help; and Portuguese is a relatively easy language. So both became fluent.

The collectors made a few excursions. They walked a few miles through tall forest to see their landlord Senhor Danin, who lived in a once-thriving cattle ranch where he now had a tile works. 'Since the political disorders [of the Cabanagem], decay had come upon this as on most other large establishments in the country. The cultivated grounds, and the roads leading to them, were now entirely overgrown with dense forest.' The lack of labourers – slave or free – added to the abandonment. Danin proved to be a courteous gentleman, who had visited England and spoke English. On one of their walks they met the Swiss consul, a Frenchman called Monsieur Borlaz who lived on a river about a mile from Belém. He introduced them to exotic fruits, and let them collect in his 'very rich' riverside vegetation.

Their most interesting journey was 19 kilometres (12 miles) to Magoari, a creek where there was a saw-mill and two mills for processing dry rice. This operation was owned by an American called Upton and managed by a very competent Canadian, Charles Leavens. William Edwards, the American author whose book had attracted Wallace and Bates to Brazil, had given introductions to these men. Wallace was fascinated by the saw-mill. It was powered by a 3-metre (10-foot) fall of water from a reservoir: this water rushed out of an aperture and spun a relatively small wheel connected to the saw with a rack-and-pinion, which made two saw thrusts with each turn. Leavens showed them different types of tree being sawn. The most interesting was the maçaranduba (*Manilkara huberi*), a magnificently tall hardwood whose crown emerges far above the forest canopy. This tree's vernacular name was 'cow tree' because it yielded quantities of juicy pulp that looked and tasted rather like cream. Local people loved it. But its milk was dangerous to drink in quantity, because it soon thickened to 'a glue which is excessively tenacious' – Leavens showed them a violin he had made with cow-tree glue. This 'wonderful tree... is one of the largest of the forest monarchs, and is peculiar in appearance on account of its deeply-scored reddish and ragged bark.' The saw-mill had made squared logs 30 metres (100 feet) long from giant maçaranduba trunks and from ipê or pau d'arco ('bow wood') trees (*Tabebuia serratifolia*). To produce cut logs that long, the living trees must have towered to between 55 and 60 metres (180 to 200 feet).

The walk to Magoari took Wallace and Bates through true virgin forest, in contrast to the secondary growths around the town. They were lucky to have their servant Isidoro with them, because he had worked in the forest and knew its trees. Normally taciturn, Isidoro 'was rather fond of displaying his knowledge on a subject of which we were in a state of the most benighted ignorance, and at the same time quite willing to learn.' His method of teaching was to address each tree that they passed, and tell its virtues and uses. There was a virola called *ucu-ubá* (*ubá* means wood in Tupi-Guarani and also a dugout canoe) that was good for sore throats – Isidoro made gargling gestures; quaruba used for house floors; putieca for paddles; nowará for fires in forges. And there were palms like the tall multi-purpose açaí (*Euterpe oleracea*) whose nutritious fruit made the favourite drink of Paraense people; inajá (*Attalea maripa*), with a thick stem and 'dense head of foliage' and whose fronds (like those of açaí) yielded the best edible palm-heart *palmitos*; and the striking paxiúba (*Socratea exorrhiza*) that stands on stilt roots above swampy areas.*

Bates was above all an entomologist. After a few weeks he was writing excitedly to Samuel Stevens about the interest and beauty of the insects. Curiously, they were not numerous – in a hard day's collecting he would get only about thirty to fifty specimens of butterfly and twenty beetles. 'But nearly all will be different species. I number now [August 1848] 460 different species of butterflies alone, and every time I go out take some new species.' He described a veritable fashion parade of Lepidoptera. Among the many about which he enthused were large yellow *Gonepteryx* and *Phoebis*, *Papilio* and other swallow-tails, a fritillary or two, orange *Cethosia*, the splendid *Philaethria dido*, long-tailed skippers, graceful heliconids with transparent wings, others with velvety black wings with green and crimson bands, countless varieties of small species, many glorious insects in marshy parts of the forest, and the dazzling blue blaze of large morphos gliding through the dense

* Modern vernacular and botanical names are given throughout, because their spelling and even their nomenclature may have changed since the mid-nineteenth century, and since the English collectors made valiant but often inaccurate transliterations of names they were told.

shady pathways of the forest. After three months in Brazil, Bates could write to his friend Edwin Brown: 'we forwarded a large chest of insects, 3,635 specimens, twelve chests of plants, &c. to England.'

Collecting was of course done on foot, but travel in Amazonia was and always had been entirely by river. On 26 August Wallace and Bates embarked on their first adventure: a voyage up the lower Tocantins, the easternmost of the great rivers that flow northwards into the Amazon. The mill manager Charles Leavens organized the trip because he hoped to get plenty of cedro logs on that river. This cedro (*Cedrela odorata*) was a favourite with carpenters – Spruce called it the 'deal' of Amazonia. Brazilians confusingly named it 'cedar' because its wood looked and smelled like that of the totally different European conifer of that name. It was an upland softwood, but because it floated easily much driftwood was cedro logs. Leavens dealt with the bureaucratic labyrinth of internal passports and licences required for even the simplest journey – complicated forms, 'signing and countersigning at different places, applications to be made and formalities to be observed'. The Englishmen were learning much more about life and travel in that world of forests and rivers.

There were boats adapted to every different river condition, but all were propelled by a mixture of sail and paddles – steam started to arrive only in the following decade. This particular journey might involve heavy seas in the expanse of open water where the Tocantins entered the Pará river, but there would be rapids and shallow draught upriver. So Leavens hired a two-masted *vigilenga*, 8 metres (27 feet) long, with a flat prow, very broad beam, decked over, and with two arched cabin awnings of palm thatch on wickerwork. Finding a crew of paddlers was usually a major problem. But on this occasion, Leavens brought three Indians from his rice-mills, headed by Alejandro, whom Bates admired as 'an intelligent and well-disposed young Tapuyo [detribalized Indian], an expert sailor, and an indefatigable hunter', thin, shortish, with fine features, 'a quiet, sensible, manly young fellow'. Bates acknowledged that without Alejandro's skill and fidelity they could hardly have accomplished anything on their journey. For their part, Wallace and Bates brought their servant Isidoro and an Indian lad called Antonio 'who had attached himself to us'.

For a century, boats going between Belém and the Tocantins had avoided the squalls, swell and rocky shoals where the mighty tributary poured into the Pará river by going up the far smaller Moju river, along an 800-metre (half-mile) canal cut through the very flat forests, and down a stream to reach the Tocantins opposite its main town, Cametá. Progress for our travellers was at first rapid thanks to a following wind and tide; but they had to tie their boat to a tree when the tide turned. At dawn it was 'beautiful in the extreme... a compact wall of rich and varied forest, resting on the surface of the stream.' To Bates it was a tapestry in which 'the dome-like, rounded shapes of exogenous trees... formed the groundwork, and the endless diversity of broad-leaved Heliconiae and palms... the rich embroidery.... The slanting beams of the early sun... lighting up the scene most gloriously.' Wallace delighted in the magical half-hour after sunset when elegant palms and towering emergents were silhouetted against the golden sky, and fruit trees around riverbank huts, 'the grassy bank, the noble river, and the background of eternal forest, all softened by the mellowed light... formed a picture indescribably beautiful.'

This hinterland of Belém was prosperous. On the lower Moju they passed the estate of a Count Brisson, with 150 slaves farming manioc to feed the town. Then there were smaller settlers' houses, villages and caboclos' riverside clearings. The Amazon contains a fifth of all the world's flowing river waters, so that its major tributaries are colossal. When the boat reached the Tocantins it was 'a grand sight – a broad expanse of dark waters dancing merrily to the breeze; the opposite shore a narrow blue line, [five or six] miles away.'

They spent the night on an island covered in two species of palm tree. Bates, Wallace and later Spruce adored palms and all three wrote about them. Those on this island were buriti (called miriti in the Amazon estuary, or moriche in Spanish; *Mauritia flexuosa*) and buçú (*Manicaria saccifera*). Buriti are glorious trees, with huge smooth stems soaring to 30 metres (100 feet) and crowned by enormous clusters of fan-shaped fronds. 'Nothing in the vegetable world could be more imposing than this grove of palms.... The crowns... densely packed together at an immense height overhead, shut out the rays of the sun; and the gloomy solitude beneath [was like] a solemn temple.' The Englishmen later learned of this palm's myriad uses, as logs for buildings, rafts, fibres for hammocks or baskets, and yielding a fruit whose yellow pulp makes nutritious porridge or (nowadays) ice cream. Early man in the Amazon planted groves of valued buriti, which would account for the dense mass on this island. The buçu palms around the edges of the island reminded Bates of vast shuttlecocks, with their stubby stems and circle of broad, 7.5-metre- (25-foot-) long dark-green leaves. These showy and durable leaves, like a banana tree's, are ideal for thatching houses or boat cabins, and pigs adore the buçu's fruit.

The two collectors were constantly learning the tribulations of river travel, and also seeing how Brazilians lived. Cametá looked good from a distance, on a high riverbank; but it was a small, straggling place. One of the Indian paddlers disappeared, doubtless in some tavern, so the boat had to sail next day without him. Because of the delay they lost a following wind, so the remaining man and boy had to row constantly 'which put them rather out of humour'.

On these thinly populated rivers travellers moved from one contact to another, and there was a culture of hospitality. In Cametá they knew a merchant, who fed them, showed them around and then accompanied them for 50 kilometres (30 miles) upriver to their next destination, Vista Alegre. This was the *sitio* (riverside farm) of a Senhor Antonio Gomez. They had a letter of introduction to him from a prominent gentleman in Belém. 'We here had an opportunity of seeing something of the arrangement and customs of a Brazilian country house.' The building itself was raised on piles, to keep it above the river at its highest. A substantial wooden pier led out to the river, with steps up to the house's verandah. This opened onto a room where visitors were received and business conducted. Another raised causeway led to a separate house where the women, children and servants resided. Bates commented that 'as usual in Brazilian houses of the middle class, we were not introduced to the female members of the family' but saw them only at a distance, and Wallace was sorry that they were 'never once honoured by the presence of the lady or her grown-up daughters'. Hospitality consisted of coffee at six in the morning; breakfast at nine was beef, dried fish and manioc, with more coffee; the main dinner was at three in the afternoon – rice or shrimp soup, meat or fresh fish, and sliced fruit (pineapples and oranges) served on saucers; then after dark at eight in the evening, tea and manioc cakes. Black and Indian boys waited at table, and were efficient at whisking off empty plates for constant washing by a woman in the background. Near the house was a crude mill driven by bullocks, for grinding sugar cane into cachaça rum. There were a few fruit trees, a neglected plantation of coffee shrubs and cacao trees, and a large shed for making manioc flour. Bates noted, disapprovingly, that ploughs and agricultural implements were virtually unknown. But land was plentiful. So, every other year, a new patch of forest was felled and the former clearing allowed to relapse into jungle. No foreigners at that time appreciated that soils under tropical forest were too weak for ploughing.

Lack of paddlers was a leitmotif for all Amazonian travellers. The Englishmen sacked their Indian boy who had 'turned lazy and disobeyed orders', so they could not proceed. Senhor Gomez tried in vain to get nearby canoe-men to enlist, and he finally lent them two of his slaves to paddle to their next destination – where they hoped that the military commander would press boatmen into their service. Exasperated, Bates thought that 'the people of these parts seem to be above working for wages. They are naturally indolent.' But he admitted that they usually had small businesses or plantations that supported them and made them independent. He also knew that local people were reluctant to work for foreigners, who were suspected of being 'strange in our habits'. Both Wallace and Bates had been sympathetic to working men in England; but in Brazil their eagerness to travel outweighed any qualms about using slaves or press-ganged Indians to propel them. Their next destination was Baião ('Big Bay'), some 40 kilometres (25 miles) up the Tocantins. There was little wind, so the men had to row. They used stout paddles attached to long poles by lianas. They stood on a raised deck of planks, above the forward thatched cabin, and they faced forward as they pushed.

There were plenty of riverside clearings on the banks and islands of the lower Tocantins. The huts found in these clearings were just an open framework, thatched by big buçu palm leaves, and resting on a high platform. Açaí palms, whose outer stem was 'hard and tough as horn', were split into narrow planks for flooring and the few walls. This was healthy country, on a clear-water river with sandy banks. 'The people seemed to be all contented and happy, but idleness and poverty were exhibited by many unmistakeable signs.' They were unconcerned by the huge rise and fall of the river. 'They seem to be almost amphibious, or as much at home on the water as on land. It was really alarming to see men and women and children, in little leaky canoes laden to the water-level with bag and baggage, crossing broad reaches of the river.' These riverbankers were polite and friendly caboclos. They lived on fish, some shellfish, manioc, and forest fruits. There were times of year when fish were scarce and people almost starved: so they had to store a little food against this lean season. When the travellers asked to buy food they would reply: 'There is none. I am so sorry but I cannot oblige you, dear heart.'

One boon to all these people was the vitamin-rich fruit of the açaí palm. Mixed with water, this made 'a thick, violet-coloured beverage which stains the lips like blackberries'. Wallace watched how a thin layer of pulp inside each fruit was soaked in warm water and then rubbed and kneaded to remove it from the stone. The resulting liquid became purple and as thick as cream. 'It is eaten with sugar and farinha [manioc flour]; with use it becomes very agreeable to the taste, something resembling nuts and cream.' Because açaí palms ripen at different times, the juice was sold constantly, all year round, in the streets of Belém. It still is, massively. Next to manioc, açaí remains a pillar of life for poorer people, who eat it with every meal. The juice is thought to be a cure for a score of ailments as well as being an energizing stimulant – although one lady told the geographer Nigel Smith that 'those who eat a lot of açaí get fat, not strong'.

Baião proved to be a village of some 400 people on a very high bluff on the east bank of the river (it is still there, but larger). Visitors climbed a rickety 120-rung ladder fixed to this bank. The commandant and most prominent citizen was José Antonio Seixas, to whom the travellers had a letter of introduction from his business partner in Belém. Wallace published a translation of this long and magnificently flowery epistle, as a fine example of Portuguese composition and politeness. It introduced Charles Leavens and his two worthy companions, and begged Senhor Seixas to give them every assistance, friendship and protection in their laborious enterprise, to alleviate the privations of 'men like them, devoted to science, and whose very aliment is Natural History, in a country like ours abounding in the most exquisite productions.' Seixas was away at his farm, but he had arranged an empty house for them. So they spent hot hours carrying their equipment up from the boat, and then settled down to a few days of collecting. Bates was amused by his first taste of 'the easy, lounging life' of people in these tropical villages. 'No sooner were we established in our rooms, than a number of lazy young fellows came to look on and make remarks, and we had to answer all sorts of questions.... In their familiarity there is nothing intentionally offensive, and it is practised simply in the desire to be civil and sociable.' The houses in Baião had only earth floors, and their doors and windows were open so that people could walk in and out as they pleased. 'There is always, however, a more secluded apartment where the female members of the families reside.' The collectors had 'very good sport' in their days at Baião. As they prepared insects or skinned birds they were watched by a crowd of boys and men. These marvelled at the patience of the whites, but they could not understand the purpose. 'They seemed to think that the English could hardly be such fools as to want to see a few parrot and pigeon skins.' So they guessed that pretty butterflies must be to provide patterns for printed calico (English cloth was a prized import in Brazil), while ugly insects could only be for medical cures. As the visitors got to know the people of this remote village, Bates was amazed by a young public clerk called Soares. This caboclo had a library that included 'a number of well-thumbed Latin classics, Virgil, Terence, Cicero's epistles, and Livy.... It was an unexpected sight, a classical library in a mud-plastered and palm-thatched hut on the banks of the Tocantins.'

When Seixas arrived, he was as kind as possible to the visitors, having them dine with him, killing an ox in their honour, and providing them with two male workers. But, as elsewhere, he did not introduce them to his family: Bates caught a fleeting glimpse of his pretty mameluca wife, tripping across the yard with her young daughter. The Englishmen slung their hammocks and worked in a room that had once been a cacao store, and which therefore swarmed with rats and cockroaches. 'The latter were running about all over the walls; now and then one would come suddenly with a whirr full at my face, and get under my shirt if I attempted to jerk it off. As to the rats, they were chasing one another by dozens all night long, over the floor, up and down the edges of the doors, and along the rafters of the open roof.' Sleep was difficult.

The land behind Baião was secondary forest, full of abandoned coffee and cotton trees. The people living in huts along the riverbanks were all poor but hard-working, the men constantly fishing, and the women planting manioc and grinding its flour, spinning and weaving cotton, and making soap of burnt cacao shells mixed with oil from andiroba-tree nuts. Bates asked them why they let their plantations run to waste. 'They said that it was useless trying to plant anything hereabout; the saúba [saúba; leafcutter] ant devoured the young coffee trees, and every one who attempted to contend against this universal ravager was sure to be defeated.' This was all too true. Yet, Bates, Wallace and all other visitors persisted in imagining that industrious Europeans or North Americans could overcome the pests, parasites and weak soils, to turn luxuriant tropical vegetation into prosperous temperate farmland.

Wallace had a different adventure. Collecting insects, he inadvertently disturbed a hanging nest of small wasps. 'They covered my face and neck, stinging me severely,' and in his haste to knock off the wasps and escape, Wallace lost his spectacles. But he was philosophical about the painful stings, which soon abated, and he had spare spectacles. He grew accustomed to a 'nightly concert' as he fell asleep in his hammock. There was the booming roar of howler monkeys, 'the shrill grating whistle of the cicadas and locusts, and the peculiar notes of the saracuras [wood rails] and other aquatic birds' together with 'the loud unpleasant hum of the mosquito in your immediate vicinity.' But the most extraordinary noises came from frogs. One was a dismal croak that you would expect from a frog. But others were 'like no animal noise that I have ever heard before. A distant railway-train approaching, and a black-smith hammering on his anvil, are what they exactly resemble.' These were 'such true imitations' that Wallace could imagine himself back in England, listening to a mail train or to boilermakers at an iron works.

After four agreeable days (but disturbed nights) at Baião, they continued up the Tocantins, with two new crewmen and fresh provisions supplied by Senhor Seixas. The broad river was full of islands, so that they occasionally ran aground in shallow channels and the men jumped overboard to push. On other days, there was a good following breeze and they set sails. They landed occasionally, either trying (unsuccessfully) to recruit more paddlers, or for collecting, or for Leavens to search for the cedro trees he had hoped to find for his saw-mill.

One stop was at rounded hills covered in light forest. Here they found an unexpected holiday camp. There were a great many hammocks slung between the trees and household possessions scattered over the ground. 'Women, old and young, some of the latter very good-looking, and a large number of children, besides pet animals, enlivened the encampment. They were all half-breeds, simple, well-disposed people.' They explained that they were from Cametá, 130 kilometres (80 miles) down the river. They had come all this way to escape the summer heat in town, and to eat plenty of fresh fish. So the men fished for their daily food, and perhaps tapped some rubber latex or gathered medicinal sarsaparilla (salsa or salsaparilha in Portuguese) and copaiba oil to sell on their return. But essentially it was just 'a three-months' pic-nic.... The weather is enjoyable the whole time, and so days and weeks pass happily away.'*

Bates disapproved of some people they encountered. A village called Nazaré dos Patos ('Nazareth of the Ducks') was typical of 'outlying settlements... the resort of idle worthless characters.' There was a festival of sorts going on there and people were fuddled with *caxiri*, fermented manioc that tasted like new beer. The local inspector or constable was a 'very slippery customer' who appeared ingratiating but probably discouraged his men from enlisting as paddlers. So Leavens had to abandon the original plan to go far upstream to the Araguaia river; but he agreed to show his friends the famous cataracts on the Tocantins. The river and its islands were becoming studded with rocks, the first that the Englishmen had seen in Brazil. They could not know, but they were approaching the Central Brazilian Plate, a mass of very ancient Precambrian geology. Each of the great southern tributaries dropped off this plate in a vortex of rapids as it flowed towards the main Amazon.

The travellers had to moor their large, flat vigilenga boat, and continue in a *montaria* that Seixas had lent them for this part of the journey. The montaria was the workhorse dinghy of the Amazon, often towed behind a larger vessel. It was a small boat made of boards, a thick broad one curved by heat for the bottom, narrow boards for the sides, and small triangular ones for stem and stern. There was usually an arched awning of woven lianas and arrowroot leaves. A montaria had no rudder: steering was by paddle. Bates was alarmed by the way in which Indians risked themselves in these 'crazy vessels'. When they and their families crossed rivers in heavily laden and leaking montarias it 'required the nicest equilibrium: a movement of a hair's breadth would send all to the bottom.' The secret was that Indians were superb swimmers. If hit by a storm they would all jump into

* This delightful campsite, Trocará, is now a small indigenous reserve.

the river and swim around until the heavy sea subsided and they could re-embark. The next-largest craft was the *ubá*, a dugout canoe hollowed from a huge tree trunk, and heavier and more stable than a montaria.

Proceeding in their little vessel, they picked up a sturdy tapuio Indian, Joaquim, as their pilot, but for various reasons they were reduced to only two other boatmen – scarcely enough to paddle against the strong currents. Wallace and Bates marvelled at the vigorous and successful way in which their three men battled against the powerful river. At times, the rushing water forced them to go close to the banks and punt with long poles. Nine days after leaving Baião, they reached the northernmost of the fifteen cataracts that block the Tocantins. Boats bringing gold and ore downstream from Minas Gerais ('General Mines') far to the south would brave all these rapids. But this took great skill and courage – many a vessel was dashed against rock outcrops and its crew drowned.

The broad river narrowed, so that the main rapids were only 400 metres (1,300 feet) wide. They landed and climbed an elevation for a stupendous view. 'A body of water rushes with terrific force down a steep slope, and boils up with deafening roar around the boulders which obstruct its course.' 'A deep and very powerful stream rushes down in an unbroken sweep of dark green waters, producing eddies and whirlpools more dangerous to canoes than the fall itself.' 'The wildness of the whole scene was very impressive. As far as the eye could reach, stretched range after range of wooded hills, scores of miles of beautiful wilderness, inhabited only by scanty tribes of wild Indians.' The toughest of these rapids was at Tucuruí, now the site of one of the world's largest hydroelectric dams.

Returning downstream, the boatmen were rewarded for their exertions. It was 'excellent fun shooting the rapids. The men seemed to delight in choosing the swiftest parts of the current; they sang and yelled in the greatest excitement, working the paddles with great force, and throwing clouds of spray above us as we bounded downwards.' At a village that evening, the admirable Joaquim rewarded himself with too much cachaça and became violent and abusive. To sedate him, the travellers plied their pilot with more rum, which had the desired effect; but next morning he was sheepish and ashamed of himself, because tapuio Indians considered it disgraceful to get drunk.

When they returned to their larger vigilenga boat, Bates spent a night ashore in Joaquim's home. He was able to see how a relatively prosperous detribalized Indian lived. The house was large and palm-thatched, but open apart from a private apartment at one end. It was raised on short stilts, and 'under this shed were placed all the household utensils; earthenware jars, pots, and kettles, hunting and fishing implements, paddles, bows and arrows, harpoons, and so forth.' A couple of chests contained the women's best clothes for religious festivals. But there was no furniture. Hammocks served as chairs or sofas, and meals were on a cloth on the floor with guests squatting around it. Hospitality and formality were the norm on these sparsely inhabited rivers. 'There were other strangers under Senhor Joaquim's roof besides myself – mulattos, mamelucos and Indians – so we formed altogether a large party.... After a frugal supper, a large wood fire was lighted in the middle of the shed, and all turned into their hammocks and began to converse.' Some guests fell asleep, but others told stories until late into the night. As with indigenous peoples throughout Amazonia their favourite topics were hunting and fishing, or *curupira* and other malign forest spirits. 'One old parchment-faced fellow, with skin the colour of mahogany, seemed to be a capital story-teller.' He enacted (doubtless for the umpteenth time) his killing of a jaguar, imitating its hoarse growl and dancing about the fire like a demon. Bates was growing increasingly fond of Indians, whose skills he admired. He later wrote that he realized that his travels and collecting depended on becoming accustomed 'to the ways of life of the humbler classes of the inhabitants'. Letters of introduction to important people were all very well. But in the interior wildernesses the boatmen were their own masters. 'The authorities cannot force them to grant passages or to hire themselves to travellers, and therefore a stranger is obliged to ingratiate himself with them in order to get conveyed from place to place.'

A different, but equally agreeable night was spent on the verandah of Senhor Seixas's plantation near Baião. His was a large, well-financed and efficient operation. For some distance around the house the undergrowth had been cleared to make way for 60,000 planted cacao trees; but since these chocolate-producers grew best in shade the large forest trees had been left. These included rubber trees, known to Brazilians as *seringa* because Indians had always used their milky latex to make little pear-shaped syringes – which indigenous peoples used to inhale snuff and hallucinogens, but modern Brazilians use for medicines. Bates rightly described the rubber tree (*Hevea brasiliensis**) as a tall but otherwise unremarkable tree, with bark and foliage similar to an English ash. On seasonally flooded islands near the mouth of the Tocantins, they saw camps of people collecting and preparing rubber. But rubber at that time became sticky in heat and rock-hard when frozen. Apart from syringes, it was used for galoshes (rubbers in North America), waterproof capes and toy animals. Demand was limited, and in 1848 rubber accounted for only

* The early botanical name for the genus was *Siphonia* (siphon or syringe), but this was changed to *Hevea* because the Tupi or lingua geral word for the gum is *hevé*. Spanish and French use *caucho* and *caoutchouc*, meaning 'tree that weeps' in the language of the Mainas people of the Pastaza on the upper Amazon. The English name 'rubber' – or 'India-rubber' because it came from American Indians – had been coined by Joseph Priestley, the discoverer of oxygen, simply because he saw the gum being sold in London to rub out pencil marks.

a quarter of Pará's meagre exports. Unknown to the English travellers, Europeans had been improving rubber production in various ways, and less than a decade earlier the Massachusetts inventor Charles Goodyear had accidently dropped rubber mixed with sulphur onto his hot stove. He called the process vulcanization, after the Roman god of fire. The result was rubber as we know it, durable, elastic, impermeable and unaffected by climate. So the great Amazon rubber boom was about to begin. Towards the end of that century the wealth it was to generate would transform this sleepy northern half of Brazil.

The voyage up the lower Tocantins was not just for Wallace and Bates to learn more about life outside the town of Belém nor to see the spectacular rapids: it was for them to collect specimens for sale. This went fairly well, despite being confined to the boat during the many hours of paddling and sailing. Wallace recalled shooting birds – waxbills, pigeons, toucans and chatterers. In his book twelve years later, Bates said that 'Mr. Wallace and I lost no chance of adding to our collections, so that... we got together a very considerable number of birds, insects, and shells....' But in a letter to Stevens at the time, he complained that butterflies were 'exceedingly scarce', as were shells and orchids, so that their collection from the five-week journey consisted almost exclusively of less showy insects.

Back in Belém at the end of September, they packed and dispatched an impressive haul of specimens (collected around the town as well as from the Tocantins) representing 400 species of butterfly, 450 species of beetle, as many species of other insects, and – as Stevens wrote in his advertisement – 'a vast number of novelties, a few shells and bird-skins.' The collection reached England in excellent condition, because the diligent young collectors had followed correct procedures in bagging their specimens, skinning, mounting and preserving them, and then packing them for the transatlantic voyage. They caught insects with butterfly nets and sugar-baited traps, but not with lighted moth traps, mist nets, or 'tree-fogging' with gas as used by modern collectors. They did not have killing-bottles using chloroform or cyanide: these were introduced only a couple of years before they left England (although such innovations may have been sent out later by Stevens). So they killed insects mostly by pinning them alive or crushing them. One of the Royal Geographical Society's most coveted possessions is a small purple pin-cushion that Bates had on his belt during those years in the Amazon. Birds could be taken only by shooting or, less damagingly, be pierced or stunned by the arrow of an Indian helper. To this day, indigenous hunters throughout Amazonia carry some arrows with blunt heads, so that they can bring down a bird without damaging the gaudy feathers they want for their headdresses. Wallace and Bates could afford only cheap shotguns. By 1848 these firearms would have used detonators rather than sparks from flintlocks, but they would still have been muzzle-loaders since breech-loading guns were only just arriving in England at that time, and ready-made cartridges came even later.

Wallace twice had mishaps with his gun during the Tocantins trip. Near Baião he went to visit a lake with their Indian Alexandre in a small two-man montaria. The lake was full of caiman (also spelled cayman, *jacaré* to Brazilians; and which the Englishmen wrongly called alligators, found only in North America). Alexandre shot one of these, of the smaller *jacaré-tinga* (*Caiman sclerops*) species that can grow to 4 metres (13 feet), as opposed to the larger black caiman (*Caiman niger*), which can reach over 6 metres (20 feet). Although menacing-looking, both species present almost no threat to humans. But Wallace got a fright when he went to gather the one shot by Alexandre: it had rolled over and looked quite dead, so he seized a leg sticking up from its grey belly. But 'dash! splash! – over he turned, and dived down under our little boat, which he half filled with water and nearly upset.' The caiman reappeared on the surface and they poked it with a stick 'to see if it was really dead or shamming'; but it was alive and swam off. A few minutes later, Wallace shot a kingfisher and was reloading his gun. Alexandre shot a coot, but the recoil threw Wallace off balance, so that he dropped his gun into the water and nearly swamped the montaria. 'I thought my shooting for this voyage was all over.' But luckily the lake was quite shallow, so they retrieved the gun and after hours of careful dismantling, drying and oiling, it worked again.

Days later, at Senhor Seixas's plantation on the return journey, Wallace wanted to shoot some birds. Standing on a jetty, he reached for his gun, which was lying loaded on top of their boat, and pulled it towards him by the muzzle. But the gun's hammer caught in a joint of the boat's boards, fell on the cap and fired the weapon. 'The charge carrying off a small piece of the under-side of my hand, near the wrist, and, passing under my arm within a few inches of my body, luckily missed a number of people who were behind me. I felt my hand violently blown away.' There was a stream of blood, which Wallace staunched with some cotton. Back in Belém, his hand was badly inflamed. A doctor put the arm in a sling and 'I remained a fortnight, unable to do anything, not even pin an insect, and consequently rather miserable.'

On a walk in the forest near the holiday campsite at Tocará they saw some of the most beautiful and rarest birds of Amazonia: hyacinthine macaws (at that time *Macrocercus hyacinthus*; but the scientific nomenclature of the macaw genus later changed to *Anodorhynchus* and is now *Ara* – after the onomatopoeic Tupi name that imitates the call of these magnificent birds). These were almost a metre (3 feet) long and 'entirely of a fine indigo-blue, with a whitish beak; but they flew very high and we could not find their feeding-place.' They did see one effortlessly crushing the very hard nuts of a mucajá palm (*Acrocomia aculeata*) in its powerful beak. Wallace wondered why this lovely bird was found only in that small area, whereas other macaws flew throughout Amazonia. The prevailing view, among scientists such as Swainson and Lyell, was that different species had been located in appropriate places by divine design. This would account for macaws thriving throughout the tropical rain forests of Amazonia – but not for the hyacinthine being in only one locality. Wallace, constantly interested in the origin of species, wondered why 'such a strong-flying bird' should have such a limited range. He guessed that it might have been because these macaws liked some food that grew only near the rocks of the rapids. In fact the hyacinth macaw has a fairly extensive range, but Wallace and Bates were at the northern edge of this. However, Wallace was precociously right to challenge the view that species were located by divine design.

As we have seen, both Wallace and Bates were exceptional students and auto-didacts. Helped by local guides, their few books, and their experience in England, they were constantly learning how to classify Brazilian fauna and flora – which was of course the basis of their new profession. They identified hundreds of specimens of insects, birds, animals and plants, giving both vernacular names and often botanical, zoological or entomological nomenclature. They learned Portuguese and some words of *lingua geral* – the Tupi-based language that the Jesuits had introduced throughout their realm, and which was still spoken by common people eighty years after the missionaries' expulsion. They were also observing and assimilating the customs of Amazonia, among all classes of inhabitant, and ranging from bureaucracy, farming and trade to travel and social mores.

In October 1848, little more than four months after their arrival in Brazil and shortly after the Tocantins journey, Wallace and Bates decided to separate. We do not know why they parted. Both were too discreet to discuss it in either their letters or in subsequent books. When Richard Spruce went out in the following year, Sir William Hooker of Kew asked how they were. Spruce replied: 'I forgot to mention that we have several times seen Mr. Wallace. He and Bates quarrelled and separated long ago.' Like a recently divorced couple, they found different houses in which to live, and they chose not to communicate when, sporadically during the coming months, they both happened to be in Belém. But it was not an acrimonious parting. They had happy rambles together when they were in Manaus at the same time in January 1850; Bates wrote to Stevens how much he missed having 'a sympathising companion'; Wallace was sorry when he just failed to meet 'Mr. Bates, whom I most wished to see' at Santarém in June 1852; and later in that decade, when Wallace was on the far side of the world in what is now Indonesia and Bates was far up the Amazon, the two men corresponded with one another.

There are various explanations for the split. Wallace's biographer Peter Raby felt that they may have had financial differences: neither had much money, so they were anxious about whether their joint collection would sell well, or undecided about how to divide its earnings. They may have felt that it was easier for one man and his equipment to get a passage on crowded river boats – although by October 1848 they had had little experience of Amazonian travel. Or it might have been temperamental. Bates was more gregarious, tolerant and relaxed; 'Wallace was more driven, impatient, competitive'. Bates's biographer Anthony Crawforth had a different explanation: that it is almost impossible for collectors to work in pairs. If one of them found a better specimen, it could arouse complicated emotions ranging from envy or even rage to a sense of inferiority. Crawforth quoted the novelist Vladimir Nabokov, who was also a passionate insect collector and who wrote that 'nothing could surpass in richness and strength the excitement of entomological exploration.... [But there was] an acute desire to be alone, since any companion, no matter how quiet, interfered with the concentrated enjoyment of my mania.' The true reason may have been a dispute over their modus operandi – whether to collect intensively from a few places, or to adventure deep into unknown forests and rivers in search of elusive prizes. In England, Bates had collected almost entirely around his native Leicester, whereas Wallace had for years roamed across-country with his brother in their land-surveying and job-seeking. So in Brazil during the coming years, Bates was destined to travel only up and down the main Amazon river. He settled for months or even years in two or three locations, and collected professionally, tirelessly and methodically in daily excursions from them. Wallace by contrast journeyed far up tributaries, sometimes beyond any previous non-Brazilian, and had many adventures in his searches for exotic species. He was an eager tourist with a delightfully enquiring mind.

It is indicative that Bates's only reference to the split in a letter to Stevens was the laconic: 'After returning from the Tocantins I remained here [in Belém] for two months, Wallace going to Marajó.' Wallace did not mention the separation at all. In his book he generally used the first person (as did Bates), but if he wrote 'we' it meant him and Bates prior to October 1848 and him and his brother Herbert after the latter arrived in mid-1849. Bates probably felt that the Tocantins trip had been an enjoyable tourist jaunt that yielded too few saleable specimens. Whereas Wallace thirsted for more adventure.

4

THE MOUTH OF THE AMAZON

After Wallace and Bates had decided to go their separate ways, Wallace made the curious decision to visit a cattle ranch belonging to one Archibald Campbell on Marajó – a flat island the size of Switzerland that sits in the mouth of the Amazon like a stopper in a bottle. There was a delay waiting for boat for the voyage; and for two weeks he had his arm in a sling from his gun accident. So he went to his French friend Monsieur Borlaz, the Swiss consul in Belém, 'who kindly offered me a room and a place at his table' at his country house Olaria ('Brickworks').

Wallace used his month at Olaria for bird watching and collecting. There were plenty of small antshrikes (*Sakesphorus luctuosus*). These had loose, long, silky feathers, with black and white bands or spots. They hopped about in the thickest bushes in search of insects, so Wallace could not see them until he had crept to within a couple of yards, by which time they were 'difficult to shoot without blowing them to pieces.' Closely related were ant-eating thrushes (*Turdidae*) whose stronger legs and short tails enabled them to walk in search of their formic prey. But 'when one is shot, it is often dangerous... to go and fetch it, for the ground generally swarms with ants, which attack an intruder most unmercifully both with stings and jaws.' Wallace often had to abandon his shot specimen and 'beat an inglorious retreat'.

Once again, the young naturalist queried an accepted theory – that each species was 'constructed and adorned' to get its favourite food in a given environment. He was observing utterly different birds seeking the same prey in the same locality. 'Thus the goatsuckers, the swallows, the tyrant flycatchers, and the jacamars, all use the same kind of food, and... all capture insects on the wing, yet how entirely different is the structure and whole appearance of these birds!' Swallows had strong wings and flew constantly; similar goatsuckers (nightjars) were weaker birds, but semi-nocturnal with appropriate eyesight, and they caught insects by short flights from the ground; flycatchers had strong legs but short wings, so swooped from perches in bare trees and caught insects in gaping broad bills; jacamars dived in the same way, but had long pointed bills almost like a kingfisher's; trogons hunted similarly, but their bills were strong and serrated; tiny hummingbirds caught insects either on flowers or in the air. Wading birds were equally diverse. 'What birds can have their bills more peculiarly [differently] formed than the ibis, the spoonbill, and the heron? Yet they may be seen side by side, picking up the same food from the shallow water on the beach; and on opening their stomachs, we find the same little crustacea and shell-fish in them all.' A similar observation applied to fruit-eating birds like pigeons, parrots, toucans and chatterers. Yet these very different families could be seen feeding together on fruit from the same tree.

Wallace's boat finally left on 3 November 1848. It was a cattle transport with only a rudimentary cabin whose two berths were far too short for the tall Englishman; so he slept in the smelly hold. Also, the boat sailed around the Atlantic Ocean side of Marajó to reach the smaller island Mexiana to the north, where Campbell had a ranch. Wallace was seasick throughout the four-day voyage. He made the most of seven weeks on Mexiana, but it was not ideal for a collector – Bates was right not to have come. The island, 40 kilometres (25 miles) long, was savannah with occasional clumps of woodland and forest along the banks of rivers. There were virtually no insects, so Wallace collected (shot) only birds – which were plentiful but 'not very rare or handsome'. He bagged hummingbirds (whose Portuguese name is the delightful *beijaflor*, 'flower-kisser'); Great Kiskadee flycatchers (*Pitangus sulphuratus*), a brown, yellow-breasted little bird whose mocking call sounds like *bem-te-ví* ('I saw you clearly') or in Spanish *Christo-fué* ('Christ lived'); cuckoos, including one whose croak sounded like a rusty hinge; black hornbill anis (*anú* in Portuguese; *Crotophaga major*), which ranchers liked because they picked ticks off cattle; tanagers (*Tangara*), a genus of small birds with brilliant lime-green or turquoise-blue plumage; a red-breasted oriole (*Icterus militaris*); and a large yellow-billed toucan (*Ramphastos toco*). He collected, in all, seventy specimens, including hawks, eagles, herons, egrets, parakeets, woodpeckers and buzzards.

On the flat and seasonally flooded island of Mexiana, Wallace saw other aspects of Amazonian life. There were only forty people living there. Half of them were black slaves who answered to Campbell's German overseer called Leonardo. They tended 1,500 cattle and 400 horses. Vampire bats were a problem, attacking the cattle; so periodical onslaughts on these blood-sucking predators had destroyed 7,000 of them during the past six months.

Wallace's attitude to slaves is interesting. One would have expected the ardent follower of Owen's socialism to deplore servitude, but he did not. Slavery of *indigenous* people in Brazil had been illegal for a century (although various forms of forced labour persisted; free Indians saw no incentive in working for derisory pay, and their numbers were reduced by disease and the ravages of the Cabanagem revolt). But slavery of *blacks* was allowed and even encouraged as an alternative to Indian slaves. We have to keep remembering that there was virtually no machinery, and scant pasture for draft animals, so that everything depended on human effort. However, relatively few African slaves reached the Amazonian part of Brazil. They were too expensive for all but the richest planters in Amazonia, all the more so in 1848 because the British Royal Navy was trying to stamp out the transatlantic traffic.*

Wallace, perhaps to his credit, reported what he saw on Mexiana without allowing his liberal conscience to colour his view. 'The slaves appeared contented and happy, as slaves generally do.' They came every morning and evening to give a formal salutation to the two white men, Leonardo and Wallace. In the evenings, they would sing improvised songs about the day's events, to the monotonous music of home-made guitars from which they got three or four notes. These extempore ballads seemed to Wallace similar to the lays of ancient bards – they made 'well-known facts interesting by being sung to music in an appropriate and enthusiastic manner.' Deeply religious, the slaves met for divine service on every Saturday evening, with the two oldest men officiating since there was of course no priest. They had 'a room fitted up as a chapel, with an altar gaily decorated with figures of the Virgin and Child and several saints' – all carved by the German Leonardo and 'painted and gilt in a most brilliant manner... and when the candles are lit... the effect is equal to that of many village churches.' Although these black slaves worked a six-day week without pay, they tended their own gardens on Sundays or could earn a little by hunting jaguars for their skins.

One day, Wallace went with three black crewmen in a small boat to the far end of the island. After a stormy sail in open water, they turned into a small meandering stream. The paddlers worked furiously to benefit from an incoming tide. Their passenger 'was much delighted with the beauty of the vegetation, which surpassed anything I had seen before: at every bend of the stream some new object presented itself – now a huge cedar hanging over the water, or a great silk-cotton tree standing like a giant above the rest of the forest. The graceful [açaí] palms occurred continually, ...sometimes raising their stems a hundred feet

* Great Britain had in 1807 passed the Slave Trade Act and established a squadron of the Royal Navy to intercept slave ships crossing the Atlantic Ocean with their tragic human cargo. In 1833 Britain abolished slavery altogether, throughout its empire. Brazil did eventually stop the slave trade, but the last individual slaves were not freed until 1889.

[30 metres] into the air, or bending in graceful curves till they almost met from the opposite banks. The majestic murutí [miriti or buriti] palm was also abundant, in straight and cylindrical stems like Grecian columns, and with its immense fan-shaped leaves and gigantic bunches of fruit, produced an imposing spectacle.... These palms were often clothed with creepers, which ran up to the summits and there put forth their blossoms. Lower down, on the water's edge, were numerous flowering shrubs, often completely covered with convolvuluses, passion-flowers, or bignonias. Every dead or half-rotten tree was clothed with parasites of singular forms, or bearing beautiful flowers, with smaller palms, curiously-shaped stems, and twisting climbers, formed a background in the interior of the forest.'

A tough walk over bare savannah brought them to lakes that teemed with caiman and fish. The slaves harpooned and lassoed a dozen great black caiman, hauled them out of the water, cautiously approached each powerful thrashing saurian, used an axe to 'cut a deep gash across the root of the tail, rendering that formidable weapon useless', and then smashed its neck to disable its mighty jaws. About eighty smaller caiman, *jacaré-tinga*, were also killed. The large black caiman yielded quantities of fat, used for lighting lamps, and the smaller ones were for eating. These lakes were also full of the delicious Amazonian fish pirarucu (*Arapaima gigas*). Wallace admired this 'splendid species, five or six feet [1.5 to 1.8 metres] long, with large scales of more than an inch [2.5 centimetres] in diameter, and beautifully marked and spotted with red.... It is a very fine-flavoured fish.' Most pirarucu were salted and dried for the market in Belém. But the fishing party ate the fish bellies, which were too fat and rich to be cured. 'This, with [manioc] farinha and some coffee, made us an excellent supper, and the [caiman's] tail, which I now tasted for the first time, was by no means to be despised.'

In January 1849 Archibald Campbell arrived with his family and friends to spend a few weeks at his ranch on Mexiana. When he left to visit his other ranch on Marajó, Wallace seized the opportunity to sail back with him. His week there was pure sightseeing. Grasslands stretched as far as the eye could see and were home to great herds of cattle. (Marajó had by then been cattle country for two centuries. The Jesuits had 'pacified' its Arawak-speaking Indians and then developed vast ranches, before their missionary order was expelled from the Portuguese empire in 1760.) Wallace admired the fine young black and mulatto cowboys, who 'led a life of alternate idleness and excitement, which they seemed to enjoy very much.' They were constantly riding, wearing only trousers and a tasselled cap, and 'displaying the fine symmetry of their bodies'. They expertly herded cattle into corrals, lassoed the ones to be slaughtered, which they hamstrung, killed with a machete thrust to the heart, and skinned and cut up, amid a crowd of dogs and vultures feasting upon the blood and entrails. 'The sight was a sickening one, and I did not care to witness it more than once.' When the time came to load a boat with wild cattle who were kicking, plunging and goring, there was a scene with all the bravery and skill of a bullfight. Each furious animal was lassoed, subdued, dragged into the water, swum towards the boat, and lifted into the air by a gantry with a hide rope around its horns, 'struggling as helplessly as a kitten held by the skin of its neck'. The animals were then lowered into the hold where, after a little disturbance, they calmed down. The cattle ranch was on the north shore of Marajó. The return voyage took a week, with various stops, and having to sail out into the Atlantic around the eastern shore of the huge island before turning up the Pará river to Belém.

While Wallace was on Mexiana and Marajó, Bates asked Archibald Campbell if he could go to collect at another of the Scot's properties, which he had acquired by marrying the daughter of a Brazilian landowner. This was a run-down *fazenda* (farm) called Caripí at Carnapijó on the shore of the vast expanse of the Pará estuary. It was reached by a voyage of some 37 kilometres (23 miles) around islands west of Belém. Bates had heard that most English and American visitors (including the author William Edwards) had visited Caripí, which 'had obtained quite a reputation for the number and beauty of the birds and insects found there'. Campbell readily agreed. So Bates bargained with a 'villainous Portuguese' for a passage on his 30-ton but deckless trading vessel, that was going past this coast. The boat's passenger list gives another glimpse of life in the region. There were three young mameluco apprentice tailors going on holiday, 'pleasant, gentle fellows' who were literate and spent the time reading a book about the geography of foreign lands, a heavily shackled runaway slave, and Bates. The crew was the skipper and his pretty mulato (white/black) mistress, a pilot and five Indian sailors. Bates was frightened about landing when the boat reached Caripí, because he had heard that an English naturalist called Graham had drowned there with his wife and child when his montaria sank in the heavy surf. He was alarmed to see that he and all his luggage had to make the landing in one trip in such a leaky little boat. 'The pile of chests with two Indians and myself sank the montaria almost to the level of the water. I was kept busy baling all the way.' But, once again, the skill of the Indian boatmen saved the day. 'They preserve the nicest equilibrium, and paddle so gently that not the slightest oscillation is perceptible.'

Caripí was a large, well-built, red-tiled bungalow surrounded by forest, on a charming little bay with a sandy beach. It had belonged to the Jesuits, expelled some ninety years earlier; so it was now abandoned and dilapidated. Campbell kept it only as a poultry farm and a hospital for sick slaves. Bates was welcomed by the manageress, an old black woman called Florinda, who 'gave me the keys, and I forthwith took possession of the rooms I required'. He was to stay at Caripí for nine weeks until mid-February 1849. It was a good decision, because he achieved plenty of serious collecting. 'I led here a solitary but not unpleasant life; for there was great charm in the loneliness of the place. The swell of the river beating on the sloping beach caused an unceasing murmur, which lulled me to sleep at night, and seemed appropriate music in those midday hours when all nature was pausing breathless under the rays of a vertical sun.'

On the day after his arrival, Bates had a stroke of luck. Two blue-eyed, red-haired boys, who spoke English, appeared at the house, and Bates soon met their father. He was a German called Petzell who had served for thirteen years in the newly independent Brazilian army, then went to the United States for seven years farming near St Louis, Missouri, where he married and had several children. But he hankered after Pará, and brought his young family there after a contorted and arduous journey. Petzell loved the pioneering life in a log cabin he had built in the Amazon forest, and admired the tapuio Indians living nearby, but his wife and sons missed the bread and other comforts of North America.

The good fortune for Bates was that Petzell and his family proved to be expert insect collectors, so he promptly employed them. He wrote to Stevens that the German was 'a capital companion for me, as he collected beetles'. Bates's daily routine was to rise at dawn (which on the equator is always at six o'clock), take a cup of coffee, sally forth after birds, breakfast at ten, then spend five hours until three in the afternoon catching insects. He sometimes went on day-long rambles with Petzell, and the settler's sons brought him 'all the quadrupeds, birds, reptiles and shells they met with' as well, of course, as insects.

With Petzell 'we took a great many beetles, ...about a hundred species of Longicornes, ten [species of] Cicindelas, two Megacephala, large Brachini, and many curious genera – such as Ctenostoma, Agra, Brenthus, Ingá, &c.' Bates was never happier than when collecting beetles and butterflies. He described them lovingly to his agent Stevens. The longicorns (of the Cerambycidae family) were 'very graceful insects, having slender bodies and long antennae, often ornamented with fringes and tufts of hair'. Common dung-beetles 'were of colossal size and beautiful colours. One kind had a long spear-shaped horn projecting from the crown of its head (Phanaeus lancifer [now *Coprophanaeus lancifer*]). A blow from this fellow, as he came flying heavily along, was never very pleasant.' [PL. 26]

Bates was constantly learning about this new world. One revelation was bats. Woken in the middle of the night by a rushing noise, he found 'vast hosts of bats sweeping about the room. The air was alive with them; they had put out the lamp, and when I relighted it the place appeared blackened with the impish multitudes that were whirling round and round.' He tried to destroy these tiny mammals by flailing at them with a stick. This seemed to cause them to retreat to their perches in the rafters, so Bates went back to sleep. But on the following night several got into his hammock. 'I seized them as they were crawling over me, and dashed them against the wall.' Ever the scientist, he managed to identify and describe four species of the little creatures as they fluttered at speed in the darkness: two were of the *Nyctinomops* genus, one a *Phyllostoma*, and one a *Glossophaga*. (There are about a hundred species of bat in Brazil – far more than the number found in Europe. Most are insectivores (farmer's friends), some fruit-eaters (to the distress of fruit growers), some large bats that skim rivers to catch fish, and a few vampires hated by ranchers and feared by all people.) One morning Bates found a wound on his hip, clearly caused by a vampire bat. 'This was rather unpleasant.' So he declared war and tried to exterminate all of them. He himself shot many as they hung from the rafters. And he got the black slaves to mount ladders to the roof outside, to rout out many hundreds of them, including young broods, from the eaves. These attacks on bats are shocking to a modern reader. But Bates's behaviour can be condoned in various ways: he was still very new to Amazonia, so did not appreciate how much good was done by bats, nor that it was futile to try to exterminate anything in that exuberant ecosystem; the concept of conservation of nature was hardly known to Europeans at that time – to most, nature was red in tooth and claw and the enemy of mankind; and Bates shared Victorians' fear of bats, identifying them with dark forces and unaware that their sonar is so acute that they never collide with human hair. Also, the bats crawling on his body were certainly vampires, so he can be forgiven for hurling these against the wall.

Another lesson that Bates learned was that a vegetarian diet was inadequate. After a few weeks he had exhausted the provisions he brought from Belém and had eaten all the chickens that the local people could spare to sell him. He wrote to his agent Stevens: 'I lived on salt fish and mandioca [manioc] root nearly two months, having mandioca and fish for breakfast, and, to vary the thing, fish and mandioca for dinner.' He felt weak and craved meat. Old Florinda asked whether he would eat *tamanduá bandeira*, giant anteater (*Myrmecophaga tridactyla*). Local people considered these unfit to eat, but Bates had heard that they were consumed elsewhere in South America. So he told Florinda that any flesh would be welcome. She got an old black man called Antonio to use his dogs to hunt an anteater, and Bates found its stewed meat to be 'very good, something like goose in flavour'. During the following weeks, Bates would send Antonio and his dogs out to get him another anteater, for a small reward. It is a lovely animal whose head and body measure 1.30 metres (4 feet 3 inches) and its flamboyant bushy tail (favoured as a bonnet by Indian chiefs) is as long again. But one evening Antonio ran in with the terrible news that his favourite dog had been caught and killed by a tamanduá. They hurried to the scene and found both animals wounded but still alive: the dog had been severely torn by the anteater's long claws (used for demolishing termite mounds) but it survived. Bates drew a vivid sketch of the two animals grappling in mortal combat. He also described four species of anteater: the giant that ambles rapidly across savannah and destroys termite mounds; and three smaller and rarer arboreal species that attack termites in their tunnels and nests on tree trunks. An Indian brought Bates one of the latter, which he kept as a pet for a day. It clung motionless to the back of a chair, but when irritated reared up from the chair and clawed out with its forepaws like a cat. Bates finally released it onto a tree. He evidently felt that English collectors would not be interested in a stuffed specimen.

There were small farms scattered in the woods near Caripí, and each had a few fruit trees. Like many Amazonian trees, these could flower at any time of the year, and they attracted masses of beautiful hummingbirds. Scores of these whirred about the blossoms, in the cooler hours of morning and evening. Bates was fascinated by their unusual motions, darting from tree to tree so swiftly that his eye could scarcely follow them, hovering before each blossom for a moment with 'their wings moving with inconceivable rapidity'. They seemed to skip about 'in the most capricious way' – unlike methodical bees. Bates delighted in the glorious colours of the different hummingbirds, so many of them with gleaming iridescent feathers. But he knew that there were fewer species in these lowland forests than on the flanks of the Andes mountains. On several occasions he shot a hummingbird hawk-moth (*Aellopos titan*) instead of a real hummingbird. The two looked uncannily alike, in size, shape of both head and wings, hovering flight, and even the way the moth probed a flower with its proboscis. It took the English naturalist many days to tell them apart. Even then, the local Indians and blacks tried to convince him that they were the same species and that the hawk-moth metamorphosed into the bird. '"Look at their feathers," they said; "their eyes are the same, and so are their tails." This belief is so deeply rooted that it was useless to reason with them on the subject.' Another of Bates's charming drawings was of the almost-identical bird and moth probing the same blossoms [PL. 24]. This heightened his interest in mimicry and camouflage – an interest that would lead to his greatest scientific legacy.

Snakes were very numerous at Caripí, with harmless ones often sliding into the rooms of the house. Out in the forest 'it was rather alarming, when entomologizing about the trunks of trees, to suddenly encounter... a pair of glittering eyes and a forked tongue within a few inches of one's head.' One of the most beautiful was the coral snake (*Micrurus* sp.), with its bands of black and vermilion separated by white rings. He did not seem to appreciate that corals are one of the most dangerous of Amazonian snakes. Although not aggressive, when they do bite their venom can be lethal and there is no serum antidote for it. Bates's collection included a few fine reptiles, but he noted that their brilliant colours faded when preserved in spirits.

In his 'solitary but not unpleasant' months at Caripí, Bates got to know local people. On one of his long walks with Petzell, he went inland through several miles of dense and gloomy virgin forest to a creek called Murucupi. This secluded area had for many generations been occupied by Indians and mestizos. Their 'open palm-thatched huts peep forth here and there' from groves of fruit trees and palms, with 'glorious vegetation piled up to an immense height' along the banks of the stream. One house was superior to the others, having whitewashed mud walls and a red-tiled roof. It was full of children 'and the aspect of the house was improved by a number of good-looking mameluco women, who were busily employed washing, spinning, and making farinha.' Two were on the verandah, sewing dresses for a forthcoming local festival. The Europeans were immediately welcomed and invited to stay for dinner. 'On our accepting the invitation a couple of fowls were killed, and a wholesome stew of seasoned rice and fowls soon put in preparation.' Bates, deprived of female company, commented wistfully that 'it is not often that the female members of a family in these retired places are familiar with strangers'. These ladies proved to be the widow and daughters of a prosperous tradesman from Belém. He had given his girls the best education he could afford, and after his death all of them, married and unmarried, had retired to his country sitio. After dinner a handsome mulatto son-in-law 'sang us some pretty songs, accompanying himself on a guitar'.

During his nine weeks at Caripí, Bates delighted in the hospitality and 'kindness of the neighbouring Indians'. One particular friend was a dignified Indian called Raimundo. Bates described him as hard-working in boat building, in which he employed two young apprentices, but poor despite his industriousness. People around Caripí subsisted, growing their own manioc, corn, cotton, coffee and sugar cane, selling any surplus 30 kilometres (20 miles) away in Belém, and with no rent or taxes to pay.

Raimundo was a legendary hunter. He kept his best hunting grounds secret, but relented when Bates begged to accompany him on a hunt. In order to start early, Bates spent the night with Raimundo and his very talkative wife. Their house was a large open shed filled with hammocks. This happened to be the holy day of São Thomé, the patron saint of Indians and mamelucos, so a procession of villagers arrived requesting alms. 'Senhor Raimundo received them with the quiet politeness which comes so natural to the Indian when occupying the position of host.' All the villagers stayed for dinner and the night. 'The fare was very scanty; a boiled fowl with rice, a slice of roasted pirarucu, farinha and bananas. Each one partook very sparingly, some of the young men contenting themselves with a plateful of rice.' In a later conversation, Raimundo impressed the young Englishman with his shrewd good nature. He reminisced about 'the Cabanagem, as the revolutionary days of 1835–1836 are popularly called', when he had wrongly been accused of being one of the Cabano rebels. He said that the Indians 'meant well to the whites, and only begged to be let alone'. It was wrong that tracts of forest were awarded to whites who had no intention or prospect of cultivating them – he himself had been evicted from his home on several occasions.

The hunting journey started in the middle of the night. After a few hours 'threading noiselessly by moonlight through winding narrow creeks, with trunks of monstrous trees slanting over, and the broad leaves of the arborescent Arums in the swamps gleaming in the moonlight', they snatched some sleep in the hard canoe. Then came the brief but dazzling Amazonian dawn. 'The change was rapid: the sky in the east suddenly assumed the loveliest azure colour, across which streaks of thin white clouds were painted. It is at such moments as this when one feels how beautiful our earth truly is!' The hunting party was Raimundo, his apprentice, Bates, and five hunting dogs. They paddled up and down a maze of creeks and channels, sometimes against the tide, sometimes helped by it. Their quarry was two species of semi-aquatic rodents: paca (*Agouti paca*), reddish with white spots, which Bates described as the size of a spaniel and 'in appearance between a hog and a hare'; and much smaller agouti (or cutia; *Dasyprocta* sp.) [PL. 72]. These rodents emerge in the early morning to feed on fallen fruit. When observed they dive into their burrows, from which the dogs extract them – then the hunters have to shoot them quickly before they disappear in the stream. They bagged half a dozen paca and one cutia. Bates then watched admiringly as Raimundo lit a fire with shavings from a bacaba palm leaf, an old file and flint, and tinder of a soft felt-like substance made by ants. The agouti was singed, prepared and roasted – with garnish of a lemon, a dozen fiery red peppers, salt and manioc. 'We breakfasted heartily when our cutia was roasted, and washed the meal down with a calabash full of pure water of the river.'

Both Wallace and Bates were unstuffy about joining into every level of Amazonian society, constantly curious and learning, admiring much that they observed, and not at all squeamish about what they ate. But they were also candid about occasionally being terrified. For the return from the hunting expedition Raimundo cut poles to form a mast and bowsprit to their canoe, rigged a sail he had brought with him, and sailed for 11 kilometres (7 miles) along the open water of the Pará estuary. The small and leaky boat was heavily laden with the three men, their catch, and the five dogs – who howled with fear in the bow and occasionally fell overboard 'causing great commotion in scrambling in again'. There was no rudder, so Raimundo and his apprentice steered with paddles, while Bates baled out water and watched the dogs. Bates thought it foolhardy in the extreme. The waves ran very high. But the old Indian's nerve and skill saved them from 'falling into the trough of the sea and being instantly swamped'. Rounding a turbulent rocky headland, 'Raimundo sat in the stern, rigid and silent; his eye steadily watching the prow of the boat. It was almost worth the risk and discomfort of the passage to witness the seamanlike ability displayed by Indians on the water.'

Not surprisingly, from encounters such as this, Bates liked and admired Indians. Back in Belém, he wrote to Stevens: 'I get on very well with the Indians, being far more at home and friendly with them than with the Brazilian and European residents. The English people here, you will be sorry to hear, have not shown a disposition to assist us in the least all along....' Of course, two young, impoverished natural history collectors who dressed as casually as caboclos had nothing to offer to the town's merchant class, who were interested only in trade, money and social status. Bates found lodgings with a family of Portuguese (themselves somewhat ostracized by both Brazilians and other foreigners), who treated him kindly and helped him procure the little things he needed. So he continued his regime of constant collecting, but now back on familiar trails near the town.

Bates returned to Belém in mid-February 1849 and Wallace finished his five-month visit to the islands at the beginning of April. Wallace had written to the shipping agent Miller, asking for his help in finding accommodation. This proved to be a little house in Nazaré, the village at the edge of Belém where they had stayed on arrival a year earlier. Wallace was pleased that he could manage without a servant because a Portuguese who had a 'sort of tavern' next door supplied his meals. Collecting was more important. The boys in the neighbourhood learned that the foreigner would buy *bichos* ('beasties'), so they brought a steady supply – particularly snakes, which Wallace preserved in spirits. Even better was finding Luiz, an elderly freed slave who was an expert hunter. The great Austrian naturalist Johann Natterer had purchased the Congo-born Luiz in Rio de Janeiro when he was a boy. For seventeen years the young slave travelled all over Brazil with 'the doctor', shooting and skinning animals and birds. When Natterer returned to Europe in 1835 he freed Luiz, who then saved enough to buy a plot of land and two slaves of his own. Wallace paid him a modest retainer and his living expenses. 'He would wander in the woods from morning to night, going a great distance, and generally bringing home some handsome bird. He soon got me several fine cardinal chatterers, red-breasted trogons, toucans, etc. He knew the haunts and habits of almost every bird, and could imitate their several notes so as to call them.' In the evenings he greatly entertained Wallace with his stories about Natterer, whom he had liked and admired.

It was strange that Wallace and Bates were both in Belém for a couple of weeks in April–May and again in early August 1849, but they apparently did not bother to meet – even though Wallace complained in his book of the monotony of life there, and we have seen how Bates was upset that the English community was not interested in either of them.

Each of the Englishmen made a second river excursion on which he witnessed another aspect of life in Brazilian Amazonia. Ever the adventurer and sightseer, Wallace went in May to witness the *pororoca*, a bore that roars up the Amazon estuary and its tributaries at spring tides. He had heard that there was an impressive pororoca on the Guamá, the river that flows past Belém just before it enters the broad Pará. For Bates, his excursion was in June, a return voyage to Cametá, the old town on the lower Tocantins.

Wallace decided that this time he needed his own boat, so he bought a small one from a Frenchman. He enlisted two motley crewmen: a lame Spaniard who said that he knew the Guamá, and a mestizo boy. The Spaniard begged an advance payment to buy clothes, but blew it on cachaça. He was too drunk to help load the boat, but became quiet and submissive when sober. The boy looked Indian but had a partly black mother, so 'of course shared her fate' of being a slave: Wallace hired him from his master, an army officer – who resembled him and was probably his father. The boy normally carried a large chain as punishment for trying to run away, which he wore concealed under his trousers but which clanked disagreeably with every step he took. This shackle was removed when he was delivered to Wallace, on his promise to be faithful and industrious. Wallace's travelling companion was his admirable collector, the freed slave Luiz. Wallace organized stores for the trip, which included a barrel and spirits for preserving fish, and collecting equipment for birds and insects.

The party paddled gently up the Guamá, southeastwards from Belém, largely propelled by each flood tide and tying up by some riverside farm or cottage during the ebb. Wallace and Luiz would go ashore to collect. The pororoca bore hit them about 50 kilometres (30 miles) upriver. It was a sudden rush of water that foamed all along the banks and lifted the boat like a great rolling ocean wave. This was all over in an instant, and was followed by a very fast current. The pororoca caused much damage to riverbanks, uprooting trees and smashing boats, but craft in mid-river were safe. Wallace characteristically tried to guess what formed this bore. He drew diagrams of how he imagined its being caused by an outcrop on the riverbed – wrongly, as is now known, since it is just a wave at the leading edge of a powerful incoming tide pushing against a river current. (Modern surfers come to this stretch of the Guamá to compete in riding the pororoca upriver.)

Further upstream, they spent a week of indifferent collecting at the village of São Domingos do Capim, eighty kilometres south-east of Belém, and then moved south up the Capim river which joins the Guamá there. All these rivers were 'prettily diversified' with sugar and rice plantations. These had solid colonial houses, each with chapels and slave quarters, 'all much superior in appearance and taste to anything erected now'.

Three days up the Capim, they reached their goal: São Jozé, the estate of a Senhor Calistro to whom Wallace had a letter of introduction. Calistro was 'a stout, good-humoured looking man' little older than the English naturalist. When he heard the purpose of Wallace's visit, he warmly invited him to stay for as long as he wished and offered every assistance. He proved to be an exceptionally generous host, helping Wallace with every aspect of his collecting, offering boats, trackers, paddlers and ample provisions. He even altered his own mealtimes to fit with the naturalist's excursions.

Once again, it is fascinating to see the Owenite and future socialist Alfred Russel Wallace react to a fully functioning slave fazenda. He was again favourably impressed. The house, rice-mill, warehouses and riverbank quay were among 'the best modern buildings I have seen' in Brazil, entirely made of stone – a rare commodity whose handling must have involved the slaves in much hard labour. Calistro had about fifty slaves of all ages and as many Indians, all of whom worked on his extensive rice and sugar fields and on his boats to take the produce to market. The sugar was distilled into profitable cachaça rum. The many workers included skilled smiths in various metals, carpenters, boat-builders, masons, shoemakers and tailors. So the estate was self-sufficient.

Calistro explained that he got more work from his Indians by having them labour alongside black slaves, because the latter had to work regular long hours and submit to regulations. Wallace admitted that 'the slaves here were treated remarkably well'. They worked a six-day week, with extra holidays on the main saints' days, on which they were rewarded with rum and a butchered ox. Each of them came at night to get the blessing of Calistro, seated in an easy chair on his verandah. In the evening they could ask their owner for tiny favours, which were granted. So they 'seemed to regard him in quite a patriarchal view, [but] at the same time that he was not to be trifled with, and was pretty severe against absolute idleness.' Any who often failed to bring in their quota of rice or cane 'were punished with a mild flogging', which intensified with persistent slacking. Calistro treated his slaves as if they were a large family of children, giving them amusement, relaxation (from grinding toil) and punishment, and providing for their needs and subsistence even when old or sick. Thus, Wallace felt that this fazenda was slavery 'seen under its most favourable aspect'. In purely physical terms these slaves could be better off than many free men. But the young Englishman still pondered the morality of slavery. Even at its best, as here, was it right to keep fellow human beings in 'a state of adult infancy – of unthinking childhood?' Wallace concluded that it was wrong, because it deprived men of the cares, challenges, aspirations, rights, duties, private property, education and intellectual pleasures available to ordinary citizens. He knew that such views were too refined for ordinary Brazilian slave-holders, who cared for their slaves' physical wants only to get the most work out of them.

Wallace's collecting at the estate on the Capim was quite rewarding. He was lent men to help him net fish or shoot birds near the fazenda. He then accompanied a few Indian hunters for several days up the Capim and its meandering creeks to enter 'wild, unbroken and uninhabited virgin forest'. Lying in his hammock in a forest clearing, after an excellent supper and coffee (which he adored), Wallace gazed up through the canopy to a sky full of bright stars and equally bright *Pyrophorus* fireflies. He caught a firefly and found that he could read a newspaper by moving it along the lines. The Indians amused themselves, as they always do, by recounting hunting adventures. Later days and nights were less agreeable, since it rained hard. One night Wallace had to sleep in drenched clothes, rolled uncomfortably into a ball in a small boat. But in the morning the rain stopped 'and a cup of hot coffee set me right'. It was a virtue of both Wallace and Bates that they made light of the occasional discomforts of life in a rain forest. Neither ever exaggerated any danger in these forests, in the way that some later 'green hell' travel-writers and adventurers did. But Wallace's week with the hunting party was disappointing from a collecting point of view. He failed to get his two goals: a large tinamou (*Tinamus major*) – a tasty partridge-like game bird – or the rare hyacinthine macaw, which they had also sought in vain on the Tocantins.

At the end of June 1849 Wallace sailed and paddled swiftly back to Belém, after seven agreeable but not particularly fruitful weeks up the Guamá and Capim rivers. He was unhappy about the unstable boat, so returned it to its French owner, but 'after much trouble and annoyance' failed to recover the £10 deposit he had paid on it. This was a serious blow to a man who had set out for Amazonia with capital of only £100. He returned to live in the small house in Nazaré.

Bates, meanwhile, had spent a further three months since his return from Caripí in intensive collecting around Belém. He wrote to his brother Frederick: 'Of course I lead rather a solitary life, only occasionally visiting English friends in the evenings; but I am always busy collecting, setting, making notes, &c.... The wet season has now set in: it rains about 1½ inches [almost 4 centimetres] per day, and the water rolls down the streets in torrents. It is finest before breakfast, and I walk for pleasure most mornings in my old haunts.' Bates welcomed the rain in one sense, because it brought out masses of butterflies 'in fine plumage'. He collected plenty of mechanites 'sailing out in their liveries of velvety black and red, with spots of bright yellow', different species of *Phoebis* and rhodocerae, and *Pieris monuste*. As always in Brazilian forests, the showiest were 'the glorious Morpho Menelaus, blazing about, flapping its huge wings of dazzling azure'. The rain also brought more flowers and made the verdure fresher. As a professional collector, Bates realized that he did not necessarily need to explore remote and exciting places. Within ten minutes' walk of his house was a splendid forest of mature trees, with the feathery heads of palms peeping out from the canopy – açaí (*Euterpe oleracea*), jupatí (*Raphia taedigera*), miriti (or buriti) (*Mauritia flexuosa*), murumuru (*Astrocaryum murumuru*) and urucuri (*Attalea phalerata*).

There were flocks of monkeys: capuchins (*Sapajus*) and coatás (spider monkeys; *Ateles*) 'playing "follow the leader" making prodigious leaps one after the other'. There was even the little midas tamarin or marmoset, called locally *sagüi*. Bates once saw these rare creatures running along a branch like three white kittens. They were sought-after as pets. He met a woman with one carried constantly in her bosom and fed from her mouth: 'a most timid and sensitive little thing' just 18 centimetres (7 inches) long and covered in long white silky hair, with a blackish tail and bare face.

Bates proudly dispatched the product of all this collecting in a huge consignment to his agent Stevens. He watched the great ocean-going ship *George Glen* sail from Belém on 8 June 1849 with his collections and letters, and wrote wistfully that the big ship was returning to his home, 'the land of civilization, cheerfulness, and activity'.

On that same day Bates himself embarked on his return visit to Cametá on the Tocantins, 'farther into this land of ignorance and barbarism'. His twinge of self-pity was misplaced, for he hugely enjoyed the outing to Cametá. The three-day voyage was a delight: 'the crew was cheerful, the weather fine, and the boat a good sailer'. As always, Bates got on well with ordinary Brazilians. Crossing the open water where the Tocantins joins the Pará, they were caught with all sail set by a severe squall that nearly brought them onto their 'beam ends'. That night they paused for the tide to turn, 'under the bank of a lovely island, clothed with forest'. It was the most beautiful starlit night that Bates had ever experienced. It was also great fun, because the boat's pilot, a handsome young mameluco called João Mendes, entertained them singing to his wire-stringed guitar. All the men sang, but Mendes was 'a capital hand at extemporaneous versifying'. After each lusty chorus, he sang a humorous verse about the journey and the people on board. This 'would sometimes set all the crew laughing, the musician himself letting fall his instrument and shrieking with delight.' One verse was about Bates – how he had come all the way from England to skin monkeys and birds and to catch insects – 'the last-mentioned employment of course giving ample scope for fun.' They sailed off, with Bates sleeping on deck wrapped in an old sail. But Mendes woke him in the early hours just 'to enjoy the sight of the little schooner tearing through the waves before a spanking breeze... with booms bent and sails stretched to the utmost.' It was a clear, cold, moonlit night. The crew were brewing herbal maté tea, and the flames from their fire 'sparkled cheerily upwards. It is at such times as these that Amazon travelling is enjoyable, and one no longer wonders at the love which many... have for this wandering life.'

Bates had spent a night at Cametá when travelling with Wallace nine months previously. He now got to know this delightful little town. It was built on a high riverside bluff with fine views over the broad river. Cametá had only 5,000 inhabitants, with their houses in three unpaved streets parallel to the Tocantins; but there were a further 20,000 living in the surrounding countryside. This was not high virgin forest, but partially felled woods with shady groves that shaded the palm-thatched cottages of small farmers. The town was named after the Camuta indigenous people, a large and sophisticated chiefdom that had generally welcomed the European colonists. Camuta women were good-looking and made excellent wives. Settlers had freely intermarried with them and during the past two centuries had produced 'a complete blending of the two races'. This was enhanced by several hundred black slaves added to the mix; and there were two or three families of white Portuguese and Brazilians. They all had a reputation for energy and perseverance, and were cheerful, quick-witted, communicative and hospitable. Bates met 'bright people' including a poet who wrote verses about the natural beauty of his homeland. So he was able to refute a common racist notion that people of mixed blood were deficient. 'It is interesting to find the mamelucos displaying talent and enterprise, for it shows that degeneracy does not necessarily result from the mixture of white and Indian blood.' The ordinary people were indolent and sensual, as elsewhere throughout Pará. They loved saints-day celebrations – which had 'about as much to do with religion as Michaelmas fair has in Leicester.' But this moral condition was 'not to be wondered at in a country where perpetual summer reigns and where the necessities of life are so easily obtained.' The river was full of fish, a small field of manioc provided a form of bread, and each family grew a little cotton for hammocks, coffee, cacao, yams, melons and a variety of fruit.

Each of the score of towns throughout the vastness of Brazilian Amazonia had a slightly different character. Cametá prided itself on being the only one that had not been captured in the Cabanagem rebellion of the previous decade. A priest called Prudêncio had helped to organize its fortification and resistance to attack by a large force of insurgents. Bates took the establishment view that the Cabanos were anarchists and half-savage revolutionaries.

As a visiting scientist Bates had entrée to many people in Cametá. They invited him into their houses, slung a hammock for him, young girls would make coffee in the back of the hut, and the host would fill a pipe of tobacco, light it and give it to his English guest. But the greatest treat for Bates was to meet the town's most distinguished citizen, Dr Ângelo Custódio Correia. This excellent man was educated in Europe, had twice been president of the province of Pará, and had served as a Liberal in the Brazilian parliament. Bates (who, growing up as a hosiery apprentice in provincial Leicester, had never met important men) regarded Correia as 'a favourable specimen of the highest class of native Brazilian.... His manners were less formal, and his goodness more thoroughly genuine, perhaps, than is the rule generally with Brazilians. He was admired and loved, as I had ample opportunity of observing, throughout all Amazonia.' This patrician was known for his kindness to strangers, and he appreciated what Bates was doing. So 'he procured me, unsolicited, a charming country house, free of rent, and hired a mulatto servant for me' – none of which Bates could have found for himself, in a place where inns were unknown. The charming little rocinha belonged to a kindly man, as stout and florid-complexioned as a typical English country gentleman. It was on a lovely sandy bay, where Bates would bathe before breakfast, tide permitting, 'and would then sit on shore on a felled palm-tree, enjoying the delicious coolness at sunrise, and looking over the expanse of blue waters dotted with palm-clad islands; but the beauty of this situation is beyond description.'

Bates spent five happy weeks at Cametá. However, surprisingly, his collecting there was less fruitful than in the tall forests near Belém. He wrote to Stevens that he was now more selective, taking only species that were new to him or that he had sent sparingly before. There were monkeys in the woods around the town, including bands of little cuxiú marmosets (*Chiropotes satanas*) that were dormant and hidden during the day but hopped through the trees at dawn and dusk.

One collected specimen was a huge spider, *Mygale avicularia* (probably what we now know as *Theraphosa blondi*), also known as the South American bird spider or trapdoor spider. Its body was 5 centimetres (2 inches) long and its legs up to 18 centimetres (7 inches). Some of these spiders live in smooth sloping underground tunnels concealed by trapdoors. This predator's speciality is to devour small birds. Bates saw (and vividly sketched) two small finches caught by his mygale: one was dead, but the other was under the spider and 'smeared with the filthy liquor or saliva exuded by the monster' [PL. 27]. The collector killed the spider and tried to rescue the half-dead finch, but it soon died. The spider had its revenge. It was covered in coarse grey and reddish hairs, which came off when Bates handled it incautiously and caused him three days of 'almost maddening irritation'.

The most interesting collecting was of Bates's beloved butterflies. He proudly wrote to Stevens that he was sending a mating couple of the glorious swallowtail *Papilio sesostris* (now *Parides sesostris*). He had actually observed five species *in copula*, and he had 'strong circumstantial evidence against four others' including one where he had observed a male (with a round green spot on its fore wings) fluttering around its mate, a 'splendid creature with an irregular greenish patch on its fore wings and a crimson band with pearly lustre on its hind wings'. The many Lepidoptera he sent included proteus, little theclas, erycinidas, and two examples of copulating mechanites, as well as butterflies that he had reared from caterpillars. He said that he had mated a great many eurygonae: he could tell the females from males by markings on their undersides so that he 'appropriated the females to their right partners'.

River travel was always a problem. When Bates returned to Belém in late July he got a passage on a cargo vessel called a *cuberta* (covered boat), the favourite craft of 'peddling traders'. This was not decked, but its bow and stern were raised and arched-over so that cargo could be piled high above the waterline. The fore arch had a planked roof, for the crew to stand and propel it with long oars if there was no wind; but normal propulsion was by fore-and-aft sails on two masts. Cooking was on a narrow deck amidships. The boat was carrying 20 tons of cacao, and it went by the Caju and Moju inland river route rather than via the dangerous open waters of the Tocantins and Pará. Bates saw why they chose the safer channels. 'I was rather alarmed to see that it was leaking at all points. The crew were all in the water diving about to feel for the holes, which they stopped with pieces of rag and clay, and an old negro was baling the water out of the hold.' The owner was very relaxed about the leaks. He admitted to Bates that he had bought the boat cheaply because it had been left to rot on a beach. But this laid-back man was very kind to Bates, taking him ashore to collect at every interesting spot on the banks of the Moju. The journey thus took five days.

Bates slept in the open, again wrapped in an old sail, but when it rained he had to lug himself and his sail 'into an oven of a cabin'. On the last lap of the voyage, where the Moju and Guamá rivers meet in front of Belém, there was a stiff wind, the vessel heeled over, the leaks burst out afresh and the owner hoisted more sail to arrive faster. But 'with an extra puff of wind, the old boat lurched alarmingly, the rigging gave way, and down fell boom and sail with a crash, encumbering us with the wreck'. The crew had to row. When they were almost there, Bates feared that 'the crazy vessel would sink before reaching port' and persuaded the owner to send him ashore in the dinghy with his most precious specimens.

On 12 July 1849, while Bates was at Cametá, the brig *Britannia* arrived at Belém carrying cargo and three passengers from England. These were the botanist Richard Spruce, his assistant Robert King, and Wallace's twenty-year-old younger brother Herbert. As we have seen, Spruce had read Bates's and Wallace's reports, which Stevens had cleverly got published in the *Zoologist* and other journals, and was encouraged by Hooker and Bentham to join the young collectors in the world's richest ecosystem. Spruce described King as tall and broad in proportion, 'a young man who had agreed to brave the wilds of the Amazon as my companion and assistant'. Herbert Wallace hoped to aid his older brother in the same way. In 1846 Alfred had brought his widowed mother, his sister Fanny, and his brothers John and Herbert to join him at Neath in Wales; but the family then scattered. Herbert sailed from Cardiff to Liverpool to attempt to make a living as a French teacher but there was little demand for his rudimentary language skills, so Alfred suggested that he come to join him in Brazil as a naturalist collector.

Wallace dispatched his collections of fish and insects to England on the brig that brought his brother. He then started taking young Herbert out for training in the tropical forest. After a few days they found some beautiful yellow and green *Conurus carolineae* parrots feeding on a tree. These were known locally as imperial parrots, much esteemed because their colours were those of the Brazilian flag; and Spix had depicted them in 'his expensive work on the birds of Brazil'. Wallace 'had long been seeking them, and was much pleased when my brother shot one.'

The Wallace brothers decided that the time had come to move westwards, deeper into Amazonia. Alfred laid in a good stock of provisions, borrowed books from English and American friends to while away the days on the river, and set off in early August. It was sixteen months since his arrival in Brazil. Like Bates, he had learned a great deal during that time, about the people, local customs, language, quirks of travel and above all about the flora and fauna and how to collect, preserve and package them. As always, it was difficult to get a passage. The Wallaces finally sailed as the only passengers on a small trading vessel returning upriver to the town of Santarém. There was no cabin, so they slung their hammocks in the hold – which smelled of its previous cargo. 'We found it very redolent of salt-fish, and some hides which still remained in it did not improve the odour.' But Wallace was now experienced in local custom: he commented philosophically that 'voyagers on the Amazon must not be fastidious'.

Richard Spruce had an introduction to James Campbell, the elder brother of Archibald who had been so kind to both Wallace and Bates. Spruce described them as colonists of long standing with extensive possessions in Pará. James Campbell cordially invited Spruce and King to stay in his house, which they did for three months. Spruce, now thirty-one and an established botanist, was soon dazzled by the magnificent vegetation, even near the town Belém and at the easternmost edge of the great Amazonian forests. He wrote that 'botany occupied me so completely' that he scarcely observed anything about the town or the Brazilians.

A mature rain forest can contain over a hundred different species of tree. 'The botanist would be struck with the wondrous diversity of forms, two trees of the same species scarcely ever growing side by side.' This was an important observation. A tropical rain forest contains scores of different tree species. Each tree devotes much energy to disperse its seeds and its pollination, in order that its offspring are scattered. To achieve this, it produces fruits or nuts or light blossoms, to appeal to its target dispersal agent – birds, bats, monkeys, insects or wind. This great effort is done to protect the tree against the myriad parasites, blights and predators that abound in that richest of ecosystems.

Spruce was immediately confronted by the problem that had baffled Humboldt half a century before him, and that still confounds ecologists to this day: the difficulty of identifying tropical forest trees at ground level. Each species can be confirmed with certainty only by examining its leaves, blossoms and seeds; but every tree rises straight up towards the sunshine and rain of the canopy far overhead. Spruce lamented that, like Humboldt, he could not find 'agile and willing Indians ever ready to run like cats or monkeys up the trees for me'; the lowest branches were far too high for blossoms to be obtained by poles and hooks; and the trunks were too thick to be 'swarmed'. So Spruce gradually realized that the best way to get blossoms and fruit was to fell the tree. 'But it was long before I could overcome a feeling of compunction at having to destroy a magnificent tree… merely for the sake of gathering its flowers.' He finally overcame his qualms about this vandalism by reasoning that it was in the interest of science and getting the tree better known in museums. More reassuring was the thought that the Amazon forest was 'practically unlimited – near three millions of square miles [8 million square kilometres] clad with trees and little else but trees' – so that local people thought no more about destroying the noblest tree than an Englishman did pulling up a groundsel or poppy from a cornfield.

Spruce of course collected flora of every description. He started with the most easily accessible. 'Having never before seen tropical plants in their homes, all were new to me and beautiful.' But he knew that these coastal plants were found throughout the tropics and therefore familiar, so that only a few of them would be of value to collectors at home. At the muddy mouth of a stream – *igarapé* or 'canoe way' – there were handsome grasses, such as tall, sedge-like cyperi that looked 'exceedingly beautiful with their umbels of polished and brown or green-and-gold spikelets'. But, like mangroves, these were so widespread that one soon tired of their monotonous abundance. Low, moist flatlands contained *Vismia guianensis*, a bush whose trunk exuded thick red juice that dried to form a substitute for sealing wax. There were several species of inga trees, legumes that each had different-shaped pods – some a metre (3 feet) long – hanging from their branches. Over the trees grew convolvuli and creepers, such as malpighiaceae with yellow or pink flowers or more showy combretaceae, one of which was *Cacoucia coccinea* with long spikes of brilliant scarlet flowers – splashes of colour in that world of greens. Most passion vines exhaled exquisite odours, apart from the common weed *Passiflora foetida* whose heavy narcotic smell was like a buzzard's nest. Various pepper plants monopolized the soil in moist places, and some 'looked like minute ferns, as they crept with thread-like stems over the trunks of trees'. In the virgin forest, every visitor is struck by the lack of flowers at gloomy ground level. But Spruce was partly consoled by the abundance of ferns and of his beloved mosses and hepatics (liverworts). One feature was new even to him. This was that in moist tropical forest 'the very leaves on the trees get covered with beautiful lichens and Hepaticae'. The lichens were usually a whitish crust with black, red or yellow shields. To the naked eye the liverworts were mere patches or slender threads of various colours. But examined under a magnifying glass they were seen to have tiny leaves in symmetrical ranks and minute flowers of various shapes.

Spruce's main excursion was in August, when he and Robert King went with Archibald Campbell (younger brother of their host James) and his family to visit his estate, Caripí. This was the dilapidated former Jesuit mission where Bates had spent such a happy time earlier that year. The large house had been closed for some time, and the room provided for Spruce had a huge pile of fresh earth, like some newly opened grave, in the middle of its floor. This was the work of leafcutter saúba ants – 'the great excavator... or navvy of the Amazon valley'. As with Bates, hordes of bats flew wildly about when light was admitted. Spruce happened to wake in the middle of one night, and saw by their dim night-light that King was sleeping with his head out of his hammock and almost touching the ground. Near it was 'a sooty imp' the size of a toad. Spruce jumped out of his hammock, grabbed his machete and sprang across the room to pin down the monster. It suddenly spread its wings to reveal that it was a young vampire bat. Thereupon 'the two parent bats sallied forth from the roof and attacked me; and when I beat them off, they flew round and round the room, attempting to strike me with their wings every time they passed me, and I them with my [machete].' Young King was now awake and 'convulsed with laughter, in which I heartily joined,' at the sight of this bizarre skirmish.

On one walk from Caripí, Spruce visited huts of hospitable Indians and was shown how they made fireproof pottery – as was used in every kind of cooking utensil throughout Amazonia. The secret was to mix fine clay with the bark of the caraipé tree, which contained a great quantity of silex or silica. When burned and crushed in a mortar it made the pottery fire-resistant. With some difficulty, they found a caraipé tree. Spruce watched delightedly as an Indian boy climbed its fairly slender trunk, with his toes in a ring of liana, clutching the trunk in his arms, and jack-knifing upwards with the ring held tightly around the tree as a form of step. The boy brought down leaves, which Spruce dried and sent to his expert patron George Bentham. 'With his vast knowledge of comparative vegetable anatomy', Bentham identified the tree as almost certainly a *Licania* of the Chrysobalanea family.

From Caripí, on the shore of the Pará estuary, Spruce and King were taken by boat to another of the properties that Archibald Campbell's wife had inherited. This was several days up the Acará river, due south of Belém. Such visits to virgin forests were the grounding of what was to become Spruce's incomparable knowledge of Amazonian plant life.

Richard Spruce later wrote a thirty-page summary of the botanical riches of what he called primeval equatorial forests. Over a century later Professor Richard Evans Schultes of Harvard University – the greatest Amazonian botanist of the twentieth century – described Spruce's summary as still the best introduction for a novice entering that environment for the first time. One of the noblest trees is the kapok or silk-cotton tree (*sumaúma* in Tupi, *Ceiba pentandra* scientifically). Spruce gazed in awe at a mighty kapok trunk, nearly 13 metres (42 feet) in circumference and just as thick 15 metres (50 feet) from the ground, where it first began to branch. He measured a fallen kapok at 49 metres (160 feet) long and was assured that others rose to fully 60 metres (200 feet), emerging from the canopy and towering above others in the forest. He loved looking up towards the dome-shaped crown of a kapok, with 'fretted leafy arches that span the space between the pillar-like trunk, and are projected on the vault of heaven beyond.'

Spruce immediately noticed ways in which Amazonian forests differed from European temperate woods. One was that the system is evergreen: 'trees are never denuded of leaves' since they grow new foliage before the old falls away, and there is no autumn or winter when all the trees are bare. Another was that 'the loftiest forest is generally the easiest to traverse' because its creepers and epiphytes are far overhead in the canopy and there is little undergrowth at ground level. Even in lower forest, the herbaceous vegetation was often little more than ferns, sedges, pretty pariana grass looking like palm fronds, and a few shrubs and bushy trees. Spruce also wrote about varieties of bark, shapes of trunks and the fact that the branches of many trees, high above the ground, are 'exceedingly regular and even geometrical, giving rise to a symmetry of outline'.

Writing about forms of leaves, he noted that these are often obscured by lianas and epiphytes. But where they are clearly visible, for instance on riverbanks, 'the general impression to a casual observer would be of massive glossy leaves, intermixed with light feather-like leaves of Mimosas and other similar plants, and with the gigantic plumes of palms.' There are surprisingly few different types of foliage, 'rendering the *tout ensemble* exceedingly monotonous; for the leaves of a large proportion of Amazon trees are ovate or lanceolate, leathery, smooth, and entire at the margins.' For Spruce it was 'a rare treat... to see deeply-divided, strongly cut, jagged or sinuated leaves' that reminded him of English trees. An exception to this monotony came from some trees of moderate size such as papayas and cecropias which 'have enormous leaves, lobed or deeply cloven into finger-like divisions'. The underside of a cecropia leaf is grey or hoary. Spruce recalled that 'when my boat has been floating lazily on the water, under a burning and dazzling sun, with not a breath of air stirring' a squall would reveal the 'glowing tints of the underside of the [cecropia] leaves, [and] waked up the scene into life and beauty.'

Some trees, including the noble kapoks, have several leaflets springing from a common stalk, as on a European horse-chestnut. There can be five, seven or nine such leaves on one stalk, but there are only three on what was later to become Amazonia's most lucrative natural product, the 'India-rubber tree', *Hevea brasiliensis*. Most of the extensive family of legumes have pinnate leaves, like European ash or walnut, although there is no northern equivalent of the bipinnate leaves of several genera of mimosas. Much of the forest undergrowth was of the order of melastomes, all of which have opposite leaves, often large and 'downy or shaggy' with ribs and veins that give them 'a remarkably neat and geometrical appearance'. These differed from leaves of 'their near allies, the myrtles', almost as abundant, but with 'smaller, glossy, ribless leaves beset with transparent dots'. Throughout a forest, 'nearly every shade of green is observable in the hues of the foliage'.

With no autumn or fall, there are no reds and browns of dying leaves, but this is almost compensated by 'the delicate hues of rose and pale yellow-green assumed by young leaves… contrasting admirably with the deep green of the rest of the tree'.

Turning to blossoms (which he always called 'flowers') Spruce cautioned his readers that a showy display of colour was a rarity, because trees did not blossom simultaneously and their flowers were often short-lived. Also, 'sooth to say, the flowers of Amazonian trees are often so inconspicuous, either from their minuteness or from their green colour assimilating them to the leaves, that none but a botanist ever would see them.' There were, however, glorious exceptions. At the mouth of the Amazon, some legumes and bignonias 'outshine all other orders in the abundance and beauty of their flowers…. The taller-growing Cassias and a Sclerobium are crowned with a profusion of golden flowers; but far more elegant are the large pure white prince's-feather-like flowers of the Bauhinias [the tortoise-ladder liana]…. [Among bignonias] lofty Tecomas [carobs] are, in their season, one mass of purple or yellow, from the abundance of foxglove-like flowers…. The showy white or red… flowers of the Clusias… are sure to attract the botanist's early attention. Some Tiliads (Molliae sp.) are studded with large star-like white flowers, as striking in their way as the gaudy stars of the passion-flowers [*Passiflora*] that spangle the liana-curtain skirting the rivers.' Spruce mentioned many other curious and glorious blossoms. But he repeated his warning that 'the great mass of the trees of the forest, and even many of the lianas, bear inconspicuous flowers' or have none at all.

'Almost the first thing that strikes the observer' in an Amazon forest is the great triangular buttresses at the bases of many trunks. Rarely more than 15 centimetres (6 inches) thick, these buttresses can extend on the ground up to 4.5 metres (15 feet) from the base of a trunk with the apex rising up to 15 metres (50 feet). There are from four to ten such roots around a tree. 'The Indians correctly call them sapopemas (*sapo*, a root; *pema*, flat)' since they are above-ground roots. They are also supporting buttresses. Spruce noted that rain forest trees spread their roots near the surface of the ground, with few deep roots. Without buttresses they could thus easily be toppled by a falling neighbour. It is now known that these shallow roots are to capture every scrap of falling nutrient, to feed each tree in its race to rise to the sun and rain of the canopy. Because, as we have seen, the resulting rain forest soil is so weak, trees need only one or two tap roots for groundwater: the rest are horizontal in the root-and-litter mat. Spruce recorded differences in size and shape of sapopemas between tree species, with some families such as Lauraceae having none. He also noticed that some trees, particularly some palms, have stilt roots – angled legs that lift the trunk above seasonally flooded ground.

Spruce marvelled at the festoons of lianas and compared them to 'the rigging and shrouds of a ship, whereof the masts and yards are represented by the trunks and branches of trees'. Called *sipó* in the Indian language Tupi, they varied in thickness from slender threads to 'huge python-like masses', round or flattened in shape, some knotted and others 'twisted with the regularity of a cable'. Local people called one type of flattened spiral 'jabutim (tortoise) ladders'. These were 'the most fantastic' lianas, with 'compressed, ribbon-like stems, wavy as if they had been moulded out of paste and while still soft indented at every few inches by pressing in the fist'. Although never more than 30 centimetres (12 inches) wide, 'they reach two or even three hundred feet [60 to 90 metres] in length, climbing to the tree-tops, passing from one tree to another, and often descending again to the ground.' Found throughout Amazonia, these particular lianas are Leguminosae (of the *Schnella* genus for Spruce, now *Bauhinia*).

Many lianas were of the bignonia family, recognized by their four-angled stems, and with swollen leaf scars every few feet. Spruce recalled 'one of the most gorgeous sights I ever saw'. Where some trees had been felled to leave a gap in the forest, a many-stemmed bignonia 'which had run lightly over [the tree tops], was left suspended between two lofty trees, 40 yards [35 metres] apart, in a graceful catenary, clad throughout its length by roseate foxglove-like flowers and large twin leaves of a deep green tinged with purple'. Spruce noted that some lianas of every species 'get up in the world by scrambling upon their more robust and self-standing neighbours. Where two or more of these vagabonds come into collision in mid-air, and find nothing else to twine upon, they twine round each other as closely as the strands of a cable, and the stronger of them generally ends by squeezing the life out of the weaker.'

Some lianas have hooks, both to help them climb and as 'formidable defensive weapons'. One of the few sought-after medicinal plants in nineteenth-century Amazonia was sarsaparilla (*Smilax syphilitica*; its roots were used in a medicinal restorative tea that – as its botanical name suggests – was thought to be a palliative for syphilis).* Spruce likened prickly sarsaparilla to 'our brambles, ramping vaguely about, but never up to a great height'. Some had 'three-cornered stems whose angles are thickly set with prickles. Sometimes they trail insidiously on the ground, where their presence is only revealed by the wounded foot that treads unwittingly upon them.' There was a thorny climber called jurupari-piná or 'devil's fish hook' because of its 'broad curved prickles in place of stipules'. One *Uncaria* was known as 'cat's claw' because of its long, tough, hooked prickles. Common on riverbank

* The Encyclopaedia Britannica of 1911 noted that, after exhaustive medical testing, sarsaparilla was regarded as 'pharmacologically inert and therapeutically useless'. So 'Salsaparilla' is now just the name of a soft drink popular in Jamaica.

forest, these were 'a serious impediment to navigation' in places where boats moving upriver had to go close to the bank to escape currents. Spruce knew of men caught by such hooks and suspended in mid-air while their craft was swept downstream. But the most dreaded riverside liana was the jacitara, a palm of the *Desmoncus* genus, whose leaves ended in rigid spines like arrow barbs. 'As the canoe shoots by or under an overhanging mass of Yacitara, woe to the unlucky wight who is caught by its claws, which infallibly tear out the piece they lay hold on, whether it be flesh or garment, or both.'

Many visitors to a rain forest are struck by a climber so 'curiously flattened to the trunks of trees' that it looks almost painted on. Its oval leaves are 'closely and symmetrically set on the stem in two rows... and of a deep velvety green beautifully netted with white veins.' This is the young state of *Monstera dubia* (*Marcgravia umbellata* in Spruce's day) and it is 'so totally unlike the mature state' that Spruce had difficulty in tracing the union of the two radically different forms of the same plant. When this monstera has climbed into the light it sends out stout branches with large green leaves that end in pointed 'drip tips' and are sculpted with so many round holes that botanists now know it as the 'Swiss cheese plant' and it is a common house plant outside Brazil [PL. 10].

'Many lianas secrete an abundance of fluid sap, usually milky and acrid.' In the many apocynaceae or asclepiad twining plants, the sap can be 'turbid and virulently poisonous in some paulliniae, but sometimes limpid, sweet and harmless'. Spruce of course drank from the famous *cipó de água* ('water liana') that has rescued many parched travellers. He explained how to get a cupful of pure water from this *Dilleniacea* (now called *Pinzona coriacea*): 'it must be cut simultaneously at two points a few feet apart, and the ends of the severed piece held at the same height; then when one end is slightly lowered the liquid runs out in a gentle stream, and may be thus conveniently drunk.'

Epiphytes are perching plants 'which roost in the forks and on the branches of trees'. They use the host tree as a support but do not harm it as parasites. Epiphyte foliage can be as abundant as that of the lianas, so that the tenant plants 'quite hide the leaves of the trees on which they sit and hang'. There are aroids that resemble the arums of English hedgerows, but with much larger leaves 'sometimes fantastically jagged or perforated, and in some instances tinged with purple or violet beneath'. Cyclanths look like aroids in having 'enormous tufts or with succulent creeping stems' and broad bifid or fan-shaped leaves. There are multitudes of bromeliads, including several species of Tillandsia and others like gigantic pineapples. The leaves of these bromeliads are viscid so that they retain rainwater: they thus become tiny ponds that form micro-environments for insects and frogs. Spruce learned that it was 'by no means agreeable to stumble against a bromeliad' because of its leaves' thorny serrature and pungent points, and because of the stings of ants living in them.

Spruce loved palms, as did both Wallace and Bates – they all later wrote books or essays on palms of the Amazon. He exclaimed: 'Intermixed with the trees, and often equal to them in altitude, grew noble palms; while other and far lovelier species of the same family, their ringed stems sometimes scarcely exceeding a finger's thickness, but bearing plume-like fronds and pendulous bunches of black or red berries, quite like those of their loftier allies, formed, along with shrubs and arbuscles of many types, a bushy [but not dense] undergrowth.' On the height of palm trees, Spruce gently corrected Alexander von Humboldt who had said that they towered above surrounding forest. In fact, this occurred only on some riverbanks or coastal shores. In virgin forest, palms rarely exceed exogenous trees and are 'often quite hidden from view until closely approached'. Years later he wrote that, gazing from an outcrop above the Rio Negro, 'I have looked over perfect oceans of forest' where palms almost never rose above the canopy.

In Belém at the end of September 1849, an 80-ton brig came down the Amazon with cargo consigned to Messrs. Campbell from their fellow Scot Captain Hislop, another long-established settler who lived at Santarém. Although small, Santarém was the largest town on the main Amazon River. It was 750 kilometres (475 miles) upriver from Belém, so it seemed to Richard Spruce a 'very desirable head-quarters for a [collecting] campaign.' He decided to get a passage on this boat on its return journey. He soon made the necessary preparations. He took letters of credit to a merchant in Santarém and a hundredweight bag (50 kilos or 112 pounds) of copper coins for small change. For food there was hard-toasted bread, manioc farinha, salted fish (strong-smelling pirarucu, which he disliked, and tastier smaller taínha), eggs, coffee and sugar. He bought a basket canteen, 'an indispensible article for a traveller' that had compartments for plates, knives and forks; also *frascos* – large square bottles that held about 2 quarts (2.3 litres) – for spirits, vinegar, molasses, etc. They sailed inland on 10 October 1849.

Henry Walter Bates had also decided that the time had come for him to leave Belém and the Atlantic seaboard and plunge into the interior. His distinguished friend Ângelo Custódio Correia in Cametá had promised him a passage in a boat belonging to his brother-in-law. Bates wrote to Stevens that he had high hopes of a fast passage in this large boat. But, as so often in Amazonia, things did not go smoothly. The entire month of

August was wasted in delays by the boat's owner, a young mestizo called João Correia. He was planning to spend several *years* on this trading voyage, so was in no hurry. The efficient Bates wrote testily that 'pleasure first and business afterwards' seemed to be his motto – and when they finally sailed, on 5 September, the first two weeks were spent revisiting Cametá to enjoy a big fiesta. Bates was by now becoming aware of the problems of movement along these rivers. Before the Cabanagem of the 1830s, rich or powerful people could travel in a fast *galiota* with a crew of a dozen Indian paddlers to augment its sails. But with the depopulation and labour shortage caused by that upheaval, almost all communication was now in privately owned trading vessels. With few paddlers available, these boats had only their sails and were of course dependent on wind. The prevailing 'trade wind' of the Amazon is easterly, up the river from the Atlantic Ocean. When this *vento geral* blew normally 'sailing vessels could get along very well; but when this failed they were obliged to remain, sometimes many days together, anchored near the shore.' This was compounded during the rainy season from January to July: not only was the east wind feebler or non-existent, but the river itself was infinitely more powerful. Swollen by rains from the Andes, the Amazon each year rises spectacularly, floods vast swathes of forest and has 'a tearing current'. (I am a feeble swimmer, but swimming downstream in the middle of the Amazon in May I seemed to move at Olympic speed.) In the mid-nineteenth century, a schooner could cover the 1,600 kilometres (1,000 miles) from Belém to Manaus at the mouth of the Rio Negro in forty days during the dry season, whereas it took twice as long, three months, to do the same voyage during the rainy months. This was a major reason why the three naturalists waited until late 1849 to move into the interior.

Bates prepared for his journey in much the same way as Wallace had done three weeks earlier. He also took a hundredweight of copper coins, as well as all provisions that would be hard to get in the interior, including ammunition, chests, storage boxes and a small library of books on natural history. One difference was that Bates also took all his household goods, cooking pots and crockery, because he planned to hire a hut somewhere, to settle down to serious collecting. He engaged a mameluco as a servant and collector: a 'short, fat, yellow-faced boy named Luco' whom he described to Stevens as a half-wild coloured youth – but an expert entomologist. All this preparation was 'troublesome and expensive', but Bates hoped that once underway his living costs would be trifling. He was looking forward to seeing the true Amazon forests, and expected that life there would be 'all unmitigated enjoyment'. But he missed being with Wallace. He wrote to Stevens, who published the letter in the *Zoologist*: 'I should have liked a sympathising companion better than being alone, but that in this barbarous country is not to be had.'

5

INTO AMAZONIA

The Wallace brothers reached Santarém at the beginning of September 1849 after a four-week journey. This was when Bates left Belém, reaching the small town on 9 October – and he sailed on upriver at dawn on the following morning. Spruce left the provincial capital Belém only on 10 October, but he had the fastest voyage: so his boat sailed into Santarém on the 27th.

The journeys up the lower Amazon were reasonably uneventful. After leaving the broad and often stormy Pará river, boats spent a week or more threading through a network of channels, *furos*, at the western end of Marajó island. These furos are attractive forested rivers. Bates marvelled at the tall walls of vegetation that turned his channel into a deep gorge. Dome-topped emergents towered above the canopy trees, with plenty of graceful palms among them – fan-leaved buriti, groups of açaí 'forming feathery pictures' amid the mass of round-leaved foliage, plenty of buçú with their vivid pale-green shuttlecock-shaped fronds, jupatí (*Raphia taedigera*) throwing its very long shaggy leaves right over the canal, stately and exceedingly elegant dark-green bacabas (*Oenocarpus bacaba*), and an infinite variety of smaller palms decorating the water's edge. Over all this, 'from the highest branches of the forest trees down to the water sweep ribbons of climbing plants of the most diverse and ornamental foliage possible.' Creeping convolvuli used lianas and hanging air-roots as ladders, and there were occasional mimosas and thick masses of inga trees with their long bean pods. The furos are still flanked by this luxurious vegetation in the twenty-first century. The largest of them are straight, wide and deep enough to take modern ocean-going liners.

In 1849 tides were of course an important factor, with daily waits for the ebb to turn to the flood that propelled boats up the lower river. But for Wallace there were days on which the wind dropped to nothing so that his boat had to resort to advancing by *espia* or cable. Such warping consisted of sending the montaria dinghy ahead with a long rope; this was fastened around a tree on the bank; the Indians then returned to the big boat and hauled it forward by hand. The laborious operation was repeated again and again, and progress was painfully slow.

Each boat eventually reached the main Amazon. The romantic Wallace gazed upon this 'mighty and far-famed river... with emotions of admiration and awe.' Bates first saw it by moonlight, looking most majestic. It was over 30 kilometres (20 miles) wide, but divided into broad channels by a series of long islands. Wallace described its waters as pale yellowish-olive; to Bates they were ochre-coloured and turbid; to Spruce, varying from dull yellow to weak chocolate. There was little human presence on this stretch of the lower Amazon – an occasional town (Gurupá, Almeirim, Monte Alegre), village or sitio in the unbroken line of forest on the north shore, but almost no life amid the desolate seasonally flooded *igapó* of the south bank.* During the first week they sailed along this drearier right bank. All three were hit by the sudden fierce squalls that occur throughout Amazonia; Bates was 'becalmed in sickening heat' for two days; his and Wallace's boats both ran aground on sandbanks; and Bates once ran into very high waves in the middle of the river, so that 'the vessel lurched fearfully, hurling everything that was not well secured from one side of the deck to the other.'

The main animation came from birds. There were numerous flocks of parrots, macaws flying in pairs and uttering their hoarse cries, many species of herons, storks, rails and ducks in the marshes, and abundant gulls and terns. Wallace was charmed by the terns who 'had a habit of sitting in a row on a floating log, sometimes a dozen or twenty side by side, and going for miles down the stream as grave and motionless as if they were on some very important business.'

All the travellers were delighted by Santarém – as people still are. Its situation was very beautiful, in the mouth of the Tapajós river, on sloping ground above a sandy beach and enjoying a balmy climate. In 1849 it was a clean and cheerful place of some 2,500 inhabitants. The town, in which 'the white and trading classes' lived, was three long streets of substantial houses, whitewashed (or yellow) with bright green woodwork and red-tiled roofs. Some were of two or three stories. However, because there were no horses or carts, the unpaved streets were covered in grass. There was a handsome church with two towers,

* There are two types of easonally inundated forests on Amazon floodplains: *igapó* and *várzea*. Although it was scarcely understood in the 1850s, the difference is that igapó is formed by flooding with clear waters (such as the Tapajós) and black waters (such as the Rio Negro), and várzeas are flooded with muddy 'white' waters (particularly those coming off the geologically young Andes). The upper Amazon-Solimões and also the Madeira (the great tributary from the south) have gigantic várzea plains. Várzeas tend to be more extensive and with more fertile soils (although not for growing manioc, which cannot stand flooding), whereas with igapó the soils tend to be sandier and the vegetation lower.

12

13

14

12 Alfred Russel Wallace, aged 25 when he embarked for Brazil.

13 Henry Walter Bates in his late thirties, soon after taking his post at the Royal Geographical Society in 1864.

14 Richard Spruce, shortly before he sailed to Brazil aged 31.

15

16

17

Men who inspired or helped the three naturalists.

15 The German polymath Baron Alexander von Humboldt, who in 1800 explored the Orinoco and Rio Negro in southern Venezuela and wrote a classic thirteen-volume account of his travels.

16 Clements Markham, who in 1859 engaged Spruce to bring cinchona trees from the Ecuadorean Andes, and who was for almost thirty years Bates's colleague in building up the Royal Geographical Society.

17 'The Three Wise Men', a modern painting imagining Charles Darwin, Sir Charles Lyell and Joseph Hooker reacting to Wallace's 1858 paper on evolution by natural selection. Darwin and the geologist Lyell befriended and helped Bates and Wallace, and Joseph Hooker and his father William (of the Royal Botanic Gardens, Kew) were admiring patrons of Spruce.

18 Carl von Martius entering a Mura hut, in 1820. The Bavarian botanist Martius and zoologist Johann Baptist von Spix were the first non-Portuguese scientists allowed on the Amazon. Bates regarded the Mura Indians as primitive; but they had been superb boatmen and fearless warriors. Martius later tried to persuade Spruce to write about palm trees for his monumental study of Brazilian flora.

19

19 Bates and his friend Antonio Cardozo watched a great turtle-egg hunt on the Solimões near Ega. One night, their sandbank camp was disturbed by a large black caiman, harmless to men but capable of eating their dog. Bates depicted himself as thin but with plenty of hair, waking from his hammock while Cardozo drove the saurian off with flaming logs.

20 Bates drew the scene when he and Cardozo caught turtles in a lagoon off the Solimões. Miranha (Witoto) Indians roar with laughter at inadvertently netting a black caiman, while Bates waits on the bank to kill it with a club. Cardozo gathers turtles into a boat in the background.

21 Wallace devised this fantasy of some of the birds he had collected. On the left is the Amazonian umbrella bird; above it a pair of tasty curassow turkeys; flying towards the artist two curl-crested toucans – the birds who mobbed Bates; standing by the water a tall trumpeter; and to the right a pair of delightful tufted coquettes, related to hummingbirds.

20

21

22 A green wood cricket (*Chlorocoelus Tanana*). Bates dissected one to find how males made a very loud twanging noise. He found that this was done by rubbing one inner wing case against a different structure in the other (as shown in the sketches beside its legs).

23 Bates was thrilled when the *Zeonia batesii* butterfly was named after him.

24 The hawk-moth (*Aellopos titan*), on the left, looked and moved uncannily like a hummingbird, right. Local people were convinced that the insect would metamorphose into the bird, and Bates could not dissuade them.

22

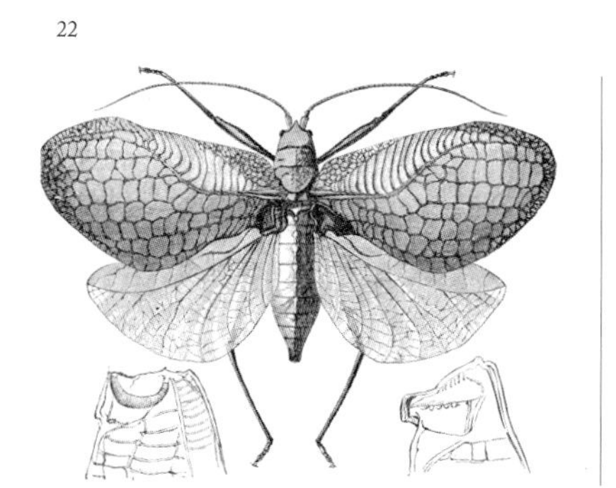

23

24

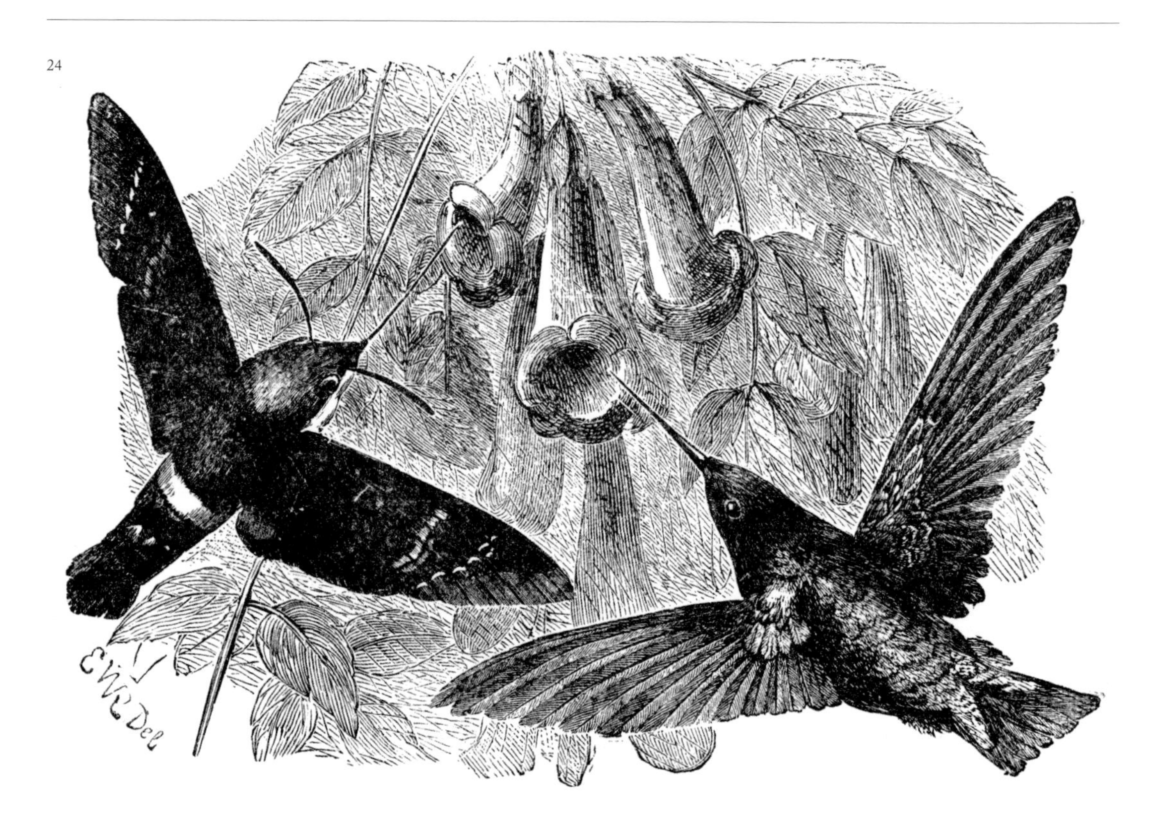

25 Bates watched a mason wasp (*Sceliphron fistularium*) build a pouch-like nest of clay pellets and then stock it with stunned spiders on which its eggs could feed.

27 Bates was horrified to find two small finches being killed by a bird-eating mygale spider, which had caught them in its dense web and was smearing them with poisonous slime. He tried in vain to rescue one bird that was still alive.

26 A tiger beetle (*Tetracha punctata*), whose larva emerges from a tunnel and grows into a chameleon-coloured horned beetle. Bates had difficulty catching them on the beach at Caripí because they were nocturnal, ran very fast, and doubled back when pursued.

26

25

27

28

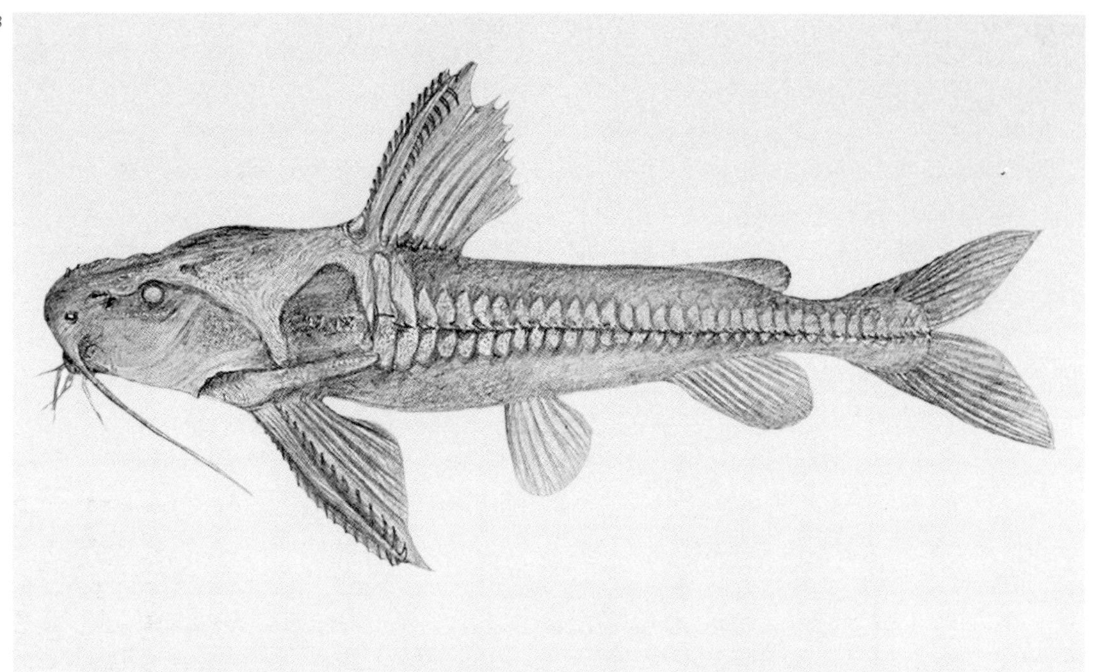

Wallace was one of the first to make beautiful and accurate drawings of freshwater fish in Amazonian rivers. He grabbed a box of these drawings from his cabin in the burning brig *Helen* and they survived days in an open boat. They are now prized possessions of the Natural History Museum and Linnean Society of London.

29 30

28 Cuyucuyú (*Hemidoras stenopeltis*), a catfish from the upper Rio Negro.

29 Wallace called this cichlid (*Pteroglossum scalara*) a 'butterfly fish'.

30 *Cichlosoma severum*, another cichlid.

and an old earthwork fort on a low hill at one end. Some of Santarém's elite led idle lives, waited on by black slaves; and the little place had a full complement of officials, including the ominously titled Comandante de Trabalhadores ('Commandant of Labourers') who superintended 'Indians engaged in [i.e. forced to perform] any public service'. Alongside the prosperous town was the *aldeia* (village) of Indians and caboclos, who lived in palm-thatched huts with mud walls or open sides. Santarém had no jetty, so visitors had to wade ashore from dinghies. But the beach was always lively, with washerwomen bleaching linen on the hot sand, and plenty of bathers – the indigenous and black children were 'quite amphibious animals'.

All the English naturalists had introductions to Captain John Hislop, 'a sturdy, rosy Scotsman' who had been living on the Amazon for no less than forty-five years. Initially a seaman, once he was resident in Santarém Hislop traded guaraná stimulant and salt up the Tapajós to the gold miners of Mato Grosso far to the south; but he now took Brazil-nuts, sarsaparilla (the best in the Amazon), manioc and salt-fish down to Belém. In the evenings, Santarém's worthies met on the verandah of Captain Hislop's house overlooking the river, 'where they would sit and smoke, take snuff, and talk politics and law for an hour or so'. Spruce was amused to find that the Captain kept a large file of old newspapers, which he read and reread even when they were months out of date, but he had only two books: the Bible and a translation of Constantin-François Volney's *Ruins of Empires* (1791). He had 'framed for himself a creed of very motley complexion' from these sources; and after a few glasses of port would regale his guests with his theories about Moses or ancient Rome. With (or despite) these oddities and his hearty bearing of a sailor, Spruce and the others found Hislop to be an amusing companion and a valued friend.

Captain Hislop was a generous host to the young naturalists. He invited Alfred and Herbert Wallace to eat with him for as long as they wished, and then found a simple house for them. This had mud floor and walls and was 'all very dusty and ruinous'; but it was the best available, so the brothers made the most of it – they had plenty of experience of rough living in England and the lower Amazon, and their voyage upriver in the smelly hold of a cargo boat had hardly been luxurious. They collected in the *campo* (savannah) country behind Santarém, including a picnic excursion with two other Englishmen called Jeffries and Golding, who had married Brazilian ladies and settled there. They went in a smart little boat – with Indians, blacks and plenty of provisions – to a delightful stream and lake surrounded by old trees and shady thickets. Here 'we amused ourselves shooting birds, catching insects, and examining the new forms of vegetation', followed by a refreshing bathe in the clear cool lake, then dinner and return by moonlight.

Wallace characteristically wanted to do more sightseeing. He had heard that there were interesting rock paintings inland from Monte Alegre (then Montealegre), a village they had passed on the voyage up the Amazon. As Spruce wrote to his mentor William Hooker, Wallace 'put it into his head to go thither'. It is extraordinary that he embarked on what proved to be a tough five-week journey, not to do profitable collecting but just in the hope of seeing 'Indian picture-writings which exist there'. Wallace had made friends with Santarém's magistrate, who lent them an excellent boat for the 100-kilometre (60-mile) trip. There was the usual week's delay finding Indian boatmen, but three men were eventually persuaded to come by the magistrate and the Comandante. After two days' sailing, just as they were about to turn into Monte Alegre's stream, they were hit by a sudden violent storm and heavy sea that almost capsized the boat and put them in 'considerable danger'; but they just made it to calmer water.

The village proved to be a miserable place. It had once been relatively prosperous and thriving, but was shattered by the Cabanagem. The open square had the ruins of a half-built sandstone church, started before the rebellion. The current church was a barn-like thatched edifice, most houses were equally poor, and there were no tidy gardens, 'nothing but weeds and rubbish on every side, with sometimes a few rotten palings round a corral for cattle'. Wallace had a letter to a French shopkeeper called Nunez – movement throughout Amazonia was always smoothed by such introductions – who found them an empty house. The problem here was mosquitoes, which bred profusely in the marshes. 'Immediately after sunset they poured in upon us in swarms, so that we found them unbearable.' The brothers kept their sleeping room tightly shut, but the mosquitoes soon found their way in through cracks and keyholes. The plague of insects became daily 'more tormenting, rendering it quite impossible for us to sit down to read or write after sunset.' Mosquito nets were unknown in the Amazon, so every house burned smoky cow dung. The Wallaces hired an Indian to cook for them. Every afternoon he was sent to gather a basketful of dung, which was readily available as there were plenty of cows about. The burning dung had 'rather an agreeable odour', and it deterred the mosquitoes sufficiently to make life 'pretty comfortable'.

The country around Monte Alegre, on the north bank of the lower Amazon, was sandy savannah, dominated by immense masses of candelabra cactus, up to 9 metres (30 feet) high and with stems as thick as a human body. There were ranches raising some cattle and horses on this weak pasture. But the main produce was brightly painted calabashes, along with cacao. Cacao trees (*Theobroma cacao*) grow naturally from the western Amazon to southern Mexico, and some had been brought down to form plantations on the lower river. The small tree grew well in cleared land on the banks of rivers. The oblong cacao fruit is about 13 centimetres (5 inches) long; green when young, then golden yellow when ripe; inside is a mass of nut seeds imbedded in white pulp. It is the seeds (called beans) that, when sweetened, yield the cocoa drink that had become a craze in Europe, and hard chocolate (an Aztec word). The white pulp is dried to make another refreshing drink. Thus, 'when an Indian can get a few thousand cacao-trees planted, he passes an idle, quiet, contented life: all he has to do is to weed under the trees two or three times a year, and to gather and dry the seeds.' (Actually, growing cacao was not as easy as Wallace imagined: 'witches' broom' fungus and other parasites attacked the trees, and processing the beans and pulp was complicated and risky.)

Wallace did some collecting in the undulating country behind Monte Alegre: trogon and jacamar birds, and 'curious ferns and mosses and pretty creeping plants' among wet rocks of springs. Shady groves around these springs yielded insects, including the beautiful *Asterope leprieuri* butterfly. But Wallace was still determined to see the 'Indian picture-writings' in the hills some 20 kilometres (12 miles) to the north. Senhor Nunez accompanied them in a montaria propelled by a heavy sail made of strips of the bark of a water plant. They spent a day sailing this top-heavy vessel up a creek, followed by hours of pushing through water-weeds so densely matted that forked punt-poles could lever against them, and then grasses taller than their heads that 'cut the hands severely if [one] merely brushed against them'. They finally reached a cattle ranch on open campo. They had the usual letter of introduction to the owner, who received them kindly. The ranch's mulatto cow-hands had 'strange accoutrements: ...curious and clumsy-looking wooden saddles, huge stirrups, long lassos, and leather ammunition-bags with long guns and powder-horns of formidable dimensions'. The travellers had an excellent dinner of freshwater turtles, all seated on a mat on the floor. They slung their hammocks among the many others that were hung across the room in every direction. It was 'rather crowded; but a Brazilian thinks nothing of that, and is used to sleep in company.'

The hills that were their destination are the southeasternmost extremity of the Guiana Shield, a belt of sandstone hills and table mountains that runs across northern South America. These are geologically Precambrian, some of the oldest rocks on earth – the formation continues in West Africa, but its two parts were separated aeons ago by plate tectonics. Wallace of course could not know this: it was established only in the late twentieth century. But it is interesting that he was so keen to see 'the cave and picture-writings' in these hills. He was one of the first to appreciate indigenous petroglyphs, which were later to become an important branch of Amazonian archaeology. The brothers and their guides spent a gruelling day scrambling over huge rocks and among innumerable chasms, all covered in coarse rigid vegetation. It was very hot, they exhausted the water in their gourd bottles, and they wasted hours reaching a small spring to slake their thirst. They climbed one hill and had a view over a wide undulating plain, uninhabited, but with low oblong hills studding it in every direction. Their 'descent was very precipitous... winding round chasms, creeping under overhanging rocks, clinging by roots and branches.' Wallace's reward was, on the return journey, to observe 'some of the picture-writings I had so much wished to see'. They were painted on a cliff in red earth-based pigment. Some showed animals, caimans, birds and household utensils, others circles and geometric figures, others 'some very complicated and fantastic forms'. They looked quite fresh and unweathered, but 'no one knows their antiquity'. On the following day the party returned and Wallace climbed to a cliff covered with different figures. He sketched and traced many of them. Then, on the third attempt, they found the cave they had been seeking – its entrance had been concealed by trees and bushes. This was a large arched chamber with a smooth sandy floor, and beyond were openings to other chambers. Without candles, they were unable to see any cave paintings or signs of human habitation. But it was almost certainly Pedra Pintada ('Painted Rock'), destined to be one of the most famous archaeological sites in the Amazon basin. This is where, almost a century and a half later, Professor Anna Roosevelt of the University of Chicago followed other archaeologists in making dramatic discoveries about early man in Amazonia. Ever curious, Wallace was ahead of his time with this pioneering interest in archaeology.

The brothers spent a few days back in Monte Alegre. There was a fiesta that attracted Indians from round about, with dancing and drinking for a night and a day, but Wallace was grumpy that their Indian servant felt it more important to play the violin for the party than to cook for them. On a collecting excursion, Herbert sprained his leg, which swelled, formed an abscess and laid him low for a fortnight – he was not proving to be as tough a naturalist as his brother. The magistrate's boat on which they had come had sailed on down the Amazon. So, after some hard bargaining, Alfred bought a small boat to take them back to Santarém and, he hoped, on upriver to Manaus. They must have travelled with a sack-full of coins and a supply of trade goods, as Bates and Spruce had done, but it is a mystery how each of them carried this heavy currency without having it stolen.

The Wallace brothers returned to Santarém towards the end of October, after five weeks on the Monte Alegre trip. They found another simple house, hired an old black woman to cook for them, and settled down to some serious collecting. They rose at dawn, six in the morning, and prepared nets and collecting-boxes, while the cook made breakfast, which they took at seven. An hour later they would walk 5 kilometres (3 miles) or so to their favourite collecting ground, while the cook went with some money to buy food for dinner – Santarém enjoyed 'an abundance of beef, fish, milk and fruits'. The brothers worked hard collecting until two or three in the afternoon, bathed in the Tapajós on the return, and when they got back there was always a refreshing watermelon ready for them. They then changed their clothes, dined, arranged their insects and other catches, and in the cool of the evening took tea and called on or received visits from Brazilian or English friends. These now included 'Mr Spruce, the botanist, who arrived here [on 27 October] shortly after we had returned from Montealegre'. Wallace later wrote that he had never enjoyed himself so much as during those weeks in Santarém. Although it was hot, 'the constant hard exercise, pure air, and good living kept us in the most perfect health'. Wallace and Spruce had met when the botanist sailed to Belém in July, but they now became firm friends. In their short time together, Wallace learned that 'Spruce was a well educated man, a most ardent botanist, and of very pleasing manners and witty conversation.' This friendship was to ripen and last throughout their lives.

Bates, meanwhile, had reached Santarém at midday on 9 October 1849 after a leisurely five-week journey in the boat of his friend João da Cunha. But he did not stay – possibly because Wallace was no longer there. Instead he continued in Cunha's boat, which sailed at dawn next morning upriver to the next town, Óbidos. Bates had heard that collecting there was good, and wanted to gather as much as he could before the rainy season set in. In the event he spent five weeks at Óbidos, from 11 October to 19 November. He narrowly missed seeing both Spruce, who arrived later in November, and Wallace who came soon after that.

It was impossible for the three Englishmen to communicate with one another. They could not even leave accurate messages, because none knew when he could get a boat or how long a journey would last. An example of this was that Bates's boat took only a day to cover the 80 kilometres (50 miles) between Santarém and Óbidos, whereas Spruce went on a large boat in mid-November but took ten days owing to lack of wind and because the owner refused to sail at night. Wallace was delayed in Santarém because he had to repair the rotten bottom of the boat he had bought in Monte Alegre (he was not a savvy buyer of boats), and because of the usual problem of finding three Indian crewmen. But the Wallace brothers then took only three days in their leaky little boat, reaching Óbidos as Spruce was still unpacking.

Óbidos is airily perched on a high bluff above the Amazon, at the only narrows during the mighty river's course from Iquitos in Peru to the Atlantic Ocean. Everything to do with the Amazon is superlative, so this 'narrows' is still 1.6 kilometres (1 mile) wide and the waters are very deep and flow with tremendous velocity. The narrowing is caused by a sandstone plate (a southern extension of the Guiana Shield) topped by layers of tough tabatinga clay on the north shore and a rock outcrop to the south.

Bates enjoyed his five weeks at Óbidos. It was an old town by Amazon standards, with 1,200 inhabitants and tile-roofed houses of 'substantial architecture'. Each town in the vastness of Amazonia had slightly different customs. Here, unusually, the townspeople spent their evenings gathered in one another's houses. They were kind and hospitable, even to the young, single Englishman. They met 'for social amusement, bachelor friends not being excluded, and the whole company, married and single, joining in simple games.' Bates attributed these 'friendly manners... and general purity of morals' at Óbidos to an excellent old vicar, Padre Raimundo de Sanchez Brito, who set a good example and insisted on strict observance of the Sabbath. Bates, brought up as an Anglican but now an agnostic, always approved of Catholic devotion, and recorded sympathetically all the local saint's-day celebrations and processions.

Some of Óbidos's old-established families had large cattle ranches or cacao plantations in the open campo country near the town. But Bates found them rather naive, because they were lazy about cattle-husbandry and farming and therefore remained poor – 'a dozen slaves and a few hundred head of cattle being considered a great fortune.' The town had been occupied by the Cabano rebels during the disturbances thirteen years earlier. An old lady from Óbidos admirably summed up that violent but unfocussed revolt: 'How we all suffered, because of men who wanted something that no one knew, not even they themselves! The Cabanagem was a scourge sent by God to punish us. It was a plague that ravaged the land where I was born. Everyone suffered from it.'

As always, Bates worked hard at his collecting. The area was rich in his favourites – butterflies and moths (*borboletas* and *mariposas* in Portuguese). There were particularly magnificent morphos, *Morpho hecuba*, with huge metallic-blue wings up to 20 centimetres (8 inches) wide. But these were difficult to catch because they glided 6 metres (20 feet) or more above the ground. Many species of heliconid, a group peculiar to tropical America, flew among lower bushes. These medium-sized butterflies have long narrow wings, generally deep black but with spots and streaks of crimson, white and bright yellow, in different patterns according to the species. 'Their elegant shape, showy colours, and slow, sailing mode of flight, make them very attractive objects', and they were so abundant that they brightened the forest and compensated for a lack of flowers below the canopy. Bates included his paintings of *Heliconius thelxiope* and *H. melpomene*, in the book of his travels. Another conspicuous genus was *Callicore*, which Bates knew as *Catagramma* – he explained that this name came from the Greek for 'a letter beneath' because below the wings are 'curious markings resembling Arabic numerals'. Their wing-tops are vermilion and black, with a velvety appearance. Bates collected 'an almost endless diversity of species', but they became even more numerous as he moved westwards towards the upper Amazon. A related butterfly was *Asterope leprieuri*, with wings of a rich dark-blue colour in a broad border of silvery green. These two groups of butterfly, *Callicore* and *Asterope*, found only near the equator in South America, 'are certainly amongst the most beautiful productions of a region where the animals and plants seem to have been fashioned in nature's choicest moulds.'

On riverbank beaches throughout Amazonia a familiar sight is vast numbers of sulphur-yellow and orange-coloured butterflies. These congregate on moist sand, particularly where an animal has urinated. They are mostly species of the *Phoebis* genus, known as sulphurs, and only males. They can form dense masses, several metres wide, with 'their wings all held in an upright position, so that the beach looked as though variegated with beds of crocuses.' These are migratory insects. During Bates's voyage up the Amazon great numbers flew across the mighty river, always from north to south, and 'the processions were uninterrupted from an early hour in the morning until sunset.' Everyone on the boat noticed these 'migrating hordes' – even though many local people were indifferent to nature. It was apparently only males who migrated, so they cannot have flown far from their females. The female sulphurs were rarer, seen only at the edges of the forest flying from tree to tree and often depositing their eggs on low, shaded mimosa.

Bates was fascinated by a pale-green wood cricket (*Thliboscelus hypericifolius*) whose males made a very loud but quite musical noise by rubbing their wings [PL. 22]. Its 'notes are certainly the loudest and most extraordinary that I ever heard produced by an orthopterous insect' (one with firm fore-wings, like a cockroach). Bates was curious. He dissected one of these 'musical crickets' to see how it made such a tremendous noise. He found that the inner joints of its two outer wings were different. One had a horny lobe, the other a row of sharp furrows like a file. Thus, a male cricket twanged the lobe of one wing against the furrows of the other; and the sound was given greater resonance by a hollow drum-like space within its parchment-like wings. The noise was, of course, a mating call – females of this insect had no furrows or lobes on their wings. Romantically, when a female approached, 'the louder notes are succeeded by a more subdued tone, whilst the successful musician caresses with his antennae the mate he has won.'

There were plenty of birds near Óbidos, but Bates was now having to be more choosy. He could no longer send Stevens common species that he had collected during the past year. One novelty was a small thrush that local people called caraxué (cocoa thrush; *Turdus fumigatus*), smaller, plainer and less musical than British thrushes. But Bates grew to love its sweet and plaintive tone, 'which harmonizes well with the wild and silent woodlands... in the mornings and evenings of sultry tropical days.'

Those woods were also full of monkeys. The Comandante Militar of Óbidos, Major da Gama, fancied the meat of coatá monkeys (*Ateles paniscus*) and used to send a man out every week to shoot one for his table. These were the largest Amazonian monkeys, up to 90 centimetres (3 feet) long, covered apart from their faces with shaggy black hair. They are known as spider monkeys because of their long, thin bodies and limbs. Bates was fascinated by the origin of species – an important reason why he and Wallace chose to go to the Amazon. He felt that these spider monkeys must be the most developed of the American primates because they had the best prehensile tails. 'The tendency of Nature here has been... to perfect those organs which adapt the species more and more completely to a purely arboreal life.' Apart from being good to eat, spider monkeys were mild creatures that became pets and would follow their masters everywhere. Indians were fond of them, and Indian women would suckle baby coatás at their breasts. Bates found that one blemish was that they were 'arrant thieves', cunning about pilfering and hiding small things. They had none of the restless vivacity of *mico* or capuchin monkeys (*Sapajus apella*) or the untameable temper of their close relatives, howler monkeys, known locally as *búgios* or *guaribas* (*Alouatta*, classified as *Mycetes* in Bates's day).

Bates left Óbidos on 19 November on a local man's ship loaded with goods to trade on the Rio Negro. The boat's owner was a kindly man who tried to make the young naturalist as comfortable as possible – giving him half the foreward *toldo* (covered space) for his hammock and boxes, and providing plenty of food and drink. But the voyage was very slow, since the owner travelled only by day and the winds were erratic.

As we have seen, Spruce and then the Wallace brothers reached Óbidos soon after Bates had left. The Wallaces wanted to press on upriver in their own leaky boat as soon as they could. Their problem, as always, was finding Indian crewmen. The three who had sailed from Santarém had contracted to come only to Óbidos, so they helped unload and then immediately found a boat to take them home. Captain Hislop had given Wallace a letter to Major da Gama – he who liked to eat spider monkeys. He found fresh Indians, but only to take them four days' journey to the next town, Vila Nova (now called Parintins). The three Englishmen had a dilemma about travel. There were no regular services. So they had either to get passages on trading boats, with all the uncertainties of departure times, conditions on board and rate of progress, or they could buy their own boat, with the problems of finding the right craft, maintaining it and above all getting Indian paddlers and pilots – difficult enough for locals and almost impossible for young foreigners.

Spruce also had an introduction to Major da Gama in Óbidos, who kindly installed him and his assistant King in a room in his own house. The only drawback was that this was the room of the Major's idle son. The rather straight-laced Spruce disapproved of this 'dissipated young man' because he divided his time between his hammock, his viola and his pipe. The two botanists spent a few weeks collecting in an interesting range of habitats near the town: some open campo, wooded streams, a marshy lake and hills with lofty primary forest. There were new varieties of mosses and ferns, which Spruce loved. But late November and early December were the height of the rainy season, a bad time for collecting. Not only were the naturalists often drenched while out in the forest, but they then had great difficulty in preserving their collections from rot and mould.

Although as passionate and methodical a collector as his two younger colleagues, Spruce – like Wallace – was also adventurous. Someone had told him that the Erepecuru river northwest of Óbidos was a splendid place for collecting, so he longed to go there.* Spruce thought that the rains meant that he would have to abandon the Erepecuru plan, but Major da Gama assured him that it should be all right: because the rains had set in early in 1849, there should be a dry spell for a month after Christmas. The Comandante even offered to lend them his own galiota boat, manned by five Indians from the Trombetas river – the northern tributary to which they were headed. Spruce of course accepted enthusiastically. There was the usual problem of finding boatmen. Even though the military commander was ordering them, only three came, and they were reluctant. Much as Spruce wanted them, he sympathized: 'They would rather, poor fellows, have been in their forest homes, hunting, working, or playing as they listed, than plying their paddles all day in the hot sun or the pelting-rain.' The boat hardly qualified as a galiota. It was an *igaraté*, a large dugout canoe, to which had been added ribs and sideboards, and floorboards at the stern 'dignified with the name of… quarter-deck'. On this stern deck was the usual toldo

* The Erepecuru is confusingly also called Paru do Oeste ('Western Paru') and Cuminá, and all three names appear on government maps even in the twenty-first century.

shelter – but because this toldo had a curved roof of boards rather than the usual palm leaves, the vessel claimed the grander appellation galiota.

Spruce and King set off on 18 December, on what was to prove a punishing adventure, with two stalwart crewmen in their thirties and a pilot aged nearer sixty. After a night at the farm of the pilot's brother, sheltering from rain and mosquitoes, they spent a couple of days ascending the lower Trombetas and then entered the scarcely explored Erepecuru. This river flows south from the Guianas, rising in the Tumucumaque hills on the Suriname border, and cascading for 350 kilometres (220 miles) before joining the Trombetas from the northeast shortly before its junction with the main Amazon. The going became tougher. At times they poled through choking luziola aquatic grass; then there were narrow channels between islands, broad stretches of river, occasionally obstructed by sandbanks, and then a canyon with waterfalls splashing off its cliffs. The current became too furious for either paddling or punting with poles. The Indians jumped ashore, cut bignonia lianas to form a rope, tied this to the boat's prow, and man-hauled along the forested shore and in the raging water, while the pilot tried to steer and wielded a pole to stop the boat smashing against rocks. They reached the first rapids, the limit of navigation, on Christmas Day. Over a cooked breakfast, Spruce and King drank to Merry Christmas and friends in England with cachaça and rapids water.

Spruce wanted to collect plants in hills called Carnaú. He could not have known that the Erepecuru is now famous for rock engravings on every one of its scores of rapids – carvings similar to the paintings Wallace had seen near Monte Alegre. But even had he known about these petroglyphs, he could not have reached them; and his purpose was botany.

Spruce's Christmas toast was his last agreeable moment. There was heavy rain and thunderstorms during the next two days. On 28 December it seemed clearer, so they set off towards the Carnaú hills. The old pilot was left to guard the boat and their shelter, so Spruce and King went with the two sturdy Indians and a cafuzo (Indian-black mestizo) called Manoel who had joined their little expedition. They decided that it would be too difficult to follow the river, so struck inland. They crossed low hills and descended into valleys choked with bamboos and murumuru palms. From time to time the Indians would climb a tall tree to try to see Carnaú hill. After six hours they paused to debate which direction to take. Then, without warning, the two Indians left to retrace their route to the boat. This was Richard Spruce's first time in unexplored virgin forest. He admitted: 'My experience of forest travelling was as yet very slight, and I knew not how essential it was to never lose sight of my Indian guides.' He, King and Manoel thought that they were near the Erepecuru, so tried to reach it down a stream. This was arduous. The igaraté was 'densely beset with bushes and lianas [and] ran through flats of entangled bamboos and cutgrass, which were passable only on our hands and knees.' It was a sultry day, but suddenly 'the solemn stillness was broken by a soughing in the forest, soon deepening into a roar, and a terrible thunderstorm burst upon us.' Robert King paused to open a pod of Brazil-nuts – no easy matter since these are extremely tough – and was left behind. It was some time before Spruce and Manoel realized that he was missing, and they failed to find him even after Manoel climbed a tree. For a while even Spruce and Manuel lost contact with one another. Reunited, at three in the afternoon 'to our very great joy, we heard King's voice.' An hour later the three finally reached the river. Being inexperienced, they tried to follow it downstream; but it was of course too dense and entangled – vegetation is always thickest beside the sunlight of a riverbank. Moving inland again, they sent Manoel ahead because he moved faster. Spruce acknowledged that this was 'another error on my part', since Manoel was the only one with a machete. Spruce and King struggled on for a while, thoroughly drenched. Night fell and they could do nothing but sit between the buttress roots of a large tree and wait for the moon to rise. They ate some cold fish and rain-soaked manioc. Then they tried to move on, but tall forest is pitch black even on a moonlit or starlit night. They had no light or blade and were rightly apprehensive because jaguars were said to prowl in forests near rapids. They scrambled on in the darkness, 'now plunging into prickly palms, then getting entangled in sipós [lianas] some of which were also prickly.' In daytime gloom creepers could trip, entangle or half-strangle people, and this was far worse at night. 'At one time we got on the back of large ants, which crowded on our legs and stung us terribly.' Bewildered and exhausted, they rested for a while on a riverbank rock, waiting for the moon to rise higher. Back under the canopy, there was just enough light for them to choose thinner parts, but not enough to see stumps or lianas. They pressed on slowly and cautiously, crossing streams by wading or finding a slippery log bridge, and finally next morning reached the boat and palm shelter. The others were calmly waiting there, but the two Englishmen were 'sadly maltreated and wayworn. The effects of this disastrous journey hung on us for a full week. Besides the rheumatic pains and stiffness brought on by the wetting, our hands, feet and legs were torn and thickly stuck with prickles, some of which produced ulcers.' They were also covered in 'bites of ticks large and small and the stings of wasps and ants'.

Spruce wrote that he was telling this adventure 'in order to give some idea of what it is to be lost or benighted in an Amazonian forest.... Let the reader try to picture to himself the vast extent of the forest-clad Amazon valley; how few and far between are the habitations of man therein; and how the vegetation is so dense that... it is rarely possible to see more than a few paces ahead.' It is all too easy to get lost, even when close to a trail or landmark. 'In the excitement of gathering new plants, or of the chase of wild animals, one often forgets to mark one's way properly; and it has several times happened to myself, when deep in the forest and quite alone, to be unable to find my track.... It is a rather painful moment when one becomes convinced that the way is irrecoverably lost....' This author agrees with everything that Spruce said. I have also experienced the panic of being lost, alone in totally unexplored forest, knowing that if I continued in the wrong direction it would be the end of me. It was imperative to find some trace of my *picada* trail. Spruce gave some good advice in this regard: that travellers should not cut saplings entirely, but break and bend them in the direction of travel. I also agree with Spruce about the bites. Glands at the tops of one's arms and legs become swollen and sore from straining poisons from a mass of bites on those limbs.

Because he had been so keen to reach the mountain, and their boat was crowded, Spruce acheived little botanical collecting. One climbing plant he did gather was *Norantea guianensis* 'an odd Guttifer [containing hard gutta latex], which shot forth from the mass of its dark green foliage as it were jets of flame – spikes two feet [60 centimetres] long, bearing each some two hundred curious pouch-like bracts of the finest rose-colour, accompanied by minute purple flowers.' He found the most curious plants growing on the constantly wet rocks of rapids and waterfalls. Also near *cachoeiras* ('rapids') was a little bird called uirá-purú ('spotted bird'), which he had been told 'played tunes for all the world like a musical snuff-box'. This was the musician wren (*Cyphorhinus arada*), famous throughout Amazonia for its melodies, sought-after as a caged bird, and thought to be a legendary bringer of good fortune. Spruce suddenly 'had the pleasure of hearing it strike up close at hand. There was no mistaking its clear bell-like tones, as accurately modulated as those of a musical instrument.' Its short phrases included an entire octave. But after repeating this perhaps twenty times, it suddenly passed to another phrase, changing key to the major fifth, and then repeated that one. 'Simple as the music was, its coming from an unseen musician in the depths of that wild wood... held me spellbound for near an hour, when it suddenly broke off.' Spruce recorded the notes of one of the little bird's songs in his journal – he had learned some music when teaching mathematics in York.*

Spruce realized that his men were cold and discontented and might desert him, so he abandoned the attempt to reach Carnaú hill. It took them only eight days to shoot

* Modern readers can hear this bird by visiting www.xeno-canto.org and searching for 'musician wren'.

down the swollen Erepecuru, and they reached the Trombetas and Amazon on 5 January 1850. Spruce's three-week adventure had been a painful but important lesson in tropical forest skills. From then on, he tended to explore along rivers rather than in forest alongside them, and he was careful not to get lost.

Spruce and King sailed swiftly back down the Amazon to Santarém. This seemed the most agreeable place to sit out the rainy season, which continued for four months of 'unrelaxing severity' with frequent violent thunderstorms, heavy rains particularly at night, and intense oppressive heat during the sunny intervals. In the event they also spent the dry season there, leaving only nine months later in October 1850.

Bates, on his slow trading vessel, also endured the sickening heat of the midday sun, which could only be escaped in stifling cabins; but this was compensated by deliciously cool evenings. The crew would cook dinner on shore, either at some plantation's landing place or in a shady nook. Just before eating they bathed in the river, then 'according to the universal custom on the Amazons' each downed a half-teacup of neat cachaça before tackling a mess of stewed pirarucu fish, beans and bacon. Bates was bored by the invariable saltfish; but he imagined strangely that the rum compensated for its lack of nourishment. The crew would then tether the boat on a long line and let it drift out into mid-river – to escape mosquitoes.

The rainy season hit them one night. There was 'a horrible uproar as a hurricane of wind swept over'. The cuberta was hurled against a clayey bank; everyone had to stow their hammocks and save the vessel from being dashed to pieces, while the pilot leapt ashore through 'drowning spray' and pushed the boat off with a pole. They were lucky not to be near a bank that was being washed away with all its trees and vegetation. After an hour the wind abated, but the deluge of rain continued, lit up by flashes of lightning, with 'the thunder pealing from side to side without interruption'. Their clothing, hammocks and belongings were thoroughly soaked by rainwater streaming through the leaky decking.

To Bates's frustration, his cuberta took an interminable nine weeks to sail the roughly 600 kilometres (370 miles) from Óbidos to Manaus. The genial owner, Senhor Penna, was in no hurry. He travelled only in daylight, and paused for long stays with friends on the banks. There was often no wind, so that the crew had to paddle against the current or even advance by laborious espia hauling. The only consolation for the gregarious and curious young traveller was to learn more about life in Amazonia.

One visit was to a typical cacao plantation – this chocolate tree was the main export crop of the middle Amazon. The mameluco family who owned this had a substantial house, whitewashed with a tiled roof, and were very relaxed in their dress and way of life. The master lay in his hammock smoking a long wooden pipe and was happy to have a long chat. Nearby on the verandah was the ever-ready coffee-pot simmering on its tripod. 'We were kindly received, as is always the case when a stranger visits these out-of-the-way habitations; the people being invariably civil and hospitable.' As they left, one of the daughters brought a basketful of oranges down to their boat. The small plantation had 10,000 to 15,000 cacao trees. Harvesting these took only a few weeks each year, and was easy work performed in shade. So it was a pleasant life. But Bates could not resist mild disapproval, as was prevalent among most European and North American visitors. He blamed 'the incorrigible nonchalance and laziness of the people' for denying themselves 'all the luxuries of a tropical country' and living on a meagre diet of fish and farinha. Like other commentators, he imagined that 'intelligent settlers from Europe' would have done far better – even though, in practice, none did. Also, he did not yet appreciate that processing the harvested cacao pods into chocolate was laborious and skilled work.

Even more agreeable was a visit, towards the end of the voyage, to a farm opposite the mouth of the Madeira river. The elderly mameluco owner, João Trinidade, was 'a model of industry': a planter, trader, fisherman and expert boat builder. He had the advantage of farming on 'rich black loam… of exuberant fertility'.* Trinidade and his wife worked their farm with four relatives, one freed slave, one or two Indians, a family of semi-nomadic Mura people, and only one female slave. It was a model of order, abundance and comfort. There were some 8,000 cacao trees along the river and, inland, large plantations of tobacco, manioc, corn, rice, melons and watermelons, as well as a great rarity: a kitchen garden of European vegetables. Bates loved his time there, partly because his hosts took an interest in his collecting.

The only other visitor was an old free black man, who had saved Trinidade's life during the Cabanagem by warning him of an attack by insurgent looters. Bates admired this old man's 'manly bearing… quiet, earnest manner, and his thoughtful and benevolent countenance.' He had noticed this, with pleasure, among many free blacks. This included 'a frank, straightforward fellow' who had brought a year's harvest of tobacco down the

* This was clearly *terra preta do índio* ('black earth of the Indians'), amazing soil that was apparently created by centuries of slow incineration of organic waste by indigenous communities. This magical terra preta was not studied by scientists until the late nineteenth century, and it still attracts much interest 150 years later. It is greatly prized by gardeners, and some researchers even imagine that it forms a living organism. One of the largest surviving stretches of terra preta is Taperinha, on the edge of the Amazon floodplain 35 kilometres (22 miles) east of Santarém, another is Açutuba/Hatahara across the Rio Negro from Manaus.

Abacaxis river in his canoe. Penna, the owner of Bates's boat, bought this crop on generous terms, thus saving the young farmer and his wife a long journey to a market town.

The Wallaces meanwhile struggled upriver in their own small boat, even though it leaked so much that they were almost too frightened to venture out in it. They hauled it out of the water and plugged its cracks with cotton dipped in hot pitch. Luckily they experienced strong following winds, so they 'dashed along furiously' by day and night, with Wallace 'rather doubtful of our rotten boat holding together'. They also had the inevitable problem about finding paddlers. Their men from Óbidos had agreed to go only for four days as far as Vila Nova. A trader was persuaded to let them have three of his Indians to continue their journey. There was an ugly incident. One of these Indians did not want to go, so was driven to the boat 'by severe lashes and at the point of a bayonet'. Indians were technically free men whom it was illegal to enslave. So this paddler was furious and sullen, and vowed to return to kill the man who had forced and struck him. Wallace treated him well, with pay, food and drink, so the Indian assured the Englishman that he bore him no ill will. But at the first night's stop, he politely said goodbye, took his bundle of clothes and walked back through the forest to get his revenge.

Both naturalists had interesting encounters that tell about life on the Amazon at that time. Wallace was welcomed on the beach of Vila Nova by the local priest, Padre Torquato de Souza. The cleric was 'a very well educated and gentlemanly man' who, like Wallace, loved riddles and word-games. Spruce, who was in this village a year later, described the fairly youthful cleric as good-looking and rosy, and exceedingly courteous – but too fond of hearing himself talk. Torquato was the most accomplished of the very few clerics active in Amazonia at that time. Seven years previously he had accompanied Prince Adalbert of Prussia up the lower Xingu river. Wallace knew that many of his readers would have read the English translation of the Prince's book about his expedition, so he commented that the priest deserved all the praise bestowed on him in it. Spruce said that the Padre seemed highly flattered that he had been mentioned by the German aristocrat.

All Amazonians loved parties. At a village called Mucambo, a couple of days' journey above Vila Nova, there was a celebration in honour of the Virgin of the Conception. Bates found this festa admirable. It was held in the house of 'a tall, well-made, civilised' Indian called Marcellino and his wiry and active elderly wife. They were assiduous hosts to a gathering of fifty or sixty Indians and mamelucos. There was a procession for the image of the Virgin, salutes by two little guns, sung hymns and litanies, and a feast of pirarucu, stewed and roasted turtle, and piles of manioc meal and banana, all consumed around a mat on terraced ground in front of the house. Then there was hard drinking of locally distilled manioc spirit. Young men played music on wire-stringed guitars, and Bates and his boat's owner Senhor Penna were invited to join the dance – a mildly erotic version of a Portuguese fandango. 'All passed off very quietly considering the amount of strong liquor drunk, and the ball was kept up until sunrise next morning.' Further up the Amazon, Bates's boat spent five days over Christmas at Itacoatiara (then also known as Serpa). This former mission, 200 kilometres (125 miles) east of Manaus, was by that time a shabby village with irregular streets full of weeds and bushes, mostly occupied by 'semi-civilised Indians living in half-finished mud hovels'. The riverbank was full of boats of people who had come for the Christmas celebrations. Bates – who was always fascinated by Catholic ceremonial – described the various processions, with women and girls dancing along in 'white gauzy chemises and showy calico print petticoats', old women holding a decorative panel of cotton and mirrors, all singing hymns in rituals devised by Jesuit missionaries a century earlier. 'In the evening good-humoured revelry prevailed on all sides.' Black people had their own celebration to the black Saint Benedito, with a night of singing and dancing to the monotonous music of a long drum called a *gambá* and a notched bamboo-tube rattle. The Indians could not organize a dance, 'for the whites and mamelucos had monopolised all the pretty coloured girls for their own ball'.

Another day's sail upriver was a former mission, now called Mura because its twenty 'slightly built mud hovels' were inhabited by that tribe. The Mura had been a powerful people, tremendous warriors and canoers who terrified settlers on rivers near Manaus until, in the late eighteenth century, they decided to make peace. They fought hard in the Cabanagem rebellion, even killing Bararoá, the cruel scourge of the rebels; so they bore the brunt of reprisal punishment after government forces regained control. Bates, who usually admired Indians, was appalled by the Mura. The women were in ragged petticoats, their skins smeared in mud against mosquitoes, and the men were surly and unfriendly. They had broad shoulders and powerful arms – from all their paddling – but short legs. 'The gloomy savagery, filth, and poverty... made me feel quite melancholy.' The Mura begged for cachaça, but offered nothing in return so got none. 'Every one spoke of them as lazy, thievish, untrustworthy, and cruel.' But this was because they were semi-nomadic fishermen who knew their streams and lakes intimately, but did not make permanent villages or farm clearings – and they hated working for whites.

The Mura loved taking the stimulating powder *paricá* (also known as *niopo* or *yopo*), made from the sun-dried and crushed beans of a species of inga tree. They puffed this hallucinogen up one another's nostrils with a reed tube. This was these naturalists' first view of hallucinogenic plants, which would later fascinate Wallace and particularly Spruce.

Bates disapproved. The Mura inhaled the snuff after drinking their own caxiri brew or, preferably, cachaça rum. They then had 'a fuddling-bout lasting many days.' The effect of paricá on usually taciturn Indians was 'wonderful: they become exceedingly talkative, sing, shout, and leap about in the wildest excitement. A reaction soon follows...'. Bates compared 'the swinish Mura', who overdosed on paricá, with the more sophisticated Tupi-speaking Mawé, who took only small quantities as a malaria palliative.*

Other disagreeable encounters for both travellers were with biting insects. On this stretch of the Amazon they saw their first pium (borrachudo in southern Brazil and jejen in Spanish), a small biting black-fly of the family known to entomologists as Simuliidae. These piums started their 'terrible scourge' here below Manaus, and infested all the Amazon's other white-water rivers. They appeared punctually at sunrise (succeeding the nocturnal mosquitoes) and in some places were in swarms so dense that they resembled clouds of smoke. The scientist Bates observed how 'they alight imperceptibly and, squatting close, fall at once to work; stretching forward their long front legs, which are constantly in motion [like] feelers, and then applying their short, broad snouts to the skin. Their abdomens soon become distended and red with blood' and they can become almost too gorged to fly. Bates dissected some piums to see how 'the little pests operate'. He found that they punctured the skin with two horny lancets and sucked blood between them. Each pium bite leaves a small red spot and short-lived itchy irritation. Bates, like any traveller with exposed skin, got several hundred of their punctures in a day. No one in the mid-nineteenth century had any inkling that insects could transmit diseases. It was later learned, of course, that *Anopheles* mosquitoes are the vectors of malaria; but black-fly transmit no disease – other than African river-blindness (onchocerciasis), which has very recently appeared on some Amazonian rivers.

All travellers suffered terribly from mosquitoes. 'They fell upon us by myriads and... came straight at our faces as thick as raindrops in a shower.' Some men crowded into the boat's cabin, or burned rags – but the smoke half-suffocated the human beings and did little to deter the insects. Wallace found the mosquitoes 'a great torture; night after night we were kept in a state of feverish irritation, unable to close our eyes for a moment.' He noted that the Indians suffered just as much as whites. They were constantly slapping their bare bodies, and tried to escape by rolling up in stifling sails.

Another pest were biting mutuca (or butuca) horse-flies of the *Tabanus* genus. It is now known that there are over 200 species of mutuca, some over 2 centimetres (almost an

* There are now about 10,000 Mawé (or Sateré-Mawé) in reserves on the Andirá and other southern tributaries of the Amazon downstream of Parintins. A sophisticated people, they are ardent converts to evangelical Protestantism.

inch) long and with a single long proboscis that Bates found to be 'a bundle of horny lancets'. They varied from bronze-black to green. Silent flyers, they pursue their prey relentlessly, alight delicately and plunge their stilettos in painlessly: a sharp pain like a hot needle comes only when they suck blood. They leave 'a wound whence the blood… oozes out in profusion.' Bates sometimes had eight or ten on his ankles simultaneously. But they were sluggish and easily killed with his fingers.

Collecting in igapó seasonally flooded woods near Vila Nova, Bates found the grasses and low forest to be swarming with carrapato ticks (of the Ixodidae family). These eight-legged arachnids lurk on plants, with a couple of legs 'stretched out so as to fasten on any passing animal'. They plunge their short thick fangs into the victim and then gradually suck blood until, after a day or two, their flat bodies are swollen to the size of coffee beans. Bates, like anyone wandering in such places, would spend an hour every day picking them off. He learned that they must be removed carefully: if the proboscis breaks off it leaves a nasty sore. He recommended tobacco juice to remove them; or it can be done with a drop of petrol or a lighted cigarette. A worse scourge is tiny *carrapato pólvora*, the dust tick, which is the larva of the common *carrapato estrela* (star tick; *Amblyomma cajennense*). Clusters of these miniscule ticks hide under leaves. Anyone brushing past finds hundreds of them swarming all over their body, with unbearable itching. Because they are so small it can take hours or even days to hunt down every last one, and even when stark naked it is difficult to see them amid the dirt of travel in a rain forest – as this author knows from itchy experience.

Unknown to one another, the Wallace brothers in their leaky little craft actually passed Bates one night, while the latter's trading boat was moored at Itacoatiara. So the Wallaces reached Manaus on the last day of 1849, and Bates not until three weeks later, on 22 January 1850. Bates did not mention it in his book, but he discovered that he had been robbed of much of his money during the voyage. He confided to his journal that he had only £11 left to pay for the voyage and his stay in Manaus. This meant that he might have to return downriver without making a good collection in the heart of the Amazon forests. In his diary he wrote: 'Shows what a booby I am at travelling.' He blamed himself for not having come with a letter of credit from a trading company in Belém – but he had not dared to ask for one because his prospects were so uncertain.

6

MANAUS

Ever since the great rubber boom at the end of the nineteenth century, Manaus has been the largest and most dynamic city in the Amazon basin. It is at the heart of the world's greatest mass of tropical forests, and close to the junctions of the Amazon's two mightiest tributaries – the Rio Negro from the north and Madeira from the south. But when Wallace and then Bates arrived in early 1850, and Spruce at the end of that year, Manaus was so insignificant that it hardly qualified as a town. It is on a bluff on the east bank of the Rio Negro, just before that river's black waters flow into the sediment-laden whitish main Amazon. It was still called Barra because its small fort controlled the bar or mouth of the river; but its name was about to change. The Brazilian government decided to separate its western Amazonian territories from the province of Pará downriver. Manaus was about to be the capital of the huge new province of Amazonas. The new name was, ironically, in memory of the Manau (or Manoa) indigenous people, who had bravely resisted Portuguese slavers but were defeated and annihilated in the mid-eighteenth century.

Wallace reckoned that Manaus had a population of 5,000 to 6,000; Bates thought only 3,000 because many Indians had left. When indigenous people became aware that the law protected them against forced service, they naturally preferred to go to live unmolested in their own farms or villages. Manaus had grown around its little fort. This was captured by the Cabanos in 1835 and then largely demolished and abandoned, so that by 1850 there was just a fragment of wall and a mound of earth. The town itself was regularly laid out, but its streets were 'quite unpaved, much undulating, and full of holes, so that walking about at night is very unpleasant. The houses are generally of one story, with red-tiled roofs, brick floors, white- and yellow-washed walls, and green doors and shutters.' There were two churches, inferior to those of Santarém.

The citizens of Manaus were all Indians or mestizos. There were virtually no pure Europeans, because the Portuguese had so completely intermarried with indigenous people in this remote place. The more prosperous townsmen were all traders. They sent Brazil-nuts, medicinal sarsaparilla and salted fish downriver, and they imported cheap cotton, inferior cutlery, beads, mirrors and other trinkets to trade with Indians and river-bank settlers up the rivers. Manaus was the main depot for this traffic. It took months to bring luxuries such as wheat flour, cheese or wine up the Amazon, so these were expensive or unobtainable.

Wallace was disgusted to find that Manaus's traders had 'literally no amusements whatever, unless drinking and gambling on a small scale can be so considered: most of them never open a book, or have any mental occupation.' But these uneducated people were obsessed about appearance. On Sundays ladies wore elegant French muslins or gauzes and carefully arranged their fine hair, which they decorated with flowers and never hid under a hat or bonnet. The men worked in shirtsleeves and slippers in their dirty warehouses during the week. But on Sundays they appeared in fine black suits, beaver-skin top hats, satin cravats and patent-leather shoes. Dressed in this finery, people went to mass and paid social calls on one another. They naturally loved to gossip. But the rather puritanical Wallace was shocked to find that in this philistine and fashion-conscious town morals were 'at the lowest ebb possible in any civilized community'. He heard rumours of goings-on in respectable families that would hardly be credited in the worst slums of London.

Bates got over being robbed during the voyage upriver. He was delighted to find that 'my companion Mr. Wallace' was in Manaus. There were also two British traders called Bradley and Williams – 'clever, good-natured fellows' – who were good to Bates, entertaining him with 'fine frolicking… almost every day', advancing him some money, and encouraging him to pursue his collecting further up the Amazon.

The saving grace for all three English naturalists was an Italian merchant called Henrique Antonij. Bates described him as a warm-hearted and never-failing friend to all travellers; and to Wallace he was a most hospitable gentleman. When Spruce arrived (after his compatriots had left), the generous Italian became his closest friend during many years in the Amazon. He invited each of the young Englishmen to stay in his house and dine at his table. Bates said that 'in the pleasant society of [other foreigners] and the family of Senhor Henrique we passed a delightful time: the miseries of our long river voyages were soon forgotten.' Spruce reported that Senhor Henrique (as he was known throughout Amazonia) was born in Livorno, and went to Belém and then to Barra

(Manaus) in 1822 when he was only sixteen. The small town was decaying at that time. But in the ensuing decades he extended its commerce, started new industries and built substantial houses, to such an extent that 'he merits indeed the title of Father of the Barra'. Now aged forty-five he was 'still young and fresh-looking, with a frank, good-humoured face of the genuine Tuscan type.' William Edwards (the American traveller whose book had inspired Wallace and Bates) had been entertained by the generous Italian three years earlier. He was struck by Senhor Henrique's exceedingly pretty young wife, who talked with strangers as freely as a lady would in his country, and also by his four charming blond children. All this appealed hugely to the young Englishmen starved of feminine and family company.

During the next few years, Senhor Henrique helped Richard Spruce in many ways, not least having him and King to stay and eat at his well-furnished table. The cuisine was excellent, and his turtle made 'splendid eating'. Manaus was in the heart of turtle country, so this host had never fewer than five turtle dishes on his menu – stewed, roasted in the shell, minced and spiced, grilled, or in soup. Not only was the food good, but there might be as many as seven different languages spoken at these cosmopolitan meals.

Spruce was later delighted to repay 'my old friend's hospitality and other virtues' by asking George Bentham if he could name the finest new genus of trees that he had discovered on the Rio Negro 'Henriquezia' after his Italian host. Bentham duly declared that it was 'with much pleasure that I can accede to [Spruce's] wish... to dedicate... an entirely new and remarkable genus of *Bignoniaceae*... to Senhor Henrique Antonij.' One of these was *Henriquezia verticillata*, a noble tree of 25 to 30 metres (80 to 100 feet) in height, 'having its branches and leaves in whorls, and bearing a profusion of magnificent purple foxglove-like flowers'.

Alfred Russel Wallace decided to spend a month collecting at a farm three days' journey up the Rio Negro. He went in his leaky little boat, without his brother Herbert who was becoming disenchanted with rain forests. Wallace had Indian paddlers that Senhor Henrique arranged for the authorities to supply, and a lad lent by his Italian friend to light his fire, boil coffee and prepare dinner – if there were any food available. He had the usual and essential letter of introduction, to a Senhor Balbino who had one of the very few two-storied houses (known as *sobrado*) outside Manaus. But Wallace and his boy were lodged in an Indian's hut a kilometre (half a mile) away, in a tiny room 'with a very steep hill for a floor'. It was an insight into how very poor Indians lived in the interior. There were three families in other parts of this hut, with the men wearing only trousers, the women only a petticoat, and the children nothing at all. Wallace was saddened to find that their food was largely variations of manioc: early in the morning a gourd of *mingau* (watery manioc porridge), at midday a cake of dry manioc farinha or a roasted yam, and in the evening more farinha, mingau or a banana. About once a week they might get a small bird or fish, but it was divided among so many people that it was little more than 'a relish to the cassava [manioc] bread'. Wallace's Indian hunter would take only a bag of dry farinha for fourteen hours of canoeing, with a gourd of mingau in the evening. Despite the meagre fare, this man was 'as stout and jolly-looking as John Bull himself'. And the Indians were hard-working, with no time for leisure. The women were forever digging up and preparing the manioc and yams, making earthen pots, or mending and washing their scanty clothing. The men were busy clearing forest (it took a man a week to fell a tree), cutting timber for boats, thatching roofs or making baskets, bows and arrows. Wallace pondered how these good people could be so poor 'in a country where food may be had almost for nothing'. He concluded that the problem was that everyone tried to do and make everything for himself, instead of specializing and trading.

Wallace's main objective in this excursion was to collect rare umbrella birds, which he had heard were to be found on the islands of the lower Rio Negro. The umbrella bird (*Cephalopterus ornatus*) is a large black cotinga – Wallace described it as the size and colour of a raven [PL. 21]. Indians called it *ueramimbé* or 'piper-bird' because of its loud, deep call. Its English name 'umbrella' was inspired by a remarkable fringe of black feathers on top of its head. 'This crest is perhaps the most fully developed and beautiful of any bird known.' Its long slender feathers can be laid back and almost invisible; or the bird can fan them out into a spectacular umbrella-like half-dome beside and in front of its head. Another ornamental appendage is a thick tassel or 'pendant plume' of glossy feathers hanging down in front of its breast. Bates later had the good fortune to see an umbrella bird make its mating call. 'It drew itself up on its perch, spread widely the umbrella-formed crest, dilated and waved its glossy breast-lappet, and then gave vent to… a singularly deep, loud, and long-sustained fluty note.' Wallace described his difficulties in collecting this rarity. The umbrella bird is shy, perches in the highest trees and, 'being very muscular, will not fall unless severely wounded'. So his Indian hunter worked very hard to locate and catch them, rising before dawn and returning late at night, with at most one or two birds. (It is notable that Wallace, Bates and Spruce always gave full credit to their local collectors. None ever pretended that he himself had got a prize when he had not.) There was also a problem in preparing specimens, because the umbrella bird's neck has 'a thick coat of muscular fat, very difficult to be cleaned away' but which, if left, would putrefy and cause the feathers to drop off. Wallace was justifiably proud of his taxidermic work with this bird, so he sent an essay about it to his agent. The ever-efficient Samuel Stevens communicated this to the Zoological Society, which published the piece in its *Proceedings* of 1850.

Back in Manaus in February, Wallace found the incessant rain depressing, and the humidity made it very difficult to preserve his specimens. 'Insects moulded and the feathers and hair dropped from the skins of birds and animals, so as to render them quite unserviceable.' But he was pleased to find that Henry Bates had arrived while he was away. Marooned in a dull town during that rainy season, the two naturalists doubtless compared their collections and might have developed their ideas about evolution. There were bright spells amid the rain, so Wallace and Bates had almost daily rambles in nearby forests. Both were struck by the abrupt contrast between the vegetation along the main Amazon and the uniform dark-green rolling forests of the Rio Negro. The former had far more palms, and masses of broad-leaved leguminous (bean-pod) trees. The latter had a gloomy, monotonous aspect, with smaller but elegantly leaved trees of the laurel, myrtle, bignonia and rubiaceae families. Both naturalists enjoyed these pleasant outings, which included visiting a waterfall that was a popular picnic destination for citizens of Manaus. This was the last time that they would collect together.

Wallace and Bates were together for the younger man's twenty-fifth birthday on 8 February 1850 (Wallace had turned twenty-seven in the previous month), but neither celebrated such milestones so far from their families. Instead, they started to plan 'further explorations' up-country, after the uninviting weather had cleared. They again decided to go in different directions. Bates wrote to their agent Stevens that he had 'agreed with Mr. Wallace to take the Solimões [the Brazilian name for the main Amazon between Manaus and the Peruvian border], leaving the Rio Negro to him'. When Bates departed on 26 March on his voyage westwards, he wrote: 'I bade adieu to my countrymen in Manaus with regret, as we had spent many pleasant weeks together.' So it is wrong to imagine that the two had quarrelled seriously. They simply had different approaches to travel and collecting, and preferred to do it alone.

Before leaving, Bates sent five boxes of insects to Stevens. He watched proudly as Senhor Henriques's 'splendid vessel [descended the river] with all sails set, with my collections and letters for home on board'. But he wrote to his agent to explain why he had collected so little in the seven months since leaving Belém. There had been a late departure 'in this lazy country [so that he] missed the right seasons for collecting', then 'a long and tedious journey' of five months broken only by five weeks' collecting at Óbidos, followed by nine weeks being 'delayed again miserably' before reaching Manaus.

His voyage up the Solimões was little better. He was in a merchant's small cuberta manned by ten stout Kocama Indians (Tupi-speakers from far up the Solimões, now almost extinct). It started well, with four days of sailing, but then the wind dropped and they progressed by painfully slow espia hauling, heaving the boat against the current with hawsers around riverbank trees. Bates joined in: 'On cool, rainy days we all bore a hand at the espia, trotting with bare feet on the sloppy deck in Indian file, to the tune of some wild boatman's chorus.... [Much of the journey] was accomplished literally by pulling our way from tree to tree.' Bates aired his woes in a letter to Stevens – it is part of an agent's duty to console his workers. 'It was the height of the rainy season: some days and nights the rain came down in torrents, pouring in upon us as we slept in our close little cabin; and in the day, when the sun came out, the heat was terrible.' There were insect pests that they all had to endure: during the day swarms of pium black-fly, and mutuca horse-flies; at night there were a few mosquitoes. Bates had embarked with good provisions, but these soon ran out because he shared them with the Indian paddlers; so 'afterwards we had nothing but rotten manioc and, when [there was] no fresh fish, semiputrid salt ditto.'

Finally, after five weeks, this 'intolerably wearisome' 650-kilometre (400-mile) journey ended when they paddled south on a 15-kilometre (10-mile) black-water channel into the great lake Tefé – actually the estuary of a river called Tefé and two others that flow into the Solimões. Bates's spirits rose. They landed on 1 May 1850 at a pretty village called Ega (now named Tefé, like the lake and river of that name). 'The little town appears quietly reposing on its green sward, encircled by a white sandy beach, on which the swell from the lake rolls with a pleasant dreamy murmur, and a line of sombre virgin forest forming the back ground.' Bates had selected this to be his base for the coming year's collecting, and he had chosen well. Ega had about a hundred houses, palm-thatched or whitewashed with red tiled roofs, each with a pallisaded garden of fruit trees. Cattle and sheep grazed on the grass streets. Behind the town was a grassy hill and, beyond that, 'the eternal forest' where the naturalist was to do months of fruitful collecting. Apart from a handful of white and black inhabitants Ega was peopled by 'indolent and peaceable' Indians, whose facial body paint indicated their tribes: Yuri, Pauxi, Ticuna, Witoto-speaking Miranha, and Tupi-speaking Kocama (in modern spellings). All these peoples survive, but only the Ticuna now flourish as a large nation. All spoke lingua geral, the lingua franca of Amazonia. Bates immediately started to learn it.

Bates wrote that his expenses at Ega were trifling. He rented a dry and spacious cottage and hired an Indian lad – all for about £6 a year. Food was wonderfully cheap: a large turtle eight pence, a large fish sixpence, beef was rare but cost only a penny for a pound, huge bunches of bananas ('a very necessary food') a few pence or given away, as were baskets of delicious oranges. The staple food was of course manioc, and here again Bates was often given it for nothing.

After his daily regime of collecting he enjoyed the company of the few non-Indian Brazilians, to whom he was introduced immediately after arrival. There was the police chief, Antonio Cardozo, 'a stout, broad-featured man' with a ruddy complexion (white, with a dash of black). 'He received us in a very cordial, winning manner' and in the coming years Bates was constantly 'astonished at the boundless good nature of this excellent fellow, whose greatest pleasure seemed to be to make sacrifices for his friends.' Originally from Belém, Cardozo was also a small-scale trader of natural produce, employing half-a-dozen Indians to gather these. Bates then met the military commandant, Commander Praia of the Brazilian army, 'a little curly-headed man… always merry and fond of practical jokes. His wife, Dona Anna, a dressy dame from Santarém, was the leader of fashion in the settlement.' The vicar, Padre Luiz Gonçalvo Gomes, was – surprisingly – an Indian from a local village who had been educated in Maranhão. Bates later saw a good deal of this vicar whom he admired (unlike most Brazilian priests) because he was 'an agreeable, sociable fellow, fond of reading and hearing about foreign countries' and refreshingly free of religious prejudices. 'I found him a thoroughly upright, sincere, and virtuous man.' Then there was a venerable merchant, Romão de Oliveira, 'a tall, corpulent, fine-looking… shrewd and able old gentleman'. He had established a range of depots, employed many people and traded extensively, because the Indians 'respected old Romão'. Once, when Bates was leaving on a trip, Oliveira placed his store at his disposal; but when Bates later tried to settle for the goods he had taken, the old merchant refused to take any payment.

Thanks to his gregarious good nature and his command of Portuguese, Bates made the most of this small group of elderly friends. 'The score or so of decent quiet families… were very sociable; their manners offered a curious mixture of naïve rusticity and formal politeness.' Although of little education, they were 'of quick intelligence [and, wishing to learn, were] civil and kind to strangers from Europe.' They, and the mamelucos and Indians, 'seemed to think it natural that strangers should collect and send abroad the beautiful birds and insects of their country', once Bates had explained about the natural history museums in every capital city. On one occasion Bates was telling all this to 'a listening circle seated on benches in the grassy street', and a prosperous tradesman exclaimed enthusiastically: 'Let us treat this stranger well, that he may stay amongst us and teach our children'. There were frequent 'social parties, with dancing and so forth'. His popularity as an eccentric was shown during a local carnival, when 'an Indian lad imitated me, to the infinite amusement of the townsfolk.' The boy had borrowed Bates's old blouse and straw hat. On the night of the performance, he was 'rigged out as an entomologist, with an insect net, hunting bag, and pincushion. To make the imitation complete, he had borrowed the frame of an old pair of spectacles, and went about with it straddled over his nose.' But the Englishman does not appear to have formed any relationship with girls (who were secluded by their families) or even made friends with young men of his age in that tiny conservative community.

Bates was initially happy at Ega because he was earning his living by doing what he loved: collecting natural history specimens. He wrote: 'I led a quiet, uneventful life... following my pursuit in the same peaceful, regular way as a naturalist might do in a European village.' For weeks on end his journal recorded little more than notes about his daily collecting. One room of his spacious cottage was his workshop and study, with a large table and his small library of reference books on shelves and in rough wooden boxes. There was excellent ventilation, with air (and sometimes rain) entering through the gap between walls and rafters. Bates dried his specimens in cages hanging from the roof beams on cords that were soaked in bitter vegetable oil to deter ants and with upturned gourds on them to stop rats and mice. Bates's routine was to rise with the sun at six in the morning, and walk down a grassy street, wet with dew, to bathe in the river from the beach. After breakfast coffee, he spent five or six hours collecting in the forest – which started close to his house. In the hot hours of the afternoon from three to six o'clock (or throughout rainy days) he prepared, dissected and labelled his specimens and made notes or drew them. He would often make short 'rambles by water' in a small montaria paddled by his Indian lad.

Bates reached Ega at the end of the rainy season of 1850. The forest was still moist and cold, so that he often missed collecting during his first six weeks there. But he wrote to Stevens in mid-June that he had already gathered 'more than forty conspicuous *new* [his italics] diurnes [day-flying butterflies]'. Butterflies and beetles were Bates's passion. He wrote excitedly to Stevens about three new species of *Papilio* (a large genus of butterflies with swallow-like points below their wings), 'one with silky-green spots on the fore wings and crimson on the hind wings, extremely beautiful'. Another unique *Papilio* was 'extraordinary, having exactly the form and style of marking of the heliconia [a less numerous family of butterflies, with more rounded wings; see also the footnote on p. 319].' (This observation about mimicry was later developed by Bates into an important scientific discovery.) Bates made another interesting observation about these particular *Papilio* butterflies: that most subspecies were very local, so that every collecting-station all the way up the Amazon yielded one or two new examples, whereas *Papilio sesostris* (now *Parides* *sesostris*) seemed to be the only species common to all localities. He sent two new species of Nymphalidae (*Adelpha*), and 'two new Haeterae, sister of Esmeralda, very beautiful, though I believe known to Europe'.* A few months later, Bates wrote to his agent about another fine collection he had sent. He was sure that there were many new Nymphalidae and Papilionidae. 'I numbered 124 species of the diurnes taken *new to me*, and now number 922 species taken by me in the province [of Amazonas].' He was also sure that the expert collector William Hewitson would appreciate his sending 'a short series of the Heliconia-like Papilio'. In a later batch, Bates included 158 specimens of micro-Lepidoptera. He reckoned that there were thousands of species of these miniature butterflies, but had scarcely bothered to catch them because collectors were not interested in these unglamorous little insects. Great blue morphos were another matter. Everyone wanted these beautiful creatures. Bates did send quite a few different species of morpho, but, as he had found in Óbidos, they were very difficult to catch. He had nothing more than a classic butterfly net: he deployed none of the armoury of mist nets, tree-fogging gases or light traps of modern entomologists. In his letters, he discussed successes and difficulties in collecting the various genera of Lepidoptera. He asked Stevens to send him the names and numbers of all the species he had sent, 'common and rare, with advice to each' so that he could recognize these in his notebook. His faithful agent also sent relevant new publications, such as 'the second part of White's 'Catalogue of Longicorns'... and the concluding volume of Boheman's 'Cassidae'.' Bates ended his letter on a positive note: 'I am still in excellent health, and ready for any kind of sport you could mention.'

Beetles were Bates's other passion. In June 1850 he sent 'a great number of Coleoptera I have not seen anywhere else'. A year later, he dispatched more 'good things' in beetles and noted that he had a further 1,600 to dispatch for sale. (He was also amassing his own private collection of every species.) There was a *Titanus* species of prionine beetle, but not the largest of these: Bates had tried very hard to collect this but, although he saw them two or three times, he had not yet taken a perfect specimen. He sent 'some new and fine things' among longicorns and curculiones, particularly of the *Gorgus* genus.

Bates explained to Stevens how a passionate collector performed – and as usual he gave due credit to others. 'I worked very hard for Coleoptera in Ega from January to March [1851], being the showery and sunny season, before the constant rains set in. Whenever I heard of beetles seen at a distance, I would get a boat and go many miles after them, and employed a man (the only one disposed for such work in the whole village) with his family,

* When Stevens published this letter in the *Zoologist* journal, he added a footnote that there was indeed one on display in a cabinet of the Natural History Museum.

who worked in some clearing in the forest, to hunt for me. Every day he brought me from ten to twenty Coleoptera, and thus I got some of my best things.' He reckoned that he had collected a pretty good representation of beetles of the upper Amazon.*

There were also some rare birds. Bates again paid tribute to his local helpers. He wrote that time and patience were needed to catch birds, 'as the hunters here, on whom you are obliged to depend, are very slow' – and he needed more funds 'from below' (i.e. from England via Belém) to pay them. He also collected birds himself. He had realized the futility of guns for doing this – their noise frightened off game, and their shot damaged specimens. So in Manaus Bates bought a zarabatana blow-gun, as used by indigenous hunters throughout the western and northern Amazon. He and Wallace got a young Yuri Indian to train them in its use. The blow-gun was, of course, silent. Bates described the enormous amount of patient and skilled labour involved in making a zarabatana of deadly accuracy. A long and straight length of wood is split, each half hollowed using a pacá-tooth tool, then bound together with coils of strips from jacitara climbing palms (*Desmoncus polyacanthos*), and smeared with black beeswax. A mouthpiece is fitted to one end of the weapon. A hunter's main difficulty was to hold this long (almost 3 metres or 10 feet) and heavy blow-gun steady. The darts were made of sharp needles from the rind of palm stalks, and winged with a small oval mass of very light silk from the seed vessels of the silk-cotton or kapok tree. A sharp puff by an expert Indian hunter propelled this straight to the target prey. The darts were smeared with curare poison, to relax a bird's or monkey's muscles to cause asphyxiation, and so that it fell to the ground rather than clinging to a branch in rigor mortis.†

Bates loved toucans, those gaudy birds whose beaks seem as long as their bodies. It was well known that toucan beaks were porous and very light, so that they did not weigh down the front of the flying bird. Bates dismissed rumours that toucans ate other birds or even fish: their food was fruit and some insects. But the young naturalist deduced a form of evolution by natural selection. Toucan beaks had become elongated, but light, so that the birds could get at fruit that 'on the crowns of the large trees of South American forests

* We now know that this was wildly optimistic. There are tens of thousands of species of beetle, so that each locality is still yielding new ones, many of them endemic to that one place.

† Alexander von Humboldt described preparation of curare, and it was the main objective of Charles Waterton's expeditions into southern British Guiana in the 1820s. On return to London, Waterton experimented with curare at the Royal Society. He apparently 'killed' a she-donkey with it. But he then showed that curare was a muscle-relaxant by having Fellows of the Society revive the animal by pumping air into its wind-pipe. The ass was named Wouralia (curare) and lived to old age on Waterton's estate near Wakefield in Yorkshire. Indigenous peoples prepare curare in many different ways. In about 1840 Robert Schomburgk showed that a frequent ingredient was the liana *Strychnos toxifera*, but there are at least twenty Amazonian plants with toxic alkaloids. Indians also mix in poisons from frogs such as the 'arrow-poison' *Dendrobates*, and from a range of other toxic worms and insects.

grow, principally, towards the end of slender twigs, which will not bear any considerable weight'. Monkeys reached for fruit with long limbs, hummingbirds by hovering, but toucans – greedy, but with heavy bodies and feeble wings – got at food by sitting on stronger nearby branches and gulping with their long beaks.

Bates had a pet toucan. He found it in the village, half-starved: the bird had apparently escaped from a house, but he could not find its owner. After a few days of care, the toucan perked up and became 'one of the most amusing pets imaginable'. Bates was struck by the 'intelligence and confiding disposition' of his 'Tocano'. He allowed him to fly freely anywhere in the house – apart from his precious work table. The toucan joined in meals, which took place on a cloth on the floor mat since the table had more important uses. The bird knew the exact time of each meal, and ate everything that the humans did: beef, turtle, fish, fruit and manioc. 'His appetite was most ravenous, and his powers of digestion quite wonderful,' but at times he became 'very impudent and troublesome'. He used to strut around on the grassy street. One day, to Bates's distress, Tocano was stolen. But two days later he stepped through the door at dinner time, 'with his old gait and sly magpie-like expression' having escaped from a house at the far end of the village.

Hunting in the forest one day, Bates shot a curl-crested toucan (*Pteroglossus beauharnaesii*). The 'curl-crest' of its name referred to black feathers on the top of its head whose ends became thin, horny, curled plates that looked like shavings of ebony wood or a wig of tightly curled hair. The bird that Bates shot fell from a tall tree, but it was only wounded. When he went to secure his prize, the curl-crested toucan 'set up a loud scream. In an instant, as if by magic, the shady nook seemed alive with these birds,' even though Bates had noticed none. 'They descended towards me, hopping from bough to bough, some of them swinging on the loops and cables of woody lianas, and all croaking and fluttering their wings like so many furies.' Bates killed the wounded bird and planned to catch other specimens, partly 'to punish the viragos for their boldness'. But as soon as the first bird ceased screaming, the others instantly disappeared. Bates drew a vivid picture of himself being mobbed by these angry toucans [PL. 1]. It is one of only three representations of him during his Amazonian years. He wears round granny glasses, has tangled hair and sports a full moustache and mutton-chop sideburns. His flat hat is broad-brimmed, his checked shirt buttoned at the neck and wrists (as is normal to protect against insects), and his trousers held up by a cord that also secures a collecting bag. In cinematic terms, Bates looks like a startled Woody Allen in a scene from Hitchcock's *The Birds*.

Four of Bates's most interesting excursions from Ega were concerned with Amazonian freshwater turtles. There are a dozen species of these famous chelonians, ranging from the great tartaruga (*Podocnemis expansa*) that can grow to almost 1 metre long by 60 centimetres wide (roughly 3 by 2 feet), medium-sized tracajás (*Podocnemis unifelis*), weird-looking long-necked matamatás (*Chelus fimbriatus*), and smaller hand-sized perema (*Rhinoclemmys punctularia*) and muçuã (*Kinosternon scorpioides*) turtles. For the first trip, Bates went in a boat paddled by ten men, who took a day to sail out into the main Solimões and to an island where turtles laid their eggs. The community of Ega had posted two sentries high in a tree, to drive off interlopers from other communities and to prevent anyone from disturbing the shy creatures, or their highly valued eggs and the oil within them. The oil, known as manteiga ('butter'), was used primarily to light lamps or fry fish. Female turtles emerged from the river 'in vast crowds' every night for a fortnight, and laid their eggs in sandbanks. 'The turtles excavate with their broad webbed paws deep holes in the fine sand: the first comer... making a pit about three feet [90 centimetres] deep, laying its eggs (about 120 in number) and covering them with sand; the next making its deposit at the top of that of its predecessor, and so on until every pit is full.' Bates had a cold night sleeping on a sandbank, and at dawn climbed to the sentinels' perch to watch the sands blackened by multitudes of turtles waddling back towards the river and tumbling headfirst down a steep sandbank.

The next adventure was a hunt for grown turtles [PL. 20]. Two boatloads of nineteen mostly Miranha Indians, Bates, his friend the police chief and trader Antonio Cardozo and a couple of others spent a day paddling down the big river against a squally east wind. They then cut a trail into forest on the north bank to reach a lagoon in the labyrinthine estuary of the Japurá river, and pushed their boats on log rollers along this picada path. The hidden lagoon was 'known only to a few practised huntsmen'. As so often, Bates lovingly described it, ringed with ferns and fine matupá grass, beyond which was a green palisade of arborescent arums, then a taller forest of palmate-leaved cecropia and slender açaí palms with their feathery heads, and 'as a background to all these airy shapes, lay the voluminous masses of ordinary forest trees, with garlands, festoons, and streamers of leafy climbers hanging from their branches.' Although the hunters had a 65-metre (70-yard) net, the Indians felt that it was more sporting to start by shooting turtles with their bows and arrows. They did this with uncanny skill. They would notice a slight movement on the surface of the water, instantly fire an arrow at the still-submerged animal, and never fail to hit it. But they preferred more distant prey, because they then shot into the sky and the arrow's trajectory curved down to hit the turtle from above where its shell was thinnest.

The skill involved in such shooting was awe-inspiring. The Indians had special arrows with a barbed metal point that broke off when it entered the turtle. The chelonian dived, but the point was attached to the floating arrow shaft by a 25-metre (30-yard) twine made from pineapple-leaf fibres. The hunter then paddled to the spot and gently raised the turtle until it could be finished with a second arrow. After his men had shot a score of turtles in this way, Cardozo ordered the net to be spread along the bottom of the shallow pool. For an hour a converging circle of Indians beat the waters to drive the turtles into the middle, 'the number of little snouts constantly popping above the surface of the water showing that all was going well.' The Indians then heaved the ends of the net together, so that the trapped turtles could easily be grabbed and tossed into the canoes. Bates jumped in and joined the melee. Cardozo remained in a boat flipping the turtles onto their backs. But he could not do this fast enough, so that many clambered out and got free again. In twenty minutes' hard work they had caught about eighty turtles. These were mostly younger animals aged between three and ten years. They measured from 15 to 45 centimetres (6 to 18 inches) long, but were very fat [PL. 20]. There was more collecting on the following two days, although with smaller catches because, according to the Indians, the turtles had got smart and ignored the beaters. The animals collected during those days became basic food. Bates and Cardozo 'lived almost exclusively on them for several months afterwards. Roasted in the shell they form a most appetizing dish.'

The third excursion was to gather the turtle eggs Bates had witnessed being laid three weeks earlier. All the people of the villages around Ega turned out for this harvest. Bates and Cardozo, in their well-manned igaraté, passed 'a large number of men, women and children in [boats] of all sizes wending their way as if to a great holiday gathering.' About 400 camped on the island, each family building itself a palm-thatched shelter. They had large copper kettles to render oil from the eggs, and hundreds of red earthenware jars were scattered about on the sand. The operation took four days. It followed a procedure laid down during Portuguese colonial rule a century earlier. An official recorded the names of each head of house, took a small payment to defray the cost of the sentinels, arranged everyone in a large circle and, at a roll of drums, all were allowed to start furiously digging. 'It was an animated sight to behold the wide circle of rival diggers throwing up clouds of sand in their energetic labours.' They rested only during the midday heat. By the end of the second day, the sandbank was exhausted. There was a pile of small round white eggs beside each family shelter, some of these around a metre and a half (4 or 5 feet) high. The heaps of eggs were then thrown into empty dugout canoes and mashed with wooden prongs or pounded by naked feet – turtle eggs are leathery but without shells. Water was poured onto the mess of crushed eggs, which was left for a few hours in the sun; the oil rose to the surface and was then scooped out with mussel-shell spoons and purified over fires in the copper kettles.

The turtle-egg harvest on the island near Ega was just a prelude. On 20 October Senhor Cardozo led another flotilla of canoes on an eleven-hour journey, for 100 kilometres (60 miles) downriver to an even greater turtle beach called Catuá. Collecting here was on an industrial scale: the beach itself was 10 kilometres (6 miles) long; hundreds a people from many communities were involved – their huts and sheds stretched for a kilometre (half a mile); and there were several large sailing ships to carry off the pillaged oil. The collectors included 'primitive Indians' from isolated forests and rivers. Among them was a family of Arawak-speaking Yumana from the lower Japurá – gentle good-natured people (now sadly extinct). These Yumana had a bluish tattoo around their mouths. Bates was smitten by their seventeen-year-old daughter, 'a real beauty' with skin of 'light tanned shade; …her figure was almost faultless, and the blue mouth, instead of being a disfigurement, gave quite a captivating finish to her appearance.' But she was extremely bashful and clung to her dignified father's side. As always, the young Englishman could admire local women only from a distance.

The egg-hunt at Catuá beach was as efficiently organized as the smaller one upriver. After days of frenzied digging and rendering there were evenings of dancing and games, to music of guitars and violins. There was much merriment, watched by 'the more staid citizens of Ega…[who] enjoyed the fun' but made sure that it was well behaved. Prodigious quantities of cachaça were drunk, which made the 'shy Indian and mameluco maidens… give way a little', but although Bates 'mixed pretty freely with the young people… [he never saw] any breach of propriety on the praias [beaches].' He was sure that this was because there were no 'low Portuguese traders, who are most certainly the inferiors of these rustics whom they despise'. These bullies would have tried to corrupt the women, get indigenous men drunk, and incite them to steal turtle oil from their masters – for many of the Indians collecting eggs were doing so for bosses like Cardozo.

Bates reckoned that with this wasteful process of extracting oil, it took about 6,000 eggs to fill a jar. Some 6,000 jars of oil were exported to Belém or consumed locally, just from this part of Brazil. So 36 million eggs, or the offspring of 400,000 turtles, were annihilated each year from the Solimões. To make matters worse, people lay in wait for any newly hatched turtles that escaped the egg massacre, since these were tasty to eat. Of course, without human hunting many young turtles would have been devoured by birds, caimans or other predators. In colonial times, Jesuit missionaries had exercised rough sustainability, by allowing only half a turtle beach to be pillaged each year. But after their expulsion and with Brazilian laissez-faire, human profligacy was unbridled. People admitted that 'formerly the waters teemed as thickly with turtles as the air now does with mosquitoes. The universal opinion of the settlers on the Upper Amazon is that the turtle has very greatly decreased in numbers, and is still annually decreasing.' Surprisingly, Bates did not condemn this slaughter as passionately as Humboldt had in 1800. Nothing was done about it, and it continued unabated every year throughout the century. Thus, both the big tartaruga and smaller tracajá turtles are now endangered species. Destruction of Amazonian turtles was the greatest environmental catastrophe of the nineteenth and early twentieth centuries. In other respects environmental damage in Amazonia was slight during this long period. This was not because people were aware of any need for conservation. It was just because they had little commercial incentive to destroy forests and none of the modern machinery (chainsaws, earthmovers) to do so easily.

On the second day of the turtle hunt in the Japurá lagoon, when the men jumped in to the closing net they found that they had also caught a large black caiman (*Caiman niger*). The Indians were not at all alarmed, apart from fear that the big animal's thrashing would tear their net. One lanky Miranha 'was thrown off balance, and then there was no end to the laughter and shouting.' Bates, standing on the bank some distance away, called to a boy to catch the caiman, which he did, grabbing it by the tail and slowly dragging 'the dangerous but cowardly beast' through the muddy water to the shore. Bates had meanwhile cut a pole from a tree and, when the caiman landed, 'gave him a smart rap with it on the crown of his head, which killed him instantly.' It was a big specimen, with jaws over 30 centimetres (12 inches) long 'and fully capable of snapping a man's leg in twain.' Bates drew the scene at the lake, with a dozen laughing Indians, three canoes, the black caiman and himself poised with his club ready to smash it. He wears the same buttoned-down checked shirt, tartan trousers and broad-brimmed hat as in his sketch of himself being mobbed by toucans. His moustache is bushy and thick sideburns come down to his chin in a nascent beard [PL. 20]. A few weeks later, when he was at the second turtle-egg harvest, there were plenty of huge black caimans feasting off the turtle offal thrown into the river. Bates amused himself by throwing chunks of meat to a group of the 'monsters', who caught the bait in their huge gaping jaws 'pretty much as dogs do'. (This author has seen black caimans taking food in the same way, at the Mamirauá research station, not far from Tefé, in the twenty-first century.) One night when Cardozo, Bates and others were asleep in their hammocks in a beach shelter, an 'intolerably impudent' black caiman crawled up onto the sandbank, passed under Bates's hammock, and hoped to devour Cardozo's beloved little dog. The pet's owner drove the saurian off with a barrage of flaming logs from their campfire. Bates recorded the incident in another of his lively drawings. This showed how much his hair, moustache, sideburns and beard had grown, and how he slept fully clothed [PL. 19].

Ega lay in the heart of the world's richest ecosystem. So, right up to the last day of his year there, Bates gathered 'an uninterrupted succession of new and curious forms in the different classes of the animal kingdom, but especially insects.' At regular intervals he dispatched his boxed collections, all the way down the Solimões and Amazon rivers, to Manaus and then on to Belém, where they were transhipped across the Atlantic by the Campbell shipping agents.

Although collecting went well and the elderly people in Ega were friendly, Bates grew lonely and depressed. He missed intellectual society and the friendship of people of his own age or nationality. To his surprise this lack of 'the varied excitement of European life... instead of becoming deadened by time, increased until it became almost insupportable.' He concluded, sadly, that contemplating nature in itself was not enough 'to fill the human heart and mind'. Twelve months elapsed without any letters or remittances, his clothes were in rags, he was 'barefoot, a great inconvenience in tropical forests; ...My servant ran away, and I was robbed of nearly all my copper money.' He was in ill health, which he imagined was caused by bad and insufficient food. In his journal in September 1850 he wrote: 'I am a prisoner here; at this season no vessels descend to the city [Manaus], on account of the trade winds blowing strong up river.... I am reduced to a low ebb.' Matters improved when, after a week's journey upriver from the turtle beach, he returned to Ega on 10 November 1850. He was greeted by a packet of letters and £40 from Stevens; and this was followed in December by two further remittances. He was able to renew his battered clothes and shoes and repay some small debts. But a letter from his father begged him to return to England, because the family hosiery business had grown and needed him. Bates wrote to Stevens that he had weighed up the prospects of collecting further upriver in Peru, but decided to accede to his father's request. In January 1851 he confided to his journal: 'Considering the unsatisfactory nature of my future prospects in this profession, I think I do better in returning to a more certain prospect of establishing myself. I have now, therefore, only one idea, that of returning to England.' So, on 21 March 1851 Bates left Ega, sailed down the Amazon to Belém, and intended to continue home to England.

After his friend Bates left Manaus at the end of March 1850, Alfred Wallace made another short excursion: for two months from May to July to Manaquiri on the Solimões. Every spring, swollen by rains in the Andes, the Amazon rises dramatically and inundates a vast floodplain known as the várzea (or igapó elsewhere). Wallace's trip started by moving through this flooded forest, to avoid the big river's current. This is a magical experience. The traveller glides above small streams, lakes and swamps 'and everywhere around him will stretch out an illimitable waste of waters, but all covered with a lofty virgin forest. For days he will travel through this forest, scraping against tree-trunks, and stooping to pass beneath the leaves of prickly palms [whose crowns are] now level with the water though raised on stems forty feet [12 metres] high' [PL. 36]. Since their boat was floating up in the dry season's canopy, they could pluck the fruits of palms such as the marajá (*Bactris brongniartii*). 'In the gloom of the forest, among the lofty cylindrical trunks rising like columns out of the deep water' they might chance upon flocks of feeding parakeets, 'or some bright-blue chatterers, or the lovely pompadour with its delicate white wings and claret-coloured plumage; now with a whirr a trogon would seize a fruit on the wing, or some clumsy toucan make the branches shake as he alighted.'*

It was uncanny how the Indians found their way through this trackless maze. They moved with unerring certainty, paddling for days on end without getting lost. They showed Wallace how endemic trees of this igapó had evolved to resist having their trunks submerged for half the year. There were also birds peculiar to this ecosystem, such as the umbrella bird (the cotinga that Wallace had collected on the islands of the Rio Negro a few months earlier) and the beautiful and rare little manakin cotinga. The Mura and other tribes had adapted to this flooding by having 'easily-movable huts' on sandbanks during the dry season and on rafts during the wet: they then slept in hammocks slung from trees above deep water and subsisted on fish and turtles but no vegetables. There was never dry land for Wallace's group to cook their supper. Back on the main river, they found a wedged log, made a fire on it, roasted their fish and boiled some coffee. But they had 'intruded on a colony of stinging ants, who... swarmed into our [boat] and made us pay for our supper in a very unpleasant manner.'

When they reached Manaquiri, Wallace stayed with Henrique Antonij's father-in-law José Antonio Brandão. This energetic seventy-year-old told Wallace how he had come from Portugal as a young man and decided to devote his life to farming. 'He built

---

* Chatterers are either waxwings (*Bombycillidae*) or various species of cotinga. The 'pompadour' was *Xipholena punicea*, also known as the red-wine cotinga. Trogons are smaller, gaudily-coloured birds often with long tail feathers decorated with horizontal black-and-white bars.

himself a country house at Manaquiri, on a lake near the main river, brought Indians from a distance to settle with him, cleared the forest, planted orange, tamarind, mango, and many other fruit-bearing trees, made pleasant avenues, gardens and pastures, stocked them well with cattle, sheep, pigs and poultry, and set himself down to the full enjoyment of a country life.' But during the Cabanagem rebellion of 1836, when Brandão himself was absent in Manaus, a nearby group of Indians 'to whom he had always been kind' burned his house and destroyed his garden, fruit trees and livestock – all because he was thought to be too Portuguese. His wife and children escaped into the forest where they spent three days living off wild fruits. Brandão remained for several years in Manaus, where he was police commissioner and where his wife died. He then returned to the Manaquiri farm and restored it with a mass of fruit trees and European cattle, but he did not bother to rebuild his house. Wallace was surprised, and perhaps wistful, to see Brandão's daughter, 'a nicely-dressed young lady sitting on a mat on a very mountainous mud-floor, and with half-a-dozen Indian girls around her engaged in making lace and needlework.' He was also impressed by the old man's remarkable intelligence, general knowledge and thirst for learning. The autodidact Wallace modestly remarked that Brandão's education had been acquired without the Mechanics' Institutes and cheap literature that he himself had enjoyed.

Two months at Manaquiri yielded good collecting. A hunter, whom Wallace had engaged in Manaus, brought him birds and monkeys, and he himself went out to shoot more and to net insects – particularly rare butterflies. One of his best catches was a curl-crested toucan – the same species that was mobbing Bates at that time, far to the west up the Solimões. Wallace worked hard, as always. He was up at 5.30 and spent hours each morning skinning birds as well as stalking and collecting in the forest. His reward was two fine meals a day: the Amazon's most delicious fish, tambaqui (*Colossoma macropomum*), or farmed chicken or pork, with rice, black beans, manioc bread and as many oranges as he wished.

Richard Spruce spent most of 1850 in and around Santarém. After his seven-week excursion to Óbidos and getting lost on the Erepecuru river, he sailed back down the Amazon to Santarém instead of continuing up to Manaus to join Bates and the Wallace brothers. He returned to the delightful town at dawn on 6 January, and 'I was struck with the beauty of its site. The newly-risen sun illuminated the lines of white houses, stretching parallel to the river, where numerous vessels of all sorts and sizes were anchored or moving about; and at the back the shrubby campos swelled into bare hills, backed by distant blue wooded ridges.'

The rains and flooding in 1850 were the worst in memory. The rains set in at Christmas and continued 'with unrelaxing severity' for four months. There were violent thunderstorms, heavy downpours particularly at night and, because the winds had dropped, oppressive heat during any 'fits of sunny weather'. Flood-marks rose to their highest ever. Rivers burst their banks, streams became lakes, and low-lying land was inundated.

Spruce made the most of this unprecedented situation. Normal collecting was obviously curtailed. So he studied a phenomenon of the floods: unusually vast 'grass-islands' floating down the main Amazon. These compact masses of living grass measured from about 50 metres (150 feet) diameter to around a hectare (several acres) in area. When struck by a sudden squall, small boats could sometimes shelter by pushing into this dense mass. But the islands, racing in a four- or five-knot current, could also carry off or even destroy boats anchored in the river.

The floating islands were generally formed of two grasses: canarana (*Echinochloa polystachya* and *E. pyramidalis*), a common pasture in the lower Amazon, and caapim (the name for grass in general, in Tupi and lingua geral). Both types spread vigorously in the shallow waters of lakes alongside the Amazon. Spruce found that when the flooding river burst into a lake it gradually washed the earth away from caapim roots, until the loosened mass was detached, whirled around, and carried off to the main stream of the world's largest river. Examining one floating island, Spruce reckoned that its mass of roots was 6 to 9 metres (20 to 30 feet) thick; and when 'after many futile attempts, I succeeded in drawing up an entire stem of the grass [it] measured 45 feet [14 metres] in length and possessed 78 nodes.' The stems bore vigorous panicles of flowers, so that the island resembled a luxuriant meadow; and there were minute plants (one new to science) growing amid the grass.

The flooding Amazon drowned riverbank cacao plantations, so that their owners took refuge in palm-thatched shelters near towns. The Englishman Jeffries had a plot of manioc bushes. Alarmed by the sudden rise of the waters, Jeffries had all his hands work for several days digging up the roots, leaching and dressing them, and baking farinha. It was midnight on the last day when they removed the last batch of farinha from an oven in the plantation. They were just in time. 'The next morning the oven and the whole of the field were laid completely under water!' Cows – a rarity in Amazonia – could find no pasture, so they wandered into the forest and thinned, starved or drowned; and Spruce was deprived of his breakfast milk and dinner beef.

Spruce studied aquatic plants as well as the islands of grasses. He found some *Victoria regia* (now *V. amazonica*), the famous water lily whose leaf was the size of a bicycle wheel. There were floating *Riccia* liverworts and *Azolla* and *Salvinia* water-ferns. But the botanist was struck by 'a curious and beautiful Euphorbiad (*Phyllanthus fluitans*)' with pale-green heart-shaped leaves, a white fascicle under each leaf, and small white flowers. Utterly different botanically from the salvinia fern, 'it was so like it in external aspect that I could hardly believe my eyes when I found it to be a flowering plant.' Spruce pondered how such widely different plants could look alike and have similar vegetative organs. One cause was doubtless 'the same conditions of existence'. But there were probably other reasons 'deeper than we have hitherto been able to penetrate' – perhaps some form of mimicry, as found among insects – that provoked these 'startling and unexpected simulations'. Spruce was groping towards evolution by natural selection among plants.

People felt that the floods made Santarém unhealthy. Many suffered attacks of influenza and 'slow fever', and Spruce himself 'did not escape'. Villages up the Tapajós river were hit by the worst kind of malaria, from which over 400 people died. Spruce called this ague, an old word for malaria derived from the French *aigu* ('acute'). It may have been falciparum malaria (*Plasmodium falciparum*) as opposed to the milder vivax (*P. vivax*) – which is still very nasty, as this author knows having had it twice. Like every scientist and intelligent traveller in the nineteenth century, Richard Spruce was baffled by malaria. Local people felt that it was caused by miasma vapours in seasons when the river was high and 'insalubrious': hence the name *mal aria* (bad air). But Spruce went further and reasoned that the Amazon rising faster than its tributaries caused dammed water, which became stagnant with a scum of yellowish-green slime called *limo*. He examined this under a miscroscope but found that it was just decomposing confervae (freshwater algae). Of course, no-one could imagine that malaria was from a *Plasmodium* parasite in the liver and blood, transmitted by the bite of a tiny female *Anopheles* mosquito. That source and vector were gradually discovered by scientists in different countries at the end of the nineteenth century.

Spruce had another affliction. One day he and his assistant King were in a ruined hut on a hill inland of the town. Each in turn trod on a rusty nail in the undergrowth. It took them three hours to limp back, in excruciating pain, and both were confined to their hammocks for three days. But, like both Wallace and Bates, Spruce was a tenacious collector. 'Neither sweltering heat nor soaking rains ever caused me to intermit my labours, and I went on collecting all through the wet season.' He found it very difficult to dry his large quantities of plant specimens. So Spruce used great piles of paper; and he made an arrangement with a Frenchman who was Santarém's baker to let him use his warm oven after he had removed each day's bread.

As the floodwaters receded, Spruce was thrilled by 'myriads of minute annuals... which spring up on the shores.' They appeared out of the sand, flowered, ripened their seeds and then withered in the sun. 'Notwithstanding their humble size and transitory existence they were all pretty things, many of them with showy white, yellow, or pink flowers; and nearly all proved quite undescribed.' He mentioned many bladderworts, including a *Utricularia* with a white flower, of simple structure, with a stem the size of a sewing needle fixed into the sand by a tiny cone of rootlets. This 'grew in such abundance that patches of sand of many yards in diameter were white with it.' Bentham named this new species *Utricularia spruceana*. With so many discoveries of all sizes of plant, Spruce exclaimed: 'Certainly vegetation is on a most gigantic scale in the Amazon valley, not only as regards the vast size attained by some of the species, but also in the range of magnitude from the enormously large to the extremely minute. Compare, for instance, the lofty Eriodendrons [silk-cotton kapoks] and Caryocars [piquiás] with these lowly Utricularias and Alismas [water-plantains].'

There were patches of seasonally-flooded igapó woods near Santarém, in which Spruce of course observed and collected profusely. One common legume was *Campsiandra laurifolia*, a low spreading tree with a profusion of blossoms, 'white within, rosy without.... On the extreme edge of the igapó it sometimes forms a continuous fringe of miles in length.' This tree's pods contain 'large flat beans, which little Indian boys find suitable for making ducks and drakes'. Their mothers grated them down to 'make passable farinha' – after they have strained and baked away its bitter narcotic component – but only when manioc farinha was in short supply. One of the most ornamental trees was a mimosa of moderate size, *Pithecellobium cauliflorum*, whose flowers grew directly from the trunk and branches. Each was a long thread-like stamen, 'lake-red above, white below', and these were so densely packed that the trunk looked as though 'enveloped in toucan's feathers', contrasting strikingly with the leafy but blossomless crown above.

Spruce was always interested in economic botany – plants that had a commercial value. One such was guaraná (*Paullinia cupana*), a stout climber that local people pruned to form a bush with entangled branches. The fruit of the guaraná – small, yellow and pear-shaped – enclosed a shiny black seed. This was roasted and pounded, then grated into water to form an aromatic drink with 'a peculiar and rather aromatic taste' that was a stimulant like coffee. The Mawé Indians were guaraná's main producers and it was (and still is) highly prized throughout Brazil. It stimulated the nervous system and was generally thought to be 'a preventive of every kind of sickness, especially of epidemics'. Spruce of course experimented with guaraná, but found that it was no panacea. He also tried the cherry-like yellow fruit of pitomba (*Sapindus cerasinus*), whose seed had 'a pleasant flavour of black currants'. His Santarém friends were horrified, telling him that this plant belonged to a poisonous family. But Spruce was skilled enough to reason that it was botanically close to guaraná and other sapindaceae lianas and therefore harmless and wholesome: it was the stems and roots that were deadly. He also ate the pleasant medlar-like fruit of genipapo (*Genipa americana*). The fruit of this common bush is used by indigenous peoples all over Amazonia. It is their ubiquitous black skin dye, smeared on as a beautiful decoration (along with bright-red annatto or urucum) that also serves as protection against insects.

The consummate botanist watched various plants, in different habitats, react to the changing seasons. At the height of the rains, there was a profusion of grasses, 'tall, rank, and succulent on the banks of rivers and in swampy ground, slender and wiry in the groves and thickets on the campo [savannah].' He rapidly gathered ninety species around the town. During the first three months of the year, little was in flower other than these grasses and sedges. Curiously, most trees here 'looked every day more and more dingy' during the rains, only shedding leaves and growing new ones at the start of the dry season.

Some excursions took Spruce to lofty primeval forest. One of the many trees he found there was the itaúba ('stone-tree'; *Mezilaurus itauba*). 'There is no timber on the Amazon equal to it' for building larger boats. But a dugout canoe of this heavy wood 'infallibly sinks when full of water, as I have found to my cost'. Spruce casually and modestly mentioned that this noble tree of the laurel family 'was undescribed until my specimens of its flowers and fruits afforded materials for its determination.'

Spruce was so busy working that he had very limited leisure time. Rather shy, he admitted that (unlike Bates) he was not fond of Brazilians as a whole. An exception was Santarém's magistrate, Dr Campos, a 'thoroughly urbane' man who knew English and French literature, shared Spruce's interest in mathematics, was uncorrupt (unlike his predecessors), and became a close friend. There was also the merry and garrulous old Captain Hislop and two other British residents. But Spruce had 'no pleasanter fellow or better friend than Abraham Bendelak, a Jew of Tangier' who often helped the English collector and went on excursions with him. Interestingly, Spruce wrote that there were a good many Moorish Jews in Amazonian towns: they came without their wives and families, to make some money and then return to Morocco. The Yorkshireman repeated nasty local generalizations about Amazonia's main racial groups. Mulattoes (white-black) were lazy and shiftless, 'apt to be proud and restive [but] tractable enough when held properly in hand'; but cafuzos or zambos (black-Indian) were dismissed as the most vicious and criminal types. By contrast, Spruce found that black slaves from Africa 'were mostly civil and humble, but merry withal and pleasant to deal with'. Indians baffled him as much as they did the other Englishmen. He admired their amazing skills as boatmen, woodsmen and hunters, but he deplored their lack of ambition, improvidence, emotional inscrutability and reluctance to work for pay. He would have liked to borrow a boat from Captain Hislop to go up the Tapajós, although he knew that it would be almost impossible to find Indian paddlers. 'Every free man of colour was in debt to the resident merchants, who would have exacted payment of the debts before allowing the men to embark on a voyage.' Curiously, Spruce seems to have blamed the caboclos for borrowing and failing to repay their debts, rather than the traders who tricked them into debt-bondage by advancing goods at exorbitant prices. He sometimes borrowed a boat to collect along creeks and lakes. But he had to ask the town's Comandante de Trabalhadores for paddlers and then wait up to a fortnight for them, because they 'were difficult to catch... for in all probability a detachment of soldiers would have to be sent into the interior, to beat them up at their sitios.' Spruce's passion to collect overcame any qualms about getting paddlers by press-gang.

When the rains finally ceased, Spruce wanted to move upriver to join the others in Manaus. But having no boat of his own, or paddlers, he had to get a passage on some passing boat. He waited for three months, during most of the dry season, before finally embarking on 8 October in the small boat of a French resident of Santarém, Monsieur Gouzennes. This proved to be a dreadful voyage. The craft was a small igaraté, half-filled by Spruce's luggage, and with a palm-thatch toldo that leaked so badly that clothes, papers and food were drenched. The boat had a crew of only three men, which proved inadequate because the weather was broken, rainy and either calm or squally. Worst of all, Gouzennes – who travelled alongside in a much larger cuberta – was on his annual trading voyage and for various business reasons advanced painfully slowly.

The naturalist made the most of the delays and discomforts. 'When slowly beating against the strong current in the strait of Óbidos, I twice swam on shore to gather a stout Mimoseous twiner that adorned the bank for miles with its spikes, a foot long, of minute pale yellow flowers.' And he revelled in the exuberance of nature. 'Rarely is there perfect silence on the banks of the Amazon. Even in the heat of the day... when birds and beasts hide themselves... there is still the hum of busy bees and gaily-coloured flies, culling sweets from flowering trees that line the shore, especially from certain Ingas...; and with fading twilight innumerable frogs... chaunt forth their Ave Marias, sometimes simulating the chirping of birds, at others the hallooing of crowds of people....' Nocturnal birds sang at intervals throughout the night. He liked the way their vernacular names were a rendering of their calls, like the owl (*Pulsatrix perspicillata*) named *murucututú* after its lugubrious song, or a pigeon whose call just before dawn seemed to say *Maria, já é dia!* ('Mary, it is already day!').

Lack of wind meant that the two boats took ten days to sail the 150 kilometres (95 miles) from Óbidos to Vila Nova (Parintins) – a flat and boring stretch of the Amazon. Here they were duly welcomed, like all travellers, by the priest Padre Torquato. He was 'exceedingly courteous in his manners, but', Spruce added tellingly, 'delighting wonderfully to hear himself talk.'

Vila Nova is at the eastern tip of Tupinambarana island, a 300-kilometre- (185-mile) long expanse of low land, with wetlands, lakes, savannahs and igapó forests. The Amazon flows to the north, but Tupinambarana is technically an island because of a network of long channels and streams to its south. Spruce was one of the few foreigners to enter the micro-habitat along these channels, which were called Urariá with the furo of Ramos at their eastern end. Gouzennes wanted to trade salted fish from riverbank settlers all along the Ramos. They expected to spend only a few days doing this, but because of delays in catching and salting the fish, it proved to be an entire month. This was made worse for Spruce because he stayed out one night, to take a star fix, and caught malaria (caused, he believed, by having his blanket soaked in dew). One of the igaraté's two Indian paddlers deserted, taking the montaria dinghy. So it became harder for Spruce to land or go on collecting excursions. But he did observe fishing on the island's lakes, with fishermen living in dormitory huts on stilts and working hard to dry and salt the catch. He also had his first sight of rubber tapping and smoke-curing of the latex. The caboclos living in the many sitios (riverside farms) on the Ramos were hard-working people. But despite the fertility of the land and the lakes abounding in fish and waterfowl, they lived from day to day in a state of comparative destitution. They rarely used money. Their only wealth was the pirarucu fish that they salted, and this was usually sold before they had even caught and cured it. In that particular year, manioc farinha was in short supply, and unusual flocks of parrots had consumed most of the plantains.

At one sitio, the elderly mameluca mistress had a beautiful twelve-year-old daughter. Spruce commented on her pale skin and learned that her father was a Spaniard but that the girl was already married. 'The old woman said she had another daughter, younger and still fairer, then at school in Óbidos, whom she should like to marry to me, as she had a great fancy for Englishmen; but as I had no fancy for a wife of ten years old, the negotiation went no further.'

It was the end of the dry season. Gouzennes had taken the cuberta onto the main Amazon, leaving the smaller boat to pursue his business along the channel. The air on the Ramos was hot and close, there was no current and even the water 'was very warm, and so thick with the slime of decomposed Confervae [freshwater algae] as to be very unwholesome'. This stagnation was because the western end of the channels was blocked. Everything changed dramatically after 18 November, when the main Amazon rose sufficiently to burst into the channels. Although the travellers were a day's journey away, they could hear the noise of this mighty breach. They decided to take on four more men to battle westwards into the rising channel. There was no wind, but by the 23rd they had advanced, painfully slowly, against the swift current to within a few kilometres of the main river. Spruce and the young captain of his boat reconnoitred by starlight and saw 'the waters of the Amazon entering with a force and a noise truly formidable, and ploughing through the sand in such a manner as to make a wall on each side of 15 feet [4.5 metres] high, from which the increasing torrent was every moment tearing huge masses and thus widening its bed.' On the following morning they fought their way forward, with Spruce at the helm, the captain Gustavo in the prow with a pole, and King and their mamaluco paddler hauling along the treacherous sandy shore. They made little headway. At noon they paused in the intense heat. Then miraculously a boat appeared with four men and two boys, friends from one of the riverbank farms. With more men hauling, the boat could go further out into the fierce current to avoid grounding on sandbanks. Spruce, at the helm, had great difficulty in keeping the boat's prow out into the stream. Had he failed it would have been dashed against the bank, swamped and buried under a mountain of sand. 'The exertion required was so great that the perspiration ran off me in streams.' Those heaving on shore suffered equally, for the sun and sand were scorching hot. But they succeeded. When they finally found themselves 'on the broad and breezy Amazon, our previous silent anxiety changed into noisy expressions of joy.'

Spruce sailed into Manaus on 10 December 1850, having taken nine weeks to cover only 650 kilometres (400 miles) from Santarém. He wrote to George Bentham that he himself (of fragile health) and his strapping assistant Robert King were ill for most of the voyage. 'If you had seen our wan and sickly looks when we landed... you would have pitied us.' The rains far exceeded those of the previous year at Santarém. 'You cannot conceive how damp everything is here, even within the houses. Everything of iron rusts, plants mould, clothes hanging up two or three days double their weight, and the effects upon myself are a feverish cough with rheumatic pains in the limbs, etc.'

As we have seen, they had introductions to the admirable Italian Henrique Antonij. This generous man immediately installed them in an upper room of his two-storey house and 'invited us to eat at his well-furnished table.' The other Englishmen had long since left Manaus: Bates on 26 March up the Solimões, and Wallace on 31 August up the Rio Negro.

Spruce immediately settled down to collecting. Despite the rains and his ill health, he told Bentham that 'we are in full work, and it is satisfactory to find oneself in the midst of a new vegetation'. Three months later, he wrote to Bentham that it had rained for all but five days. 'For three weeks together I have not once stirred out without getting a thorough soaking. I have certainly not shrunk from exposing myself, and hitherto I have not felt any ill effect from it.' Bentham had written to him about his first collection from Santarém, and Spruce was delighted to learn that it had included many new species. But, despite the difficulty of collecting and preserving in the rainy season, he had already collected over 300 *species* of plant – which must have represented thousands of sheets of dried herbarium specimens – and 'I have no doubt my Barra [Manaus] collection includes more variety and novelty than any previous one.' At the end of April, he dispatched two very large cases of plants, of between 300 and 400 species. These went down to Belém on Senhor Henrique's large cutter and, since that ship had come from the Solimões, it probably also carried a consignment from Bates in Ega. Robert King was also on board. Although bigger and healthier than Spruce, this young man did not share his passion for natural history and could no longer endure the hardships of Amazonian forests.

Spruce spent eleven months in and around Manaus. Years later, his friend Alfred Russel Wallace wrote that 'rarely has a small tract of tropical forest... been so well explored botanically, in so limited a time, and with the constant drawbacks of an excessively wet climate and very restricted means.' Spruce was meticulous in recording his movements and exact collecting sites. He also noted the species of every plant, with its genus and natural order and if possible its vernacular name, giving a very detailed botanical description of its leaves, bracts, blossoms, fruits and appearance as well as any remarkable characteristics or properties. 'On the average he went out collecting every other day, the intervening day being occupied in preparing and drying, and describing and cataloguing the specimens. Every road and path, every clearing, farm, or swamp, every stream or hill within reach were visited at intervals as the various trees, shrubs, or other plants came into flower. ...The results of this assiduous work were very gratifying from a botanical point of view.... The eleven months spent at the mouth of the Rio Negro added no less than 750 additional species' to the 1,100 he had already collected on the lower Amazon.

Spruce made various excursions, to collect in different habitats. In late January 1851 he crossed the Rio Negro to Jauauarí, a patch of campo savannah and marshes in the promontory where the Negro joins the Solimões. The ubiquitous Senhor Henrique had a run-down cattle ranch there. Spruce decided to occupy an abandoned mud and palm-thatched hut – its 'very indolent herdsman' had left it, to live in a crowded oven house for roasting manioc. The hut was surrounded by mud and water, accessible by a plank at one corner, with two rooms under water, and a third with a dry floor but two door openings 'through which, during squalls, the wind swept furiously'. Spruce spent a week in this, with Pedro, a young mameluco who cooked his meals. It rained every day. On some mornings there was enough sunshine for Spruce to dry his pressing paper; but, if not, they had to paddle up a forested stream to dry the papers in the oven house. Spruce was still constantly learning local skills. He had never before steered a canoe with a paddle, and he often rammed the boat plump into the bushes. He was also learning lingua geral, the Tupi-based language spoken throughout Amazonia. So he overheard his paddler telling his sister: 'This man knows nothing – I doubt if he could even shoot a bird with an arrow!' He later became a proficient paddler – he boasted in a letter to Sir William Hooker of Kew Gardens that he doubted whether the most eminent botanists in Europe could have steered a canoe any better than he did. But he could never match Indians in archery.

Spruce was just as interested in small and insignificant plants as in large showy ones. He revelled in the bladderworts, small orchids, arums, grasses and sedges – 'one of the latter was an abominable "cut grass" by walking among which my ankles were completely tattooed'. Decayed ant hills formed islands in the marsh, with their own flora. One part of the campo seemed more fertile, with profuse and high grasses and some glorious shrubs, including a *Melastoma* 'completely clad with large purple flowers; it is quite new to me.' Another stretch of campo was slightly higher and also very different 'in every respect'. This was on loose white sand, with masses of a *Humiria* shrub peculiar to tropical America. Parts of it had patches of mosses and ferns on burning sand that reminded Spruce of an English heath; but then there were curious *Schizaea* ferns, two new species of grass, and orchids – all of which looked decidedly tropical.

There was a small plantation of coca (*Erythroxylum coca*), known locally as *ipadu*. This was Spruce's first experience of a narcotic plant, a branch of botany that was later to fascinate him and in which he became a pioneering authority. He knew, of course, that coca was chewed throughout Amazonia, and that it had for centuries been central to the society of the Incas and other Andean peoples. He reminded Hooker that an Indian with a wad of ipadu in his cheek could go for several days without needing food or sleep. He sent Hooker powdered ipadu and flowering specimens of the plant. But he failed to persuade any Indian to trade a coca-pounding mortar, so that he could also send that to Kew. The Indians were understandably unwilling to part with an artifact that was so useful and had taken so much time and skill to make.*

The skilled botanist was delighted to find utterly different habitats surrounding Manaus. Some 13 kilometres (8 miles) to the southeast was the Lajes ('Ledges') so-called because of layers of sandstone at the riverbank. This Lajes overlooked the famous 'meeting of the waters' – where the tannin-rich black waters of the Negro clash with the sediment-laden brown of the turbulent Amazon. The two streams flow, sharply separated, for several kilometres before eventually blending. Spruce loved to climb low hills at the Lajes for a stupendous view. Woods, river and sky stretched eastwards towards the mouth of the Madeira in a great curve, studded with islands and fringed with lakes; southwestwards the vista was up the Solimões towards the Purus. 'It is impossible to behold such immense masses of water in the centre of a vast continent, rolling onwards to the ocean, without feeling the highest admiration.' He loved to gaze at this 'truly grand' view at sunset, when the dazzling pyrotechnics of pink and golden clouds deepen into the gloom of night. He wrote to his Yorkshire friend John Teasdale that it filled him with a mixture of tenderness and awe. Collecting at the Lajes was good: Spruce returned several times. On another excursion, to the distant southern shore of the Amazon opposite Lajes, the vegetation was totally different. He doubted whether any other collectors would work there because 'such a place for snakes and ants in the trees I never met with'.

Spruce's most adventurous trip was in June 1851, some 100 kilometres (60 miles) up the Solimões to Manaquiri, the village up an inlet on the south bank where Wallace had had a fruitful visit in the previous year. This journey should have taken three days but, although Spruce had four paddlers, it took a week, such was the strength of the current and lack of wind. The river was at its highest, flooding vast areas of forest in the várzea. Collecting was difficult. Spruce wrote to Bentham: '*we never saw land*' [his italics] – just the trunks of igapó trees that were conditioned to having their bases under water for months on end. When they hugged the bank Spruce stood in the prow and occasionally

* What Spruce did not know was that at the end of that decade the German chemist Albert Niemann would isolate the alkaloid cocaine. This forms only 0.4 per cent of a coca leaf and is difficult to extract. Cocaine is a hallucinogen that was for a while hailed as a wonder drug, but is now the basis of an international drug traffic.

gaffed flowering lianas with a long hooked pole; and he did some collecting in the waterlogged forest during brief moorings. Drying and pressing plants had to be done on board. But he had trouble drying his paper because of the rain – 'the weather was dreadful, being the fag-end of the wet season'. When there was a wind, drying-papers blew away, and he could not spread them without getting in the way of the paddlers.

These tribulations were insignificant for a botanist in the heart of a great rain forest. It was now that he wrote excitedly to his botanical friend Matthew Slater: 'The largest river in the world runs through the largest forest. Fancy if you can *two millions of square miles of forest* [5 million square kilometres], uninterrupted save by the streams that traverse it.' To a Dr Semann he wrote how much he was enjoying himself. 'Here are only trees – trees – trees! Flowering... all the year round, [but] never so many blooming at one time as to cause me any excess of work in preserving them.' In his journal he revelled in the forest's beauty. 'The banks of inland rivers should be seen early in the morning, before or after sunrise. In passing along one of these at six in the morning, when the trees had mostly acquired their new foliage, some of fine pale green, others of pink or red..., standing out from deep dark recesses, occasionally varied by the finely divided tremulous foliage of a graceful Acacia and the large white star-like leaves of a Cecropia, while here and there hang festoons of some purple-flowered Bignoniacea, white- or red-flowered climbing Polygaleas often exhaling a most delicious odour, while lower shrubs [on the waterline] are bedecked with countless flowers of various Convolvulaceae, chiefly of the species of Batatas, mixed here and there with two or three Phaseolae, some yellow-, others purple-flowered.'

His excursions from Manaus also gave Spruce glimpses of different strata of Amazonian society. At the top were the few white plantation owners. At Manaquiri he stayed with his friend Senhor Henrique's father-in-law, José Antonio Brandão, who had been so good to Wallace. Although in his seventies, Brandão was hale and hearty – Spruce attributed this to his hard work as a farmer, unlike most Brazilians who chose an easy way of life and became ailing and corpulent. The farm reminded him (as it had Wallace) of home, with horses, cattle, sheep and pigs grazing on weak campo grasses, tropical fruit trees, and an ox-driven mill to make molasses and cachaça from sugar cane.

Spruce did not comment that this handsome farm was partly worked by African slave labour. Brandão was one of the few rich enough to buy such workers. Spruce also visited the coffee groves of a plantation owner called Zany, the son of another successful Italian immigrant called 'Captain' Francisco Ricardo Zany. Thirty years previously, the Bavarian scientists Spix and Martius (the first foreigners permitted to travel up the

Brazilian Amazon) had been entertained by the elder Zany. Those visitors were impressed by his many farms, worked by scores of apparently contented Indians as well as black slaves. But when Zany took Martius up the remote Japurá river, the German was shocked to see forced labour by Indians on fictitious 'public works' and rampant human trafficking – thinly-disguised illegal slavery.

As Spruce was approaching Manaus at the end of the previous year, 1850, he had visited a sugar plantation at Itacoatiara (then Serpa) that was being created by a 'good-looking, muscular, certainly enterprising, thoughtful, clear-headed' Scot called M'Culloch. This pioneer had saved money in eleven years of work in Belém, but had lost it all in a series of setbacks when trying to establish a water-driven saw-mill, which was thwarted by local opposition, official bureaucracy, legal action and eventually arson. So M'Culloch started on the sugar plantation to make cachaça, in partnership with the ubiquitous Senhor Henrique. The Scot himself worked tirelessly. He sometimes used a band of Mura Indians (the people despised by Bates) who would work when they felt inclined and expected nothing more than cachaça and a 'hatful' of manioc; 'but the only workmen on whom he could rely were four [African] slaves of Henrique's.' Back in Manaus in August 1851, Spruce's friend Henrique arranged lodgings for him, in a house belonging to another slave-owner. Five of this man's slaves ran away but were hunted down by the police and returned to their owner. The 'refractory' leader of the fugitives was chained to a post in the yard; as his master passed one night the man tried to stab him with a hidden knife, but made only a shallow wound; so the slave fixed the knife to his post and 'with desperate resolution thrust it into his own stomach.' Next morning his fellow-slaves carried his body to the river, sewed into a sack, but they were laughing and joking as if carrying a dead dog. Spruce's laconic comment to his friend John Teasdale was: 'So much for the "beauties" of the slave system!'

Spruce was at Manaquiri in June, so he joined local people celebrating one of Brazil's great annual festivals, the Eve of São João's (Saint John's) Day. He went with one of Senhor Brandão's sons and another young man from Rio de Janeiro. They paddled through several miles of flooded forest to a brightly lit house, in which one room was fitted as a chapel to Saint John. There were rockets, firing of muskets brimming with gunpowder, and singing to bamboo flutes, 'the hammering of a crazy drum' and tambourines. After the saint's mass, there was delicious papaya jelly and tapioca on a long table covered in a white cloth, coffee, and cachaça 'unfortunately in too great abundance'. There was a pecking order for this meal. The white men ate first, followed by a tapuio Indian, then mulattoes, and lastly 'various shades' of mestizos. There were ritual dances around a circuit of bonfires, musical pranks by a lad wearing a real ox's head, and comic dances by two Indians in tall masks (evidently Ticuna from higher up the Solimões).

Spruce described it all to his friend John Teasdale. The verandah was cleared for the ball, to music of a fiddle and two guitars. Every such celebration was conducted by a Juiz and Juíza – a 'judge' chosen for his ability to pay for part of the entertainment, and a 'lady judge' for her attractiveness. The Juiz insisted that Spruce should have the honour of opening the dance with the Juíza. So the tall, gangly Yorkshireman shed his coat and shoes, to be more like the others, and led the pretty lady out. 'We got through the dance triumphantly, and at its close there was a general *viva* and clapping of hands for "the good white man who did not despise other people's customs!" Once "in for it" I danced all night.' There was an interruption by a knife fight between two mulattoes, but Spruce was whisked away in the canoe, to the Juíza's house. There the dancing continued, but demurely in rings like English country dances. In one, men and women imitated woodpeckers, hopping, singing, clapping and snapping. 'I know not when I have laughed so much, especially at the hopping.' In another, called açaí after the palm tree, the dancers whirled round and at one point caught whoever was near them. 'The ladies were very fond of this dance, especially the hugging part of it, and I had often some difficulty in extricating myself from their embrace.' The lonely Spruce clearly loved it all, since among the dancers were 'two very pretty Mameluco girls', almost white, but 'the rest were only so-so. During the course of the night I danced with every one.'

On a return visit to the Lages, Spruce also enjoyed a couple of weeks among pure but detribalized Indians. These were better-off than most indigenous people. They were of course skilled boatmen and fishermen, and each family was self-sufficient with its own roça clearing of manioc, a few coffee shrubs, plenty of fruit trees and with tobacco grown communally. He contrasted the Indians' casual clothes to the ridiculous pomposity of citizens of Manaus, toiling on Sundays under black coats and top hats. Spuce himself wore only a light cotton jacket and a pair of pantaloons – with no shirt, hat, shoes or stockings. The Indians wore even less: the men only trousers, and the women a blouse 'descending below the breasts' and a skirt, often with a bare midriff. Young, unmarried girls wore only one or other of these garments; and when a white stranger appeared, they would lift it to shield their eyes from his gaze. Without his assistant King, Spruce relied on Indians to help him collect. He felt that he had learned how to manage them better than Humboldt had done. Spruce's secret was never to 'ask them to do anything *as a task*', even for handsome payment. Instead, he would suggest, in lingua geral, 'Let us go for a walk', set off up a creek in a dinghy, and 'when we reach the heart of the forest they are all alacrity to climb or cut down the trees, the gathering of the flowers [blossoms] being all the while represented as a mere matter of amusement.'

As we have seen abundantly, all three Englishmen were outstanding and tireless collectors. But they operated in different ways. Henry Bates felt that it was best to work intensively in a few locations, along the main Solimões-Amazon river. Wallace and Spruce were more adventurous. In the lower Amazon Wallace had charged off to Mexiana island, up the Guamá to watch the tidal bore, and to Monte Alegre in pursuit of rock paintings; Spruce had got lost up the Erepecuru, and made tough excursions from Santarém, Óbidos and Manaus.

In March 1851, two Englishmen came down the Rio Negro to Manaus, bearing a letter from Wallace saying that he was far to the north, on a tributary of the Orinoco in Venezuela, 'enjoying himself amazingly in a romantic and unexplored country.' So Spruce was determined to follow his friend up the Negro. He was annoyed that he had wasted an entire summer at Santarém trying to get passage to Manaus, and when he finally found one, that the boat had taken a miserable sixty-three days over the voyage. So he decided that he must have his own boat. He finally bought a nearly new *batalão*, 6 or 7 tons burthen, for the considerable sum of £10; and he had to spend almost as much again to fit it out. This was one reason that took him to the Lages, because it was home to an Indian master-carpenter who made tolda cabins. Spruce ordered one at the stern of his boat for him to sleep and another in the bow to keep his goods dry. Then there was the perennial problem of finding a crew. He wrote to Bentham that 'there is only *forced labour* here – no sum of money in the world would induce a Tapuya [tapuio] to work voluntarily.' Everything in the Amazon depended on letters of introduction. But, because there had been no British consul in Santarém while he was there, he had no letter to the Comandante de Trabalhadores in Manaus. His friend Senhor Henrique came to the rescue, sending for five Indians from São Gabriel da Cachoeira ('Saint Gabriel of the Rapids') halfway up the Rio Negro. After several weeks' wait, to Spruce's amazement, the five men suddenly arrived: stout fellows who were added to two Indians whom he had somehow 'arranged' to get in Manaus. Buying food for the journey was equally difficult. Spruce and Henrique together purchased a young bullock, and he had his half salted. Because the rivers were high, there were no pirarucu fish to be had, but Spruce bought many turtles. There was no point in taking money up the Rio Negro, other than some copper coins, so Spruce spent his modest savings on 'prints and other fabrics of cotton, axes, cutlasses [machetes], fish-hooks, beads, looking-glasses, and a host of sundries. The trafficking of these involves a serious loss of time, but there is no alternative.' He then packed his splendid collection and shipped it off to his mentor George Bentham. There had been a setback, however, in Spruce's chain of transportation: Daniel Miller in Belém had suddenly died of 'a severe chill... aggravated into brain-fever.... Poor Miller was a very fine young man, and his loss to me is irreparable, as he was so ready to do anything I needed, even to putting himself to inconvenience.' Miller had also previously assisted Wallace and Bates when they first arrived in his town. It was extraordinary how many helpful people maintained the tenuous links between the isolated naturalists and their home country.

On 14 November 1851 Richard Spruce sailed from Manaus up the Rio Negro. Although alone he was in high spirits, with a fine crew and in his own boat, which sailed well and had a comfortable cabin.

7

WALLACE ON THE RIO NEGRO

By the time Spruce started up the Rio Negro at the end of 1851, his adventurous friend Alfred Russel Wallace had already spent a year and a quarter far up that great river.

In July 1850, Wallace had hurried back from Manaquiri to Manaus because he heard that a boat belonging to an Irish merchant, Neill Bradley, was bringing him letters and all-important financial remittances. After several 'weary weeks' the boat finally arrived. It brought 'a long arrears of letters from Pará [Belém], from England, from California [where his older brother John had gone for the '49 gold rush], and Australia, some twenty in number, and several dated more than a year back.' The three collectors had to worry about communication with far-away England as well as boat sailings in Amazonia, and Wallace stayed up half the night replying to the more important correspondence, because his boat for the Rio Negro was leaving on the very next day. He completed his 'box for England' of precious collections. His much-loved younger brother Herbert had no stomach or aptitude for rain forest collecting, so Alfred arranged for him to stay for six months in Manaus and then return to England. He also made final purchases for his own year's journey. The friendly local Juiz de Direito (magistrate) gave him a parting present of a turkey and a suckling pig – he kept the former alive but roasted the latter to provide 'a stock of provisions to commence the voyage'.

Wallace had wisely abandoned his own leaky little vessel – he does not reveal whether he managed to sell it. He travelled instead on the tolerably roomy boat of a Portuguese trader, João Antonio de Lima, leaving Manaus on 31 August 1850. This craft was 10.5 metres (35 feet) long by 2 metres (7 feet) beam; aft was a rough deck of split palm stems covered in a curved palm-thatch tolda 'high enough to sit up comfortably within it'; while in the bow was a flat 'raft' on which the Indians stood while rowing with oars of paddle-blades fixed to long poles. Senhor Lima was 'a middle-sized, wiry, grizzly man', very polite and agreeable. He was a typical Amazonian trader known pejoratively as *regatão* ('haggler' or 'huckster').* His boat was full of 'all the articles most desired by the semicivilized and savage inhabitants of the Upper Rio Negro.' These included: bales of coarse cotton and cheap calico, flimsy but brightly coloured prints of checked or striped cottons, blue and red handkerchiefs, axes, machetes, masses of coarse pointed knives, thousands of fish hooks, flints, gunpowder, shot, quantities of beads (blue, black and white), countless small mirrors, needles, thread, buttons and tape. There was also plenty of cachaça, wine and food. The tolda contained boxes of Lima's and Wallace's personal baggage, but there was still plenty of room for them to sit or sleep comfortably.

Sailing up the Rio Negro was delightful. Although a tributary of the Amazon, the Negro is one of the world's largest rivers in its own right. It is 2,230 kilometres (1,390 miles) long and its basin of 691,000 square kilometres (267,000 square miles) is greater than many countries. It is amazingly wide. For the first week's sailing north from Manaus it is broken by an archipelago of islands known as Anavilhanas (now a protected national park), but beyond these for a great distance the Negro is so broad that the two banks cannot be seen at once: Wallace guessed that they were from 15 to 40 kilometres (10 to 25 miles) apart. Its waters flow more gently than those of the main Amazon. But their tranquillity is often interrupted by sudden, violent squalls with ferocious gusts of wind and downpours.

The daily routine on Lima's boat was for the boatmen to start sailing and paddling soon after midnight (perhaps guided by the moon or starlight); stop on the bank at day-break to light a fire for breakfast of coffee with buttered biscuit; stop again at ten or eleven in the morning for the Indians' main meal of cooked fowl and any fish caught during the night; moor again at six o'clock nightfall to prepare supper with more coffee, which was taken on board while proceeding upriver; and then moor at eight or nine at night, sling the hammocks in the riverside forest, and sleep for a few hours. During the day they sometimes landed at a riverbank hut in the swathes of unbroken forest, to buy a fowl, eggs or bananas. Or if they saw a clearing they might stop to try to shoot a bird or catch some fish. 'In the cool of the morning and evening we stood upon the plank at [the boat's] mouth, or sat upon its top, enjoying the fresh air and the cool prospect of dark waters around us.' Wallace and Lima would drink coffee together, admiring the dazzling pyro-technic display of the sunset – illuminating the clouds from silver to pink, scarlet, purple

* When the rubber boom gathered momentum later in the nineteenth century, governors of the province of Amazonas wrote annual reports that often fulminated about rascally regatão traders. But although these peddlers drove usuriously hard bargains, the Indians and riverbankers of the upper rivers depended on them to barter their meagre produce for manufactured goods.

and gold – or watching the little goatsucker nightjars (*Caprimulgus*) pirouetting above the water in search of insects.

All travellers on the Rio Negro commented on its black waters – which of course inspired its name, given by the first Spaniards who descended the Amazon. From above, the river looks black, but when swimming its water is the colour of strong tea. No one in the nineteenth or early twentieth century knew the reason for this phenomenon. It is now thought to be caused by tannin or the carbolic acid phenol from decaying vegetation. This content of humic acids is visible because the Negro is very pure: it rises in the ancient and totally weathered sandstones of the Guiana Shield, whereas the Amazon-Solimões is muddy-white from sediment off the geologically younger and crumbling Andes. Wallace also noted the 'great luxury' that the Negro did not have mosquitoes. He did not guess that their absence is because mosquito larvae cannot breed in the black waters. Another bonus was that the black river was not plagued by malaria. But two and two were not put together: it was too extraordinary to imagine that a disease that prostrated or even killed big men could be transmitted by the bite of a tiny insect.

With fewer aquatic resources, the Rio Negro had never been as densely inhabited as the main Amazon. But by the mid-nineteenth century it was almost totally empty – denuded by the massive depredations of colonial slavers, alien diseases and the slaughter of the Cabanagem rebellion. The region was in economic decline because it yielded almost no commercial product. All the villages were desolate and half-deserted. At Carvoeiro there were only two families – those of a blacksmith and of a drunken captain; at Barcellos (where a century earlier there had been a doomed attempt to create a provincial capital), the once-proud riverside houses were 'ruinous mud-huts' and the streets were just 'paths through a jungle'; Santa Isabel was 'a miserable village overgrown with weeds and thickets' whose only inhabitant was a decent Portuguese chap who did not mind the hardships and privations of extreme poverty.

Between the derelict villages, there were a few riverside sitios and plantations. These gave some insight into how trade worked on such remote rivers. The mulatto owner of one took much of Senhor Lima's stock, on credit, promising to provide plenty of sarsaparilla and other produce on the return journey. In contrast to Lima's credit arrangement 'which is the curse of this country', Wallace admired another Portuguese planter-trader who had grown rich because he himself took a boatload of goods right down to Belém at the mouth of the Amazon and doubled his investment by bringing back desirable merchandise. This prosperous trader was honest and fair, but was disliked 'because he does not enter into the extravagance and debauchery which it is thought he can well afford.' Wallace prudishly commented that, with food so easily available, life in this country cost nothing – unless someone drank or gambled.

Wallace enjoyed the placid voyage up the Rio Negro. With his habitual curiosity and enthusiasm, he investigated two interesting lines of research: exotic fish and rock engravings. He caught, described and drew many beautiful fish, for the Rio Negro abounds in endemic species (and it is still a source of gaudy specimens for the world's aquaria). As well as collecting brilliantly coloured small fish, he enjoyed eating those that were full of fat and made tasty chowder. He was also ahead of his time in his interest in indigenous rock art. As we saw, a year earlier he had made a great effort to see painted cliffs and caves near Monte Alegre. He now drew engraved picture-writing of animals and men found on an island near the mouth of the Rio Branco, and collected some of the 'picturesque granite rocks' on which they were engraved. Tragically, as we shall see, these drawings and many of his specimens were later lost.

The Amazon basin is so flat that the main river and most of the mighty Rio Negro is free of rapids. But a third of the way down the Negro's long course it falls off the Guiana Shield in a series of turbulent rapids and waterfalls. These are at São Gabriel da Cachoeira. Although Lima's boat had sailed and paddled with few stops and with its rowers sleeping for only a few hours each night, it was not until 19 October 1850 – seven weeks after leaving Manaus – that they reached 'the celebrated Falls of the Rio Negro'. They had to abandon the large boat and hire a smaller one in which to force their way upriver. 'Small rocky islands and masses of bare rock now began to fill the river in every part. The stream flowed rapidly round projecting points, and the main channel was full of foam and eddies.... Beds and ledges of rock spread all across the river, while through the openings between them the water rushed with terrific violence, forming dangerous whirlpools and breakers below.... We dashed into the current, were rapidly carried down, got among the boiling waves, then passed suddenly into still water under shelter of an island.' They then had to clamber onto rocks while the Indians hauled the boat up 'a great rush of water'. They wove back and forth across the river, to find channels and avoid jagged rocks. Wallace was awed by the boatmen's skill and courage. 'The Indians, all naked, with their trowsers tied round their loins, plunged about in the water like fishes.' To fix a rope over a projecting crag, an Indian would leap into the seething river, dive down to stiller current, then try to clamber up onto the slippery rock. Two or three of them would fail, falling back exhausted and floating down to the boat 'amid the mirth and laughter of his comrades'. Later, they sought to reach the lee of an islet. 'In a moment we are in a stream running like a mill-race: "Pull away, boys!" shouts Senhor Lima. We are falling swiftly down the river. There is a strong rapid carrying us, and we shall be dashed against those black masses just rising above the foaming waters. "All right, boys!" cries Senhor L.; and just as we seemed in the greatest danger, the boat wheels round in an eddy, and we are safe under the shelter of a rock. We are in still water, but close on each side of us it rages and bubbles, and we must cross again.' So it continued, for hour after hour. They paused for the night at an island. The Indians got some sleep, to prepare for another day of fighting the falls. Unable to help, Wallace enjoyed the scenery. 'The brilliant sun, the sparkling waters, the strange fantastic rocks, and broken woody islands, were a constant source of interest and enjoyment for me.' They reached São Gabriel on the afternoon of the second day. The final obstacle was a great sloping incline over which the river rolled in a tremendous flood and 'below, the water boils up in great rolling breakers, and a little further down forms dangerous eddies and whirlpools.' Here the boat had to be unloaded and pulled up through the foaming water, as near as possible to the shore. Lima and Wallace wore little because of the drenching spray and frequent wading in the rapids. They now got dressed and walked up to the village to pay their respects to the military commander – they had to get his permission to continue upriver. He was a friend of Lima, and Wallace had a letter of introduction to him, so they had a gossip over a cup of coffee, and then spent the night in a trader friend's house.

Above São Gabriel there was a further two days of lesser rapids, with the boat weaving across the river from island to island and rock to rock. From then on, the Rio Negro was plain sailing, across a vast plain of tall rain forests and lower *caatinga* (thorny brush woods). They passed the mouth of the large Uaupés river, after which the Negro was calm and placid, still very wide and blacker than ever. On 24 October they reached Nossa Senhora de Guia ('Our Lady of Guidance'), once a 'very populous and decent place' but now just a dozen huts, as miserable as all the other villages along the river. Most of Guia's Indian inhabitants were off in their sitio riverbank farms, but they returned to welcome Lima, for this was his home. He introduced Wallace to his partner, a good-looking mameluca of about thirty, and their five children. Lima explained that he did not believe in marriage. He had had a number of older children by another woman, but he had evicted her because she was an Indian who could not teach her offspring Portuguese or educate them. She had once saved Lima's life by nursing him during an eighteen-month illness, but he had no qualms about 'turning her out of doors'; and the poor woman had since died 'of passion'. The crew of Lima's boat also came from Guia, so several days of non-stop drinking and dancing ensued.

An endearing characteristic of Wallace was his enthusiasms. If he heard about some wonder he would pursue it with single-minded boyish determination – this may have been why Bates preferred not to travel with him. In the lower Amazon it had been the islands in the river's mouth, then the pororoca tidal bore; moving upriver it was the rock art near Monte Alegre; then the umbrella bird on the lower Negro. Now his passion was to find and collect cocks-of-the-rocks. This medium-sized bird, *Rupicola rupicola*, is a cotinga similar to the umbrella bird, except that it is a brilliant orange-red whereas the latter is black, and its comb is a semicircular fan of feathers over its head and beak where the other cotinga has an umbrella-shaped lid of feathers.

The search for the cock-of-the-rock began at the large Içana river, which enters the Negro from the west just below Guia. Wallace set off in a small canoe, with two Indian paddlers and plenty of food and ammunition. After a few hours up the Içana they started up the Cubaté, a smaller river that joins it from the west. The Cubaté's waters were very black, and it meandered tortuously. For the first two days they moved through typical caatinga with dull and monotonous vegetation. Then the forest became loftier and they could see the rocky Cubaté hills in the distance.

They reached a village of Arawak-speaking Baniwa. This was Wallace's first sight of an indigenous village, albeit of semi-acculturated people. There were six huts embedded in the forest, consisting only of thatched roofs supported on tree-trunk columns – without walls, apart from a few that had a palm-leaf partition to form a sleeping apartment. These sheds (as Wallace called them) were full of naked children and their almost naked parents. Wallace offered good pay for every cock-of-the-rock caught, so most Baniwa men and plenty of boys joined his expedition next morning. They had to walk for 15 kilometres (10 miles) through dense forest. After an hour they came to the last house before the hills. They were greeted by a wrinkled, white-haired old Indian woman; but with her was a young mameluca, 'very fair and handsome, and of a particularly intelligent expression of countenance'. The moment Wallace saw this beauty, he was sure that she was someone Lima had told him about: the daughter of the great Austrian naturalist Johann Natterer and an Indian woman.*

---

* Two years previously, in Belém and on the excursion to see the bore on the Guamá river, Wallace had employed the admirable Congo-born Luiz – who had been freed from slavery by Natterer. Luiz thought very highly of the Austrian. Natterer had reached Brazil in 1817 in the entourage of the Habsburg Archduchess Leopoldina, and then spent a remarkable eighteen years travelling throughout the country. He was on the upper Negro in 1830–1832 when in his mid-forties. He married a Brazilian woman at Barcellos on the middle Negro and had a daughter by her – in addition to the Baniwa girl that Wallace met (Natterer was with the Baniwa on the Içana in June 1831, so if he did indeed father a girl there, she would have been nineteen at the time of Wallace's visit). In addition to zoological and entomological collections, Natterer compiled an amazing *sixty* indigenous vocabularies and collected artifacts from *seventy-two* tribes. But he sadly never published an account of his travels or any academic studies. He himself died of a lung disease in Vienna in 1843; most of his papers were burned in a fire during the 1848 revolution, and his only legacy (apart from his daughters) was superb collections in Vienna's ethnographic and natural history museums.

Beyond this final house, the going got very tough. They walked at top speed, as Indians always do. The Baniwa were all naked, while Wallace deplored 'the uselessness and bad consequences of [his] wearing clothes upon a forest journey.... Hard roots rose up in ridges along our path, swamp and mud alternated with quartz pebbles and rotten leaves; and as I floundered along in the barefoot enjoyment of these, some overhanging bough would knock the cap from my head or the gun from my hand; or the hooked spines of climbing palms would catch in my shirt-sleeve.' His bare feet must have suffered painfully from the quartz and the spines. They started to climb over very rough, uneven ground, clambering up steep ascents and over the rotting logs of fallen trees, or dropping down into gullies.

They had a stroke of luck when the men killed a peccary. It must have been an animal foraging by itself (as this author has shot), probably a caitetu or collared peccary (*Tayassu tajacu*; 'collared' from a ring of whitish bristles around its neck). These are more likely to be alone than the slightly smaller queixada (*Tayassu pecari*) or white-lipped peccary (with white hairs on its lower jaw), which move in large herds. A band of peccary of either species can be dangerous. Once aroused, they attack en masse and are aggressive, vengeful and fearless. Even jaguars are wary of them, and an attacked man can do nothing but try to climb out of reach. Wallace watched as the Baniwa swiftly butchered their kill. The priority was to remove the scent gland on its back. This gland is the main difference between peccary and pigs and boars; if not fully excised, it gives a horrible piggy taste to the meat. The animal was then swiftly skinned, part of it boiled and eaten for dinner, and the rest smoked over the fire and packed in palm leaves for carrying. Replete with this delicious pork, they spent a reasonably comfortable night in a cave at the base of a mountain.

The following day's march was far worse. 'We commenced the ascent up rocky gorges, over huge fragments, and through gloomy caverns, all mixed together in the most extraordinary confusion. Sometimes we had to climb up precipices by [clinging to] roots and creepers, then to crawl over a surface formed by angular rocks, varying from the size of a wheelbarrow to that of a house.' Wallace was rewarded by shooting a cock-of-the-rock, which his men retrieved from a deep gully. The thirteen Indian men pressed on to a tougher area, in search of more birds, while Wallace decided to return to the cave with the Indian boys. But they got lost. 'We descended deep chasms in the rocks, climbed up steep precipices, descended again and again, and passed through caverns with huge masses of rocks piled above our heads. Still we seemed not to get out of the mountain, but fresh ridges rose before us, and more fearful fissures were to be passed. We toiled on, now climbing by roots and creepers up perpendicular walls, now creeping along a narrow ledge with a yawning chasm on each side of us. I could not have imagined such serrated rocks to exist.... Fissures and ravines [were] fifty to a hundred feet [15 to 30 metres] deep.' The boys finally admitted that they were lost, in the dense forest and matted undergrowth that smothered the jagged rocks. There was no alternative but to retrace their route. 'It was a weary task. I was already fatigued enough, and the prospect of another climb over these fearful ridges, and hazardous descent into those gloomy chasms, was by no means agreeable.'

When they finally found the cave, the Baniwa men were already there. Although he must have been utterly exhausted, Wallace skinned his cock-of-the-rock, 'lost in admiration of the dazzling brilliancy of its soft downy feathers.' Two of the men spoke some Portuguese, and Wallace chatted to them in the gloom of the cave. They asked him about the white men's world, then more metaphysical questions 'about where the wind comes from, and the rain, and how the sun and moon got back to their places again after disappearing from us'. He did his best to satisfy their curiosity, even though he often knew little more than they did. In return, the Indians told him 'forest tales' about jaguars, fierce peccary, 'and of the dreaded curupira, the demon of the woods, and of the wild man with a long tail found far in the centre of the forest.' They talked into the night, until they fell asleep.

Everyone was up at crack of dawn because the Indians, with no covering, felt the chill morning air. The first task was to stoke their fires and brew mingau manioc gruel. No more cocks were caught on the following day, so they decided to try the far side of the mountain. 'If our former path was bad enough, this was detestable. It was principally through second-growth woods, which are much thicker than the virgin forest, full of prickly plants, entangled creepers, and alternations of soft mud and quartz pebbles under foot.' This ordeal led them to a clearing with an abandoned Indian hut, in which they camped for four days. There were melastome trees (*Leandra micropetala*) nearby whose blue berries on red stalks are favourites of fruit-eating birds, including cocks-of-the-rocks. There was also plenty of food for the hunting party – monkeys and the two tastiest game birds, turkey-hen guans (*Penelope*) and big black *mutum* curassows (*Crax alector*). So when Wallace returned to the Baniwa village after nine days in the forest he had a good haul – twelve cocks-of-the-rocks, 'two fine trogons, several little blue-capped manakins [*Pipra coronata*, a small black bird with a pale-blue top to its head], and some curious barbets [brightly coloured birds, like toucans but with smaller beaks], and ant-thrushes [Formicarius, tawny-red birds of medium size].' He spent a further fortnight in the Indian village, collecting many small birds and the skin of a black agouti rodent, and drawing curious fish.

Back at Guia, on the Negro near the mouth of the Içana, Wallace met some of the nefarious characters who operated in that remote region – but more about them later. His trader friend Lima lent Wallace a boat with four Indian paddlers, and they finally set off up the Rio Negro at the end of January 1851. Each Indian's baggage consisted of a zarabatana blow-gun with a quiver of curare-tipped darts, a pair of trousers and a shirt, a paddle, knife, hammock, and tinder-box for lighting fires. Wallace himself had more: 'my watch, sextant and compass [he was a trained surveyor], insect- and bird-boxes, gun and ammunition, with salt, beads, fish-hooks, calico, and coarse cotton cloth for the Indians.' They soon reached Brazil's frontier fort Marabitanas – no more than a ruined mud rampart manned by a few soldiers – and the following day passed a great outcrop, Cucui. This 'granite rock, very precipitous and forming nearly a square frustum of a prism, about a thousand feet [300 metres] high' towers out of dense forest and has its own crown of trees on top. Cucui (Cocuy in Spanish) has not changed: it is still an isolated landmark, jutting above an unbroken canopy of forest that stretches to the horizon in all directions [PL. 9]. The crag of Cucui marked the actual frontier between Brazil, Venezuela and Colombia (which has a dog-leg of land down the west bank of the upper Rio Negro). From now onwards Wallace was moving up the Negro with Venezuela to the east and Colombia to the west – although neither Spanish-speaking republic had any presence in these almost uninhabited forests.

After four days they reached San Carlos de Río Negro, the main town in southern Venezuela. It was a neat little place, with a large plaza typical of Spanish colonial towns, and a grid of streets with whitewashed houses. Wallace dined with the comisario, an elderly Baré Indian. This official occupied a former mission convent and himself conducted a daily service with hymns for the Baré and Baniwa townsfolk.

San Carlos is close to the southern end of the Casiquiare Canal, a meandering river that joins the Orinoco and Negro in this flat landscape. Wallace, like most travellers, chose a straighter link between the two great rivers. He continued north up the Negro (which changes its name to Guainía) and took a portage to the Atabapo river, which flows north to the middle Orinoco. Half a century earlier, in 1800, Humboldt and Bonpland had taken this route south. Wallace finally had to park his little boat because there had been so little rain that the Guainía – although still a mile wide – was so shallow that he often ran aground. He continued in an ubá, a dugout canoe of a single tree trunk, that could resist scraping on rocks. This canoe was full to the gunnels with all his baggage, so they often had to unload and drag it over rock ledges or reedy shallows. But on 10 February he and his two Indian paddlers reached a village called Tomo and then, after two more days up a small headwater, Pimichin the southern end of the watershed portage.

There was a curious burst of commercial activity among the Christian Indians of this remote region. One speciality was building large boats and schooners for traders on the Negro and Amazon rivers. At Tomo there was a Portuguese, Antonio Dias, who welcomed Wallace to stay in his little shed and who was building a fine 100-ton-burden boat. These craft were roughly made, of poor-quality wood; but they were the workhorses of river commerce. All inhabitants of the nearby villages were hard at work building boats, for throughout the Brazilian Rio Negro there was demand for these 'Spanish vessels' as they were known. The Indians had learned their craft in colonial days and worked without plans or drawings – just axes, adzes and hammers, with hand and eye measurements. (This author has had illiterate men build a dugout canoe in this way in the late twentieth century. They also used only axes, with an adze for finishing touches; and their measurements were by hand-spans, knuckle joints (inches), and forearms (cubits). It took five men a week to fell a tree and hollow-out a superb canoe in this way.) The big boats built on the upper Negro had to await the rainy season to be sailed downriver; they could shoot the mighty rapids at São Gabriel only when the waters were at their highest.

The new vessels carried bulky cargoes of piaçaba, vegetable pitch and manioc flour; and smaller boats returned with iron and cotton goods. Piaçaba (*Attalea funifera*) is a curious palm tree, rare elsewhere, and therefore providing another local industry. The greater part of the population – men, women and children – went out to gather piaçaba fibre, which hangs down from the palm trunks like a thick matting of coarse hair, and bristles that grow from the bases of the fronds. The fibre was for making ropes and cables, and the bristles for brooms and brushes [PL. 57].

Antonio Dias at Tomo had another skill: he made fancy hammocks decorated with his own border designs. The decorations were of exquisitely coloured feathers from 'cock-of-the-rock, white egrets, roseate spoonbills, golden jacamars, metallic trogons, little seven-coloured tanagers, many gay parrots, and other beautiful birds.' Wallace was shocked that this Portuguese 'was rather notorious, even in this country of loose morals, for his patriarchal propensities, his harem consisting of a mother and daughter and two Indian girls, all of whom he keeps employed at feather-work, which they do with great skill.'

As Wallace finally approached the portage between the Amazon and Orinoco basins, he and his men were hungry: for a few days his meals were only manioc mingau and a cup of coffee. At one overgrown hamlet, the only inhabitant was a Portuguese deserter, 'a very civil fellow, who gave me the only eatable thing he had in the house, which was a piece of smoke-dried fish, as hard as a board and as tough as leather.' Wallace nobly gave this to his Indians, and contented himself with a cup of unsweetened coffee with the down-and-out. It is remarkable how generous people were in these remote places. Indian comisario headmen invariably found accommodation for the young traveller, usually in what had once been the house of a missionary. Wallace sent his Indians off to fish with bows and arrows, but they had no success. So he spent a day with them using juice from roots of the timbo liana in a small stream. Timbo is poisonous enough to stun fish, which then float to the surface; but it does not make them inedible. 'The Indians were in the stream with baskets, hooking out all that came in the way, and diving and swimming after any larger one that appeared affected.' In an hour they had a basketful of mostly small fish. Wallace selected half-a-dozen of the most novel and interesting, which he described and painted; the rest went into the pot for 'a rather better supper than we had had for some days past.'

Wallace walked from Pimichin to Javita at the northern end of the portage in order to find porters. The 16-kilometre (10-mile) portage road was straight and level, with tree-trunk boarding where it crossed numerous streams or marshes. With help from a comisario he hired one man and eight or ten women and girls, and then did the three-hour walk back to prepare loads for them. There was a lot to be carried: a 45-kilo (100-pound) basket of salt, four baskets of manioc farinha, a jar of oil, a demijohn of molasses, a portable cupboard, boxes, baskets and numerous other articles. Wallace himself was 'pretty well loaded, with gun, ammunition, insect boxes, etc.' All this took time, so that night fell during the walk back to Javita. They had no lantern and there was no moon or starlight. 'I was barefoot, and every minute stepped on some projecting root or stone, or trod sideways upon something which almost dislocated my ankle. It was now pitch-dark... and the road now totally invisible.' He often fell into water (mercifully shallow), or had to grope to find bridges or causeways. 'To walk along a trunk four inches [10 centimetres] wide under such circumstances was rather a nervous matter.' He was alone – the sure-footed porters had trotted ahead into the darkness, and his own Indians had spent the day fishing and decided to follow him north the next day. So Wallace admitted fear. Strolling out the previous evening he had come face-to-face with a black jaguar; he had then been thrilled to see 'the rarest variety of the most powerful and dangerous animal inhabiting the American continent'. He had thought of shooting it, but his gun was loaded with small shot. So he simply admired the magnificent animal and was too awe-struck to feel fear. But now he was terrified because he knew that jaguars abounded in the area and could be fierce and cunning.

He had also seen the deadly viper jararaca (*Bothrops jararaca*) and knew that they, too, were nocturnal and plentiful.

The ordeal finally ended when they reached Javita. Wallace was again lodged in a former priest's house. Next day his porters came for payment. 'All wanted salt, and I gave them a basinful each and a few fish-hooks, for carrying a heavy load ten miles: this is about their regular payment.' Wallace observed that 'Salt here is money': it was local people's only imported luxury, and Wallace used it to pay for everything. No one stole from him, even though he was alone, remote and vulnerable. The Indians heaving his salt could easily have taken some of it, but they did not.

Carrying loads along the portage road was the speciality of the people of Javita – about 200 in number, all pure Baniwa Indians. Lifelong practice meant that they sometimes did two portages in a day, walking 65 kilometres (40 miles), half of this carrying heavy loads. This was far more than unskilled porters could possibly do. They went at a trot and hardly ever rested, and they crossed the log bridges with perfect balance. Twice a year the entire population went out to tidy their half of the road, cutting back and removing trees and undergrowth, weeding, and sweeping all the dead leaves and twigs. (Villagers from Tomo and other places to the south cleared the other half of the portage.) The men often had to go some distance to find logs to repair the bridges, dragging them on rollers with liana cords. Wallace marvelled that 'to clear up a road five miles in length in this manner was no trifle, but they accomplished it easily and very thoroughly in two days.' There was no pay for this communal work, which was done cheerfully. The other main livelihood of Javita's Indians was gathering piaçaba fibres and twisting them into rope.

Wallace liked Javita, a pretty village of clean streets lined with palms, fruit and other trees, and houses of whitewashed mud and palm-thatch roofs tied down with cords. The Baniwa-speaking people were devout Christians who, having no priest, conducted their own masses – merry fiestas, with 'peculiar monotonous dances' in a ring, shouting and singing to music of reed pipes and small drums. Wallace even composed a doggerel poem about Javita, praising its simple, peaceful way of life, with no lust for riches and a rural existence of contented hunting and fishing. He particularly admired the village maids, with flowers and ribbons in their beautifully combed hair, and graceful bodies unimpeded by straps or bands: 'simple food, free air, and daily baths and exercise give all that nature asks to mould a beautiful and healthy frame.' With poetic licence, he mused: 'I'd be an Indian here, and live content to fish, and hunt, and paddle my canoe, and see my children grow, like wild young fawns, in health of body and peace of mind, rich without wealth and happy without gold!' But Wallace was no 'noble savage' romantic. Like Bates, he missed the intellectual conversation and pursuits of his own country. He soon ran out of conversation with the Indian comisario, and had 'dull and dreary' evenings because no one could understand his mixture of Portuguese and a little Spanish.

Wallace wrote urging other collectors to work in this rich ecosystem on the Amazon-Orinoco watershed, where the cost of living was trifling. (Spruce was to heed this advice in the following year.) But Wallace himself had bad luck: the rainy season started immediately after his arrival, in March 1851, a month earlier than usual. 'Day after day the rain poured down; every afternoon or night was wet....' He tried to make the most of the unforeseen bad weather. He sent his Indians with their zarabatana blow-guns to seek 'splendid trogons, monkeys, and other curious birds and animals in the forest'. He himself was out every day. His Indians also fished, for them to eat. But Wallace examined their catch to find new tropical species: when he spotted a gaudy new one, he had to paint it immediately before night fell. This exposed him to the pest of tiny sand-fly midges (*Phlebotomus*) that swarmed in millions every afternoon: his hands and face would be 'rough and red as a boiled lobster, and violently inflamed'. But Wallace was philosophical: the swelling and itching soon subsided, and these sand-flies were far preferable to pium black-flies, mosquitoes or mutuca horse-flies. The days and nights of incessant rain made it almost impossible to collect small insects. 'In the drying-box they got destroyed by mould; and if placed in the open air and exposed to the [brief morning] sun minute flies laid eggs upon them and they were soon eaten up by maggots.' Beetles were plentiful and often new, since there are tens of thousands of species of Coleoptera in Amazonia. Bates paid Indian boys a fish hook for every beetle they brought him. The great blue butterflies, *Morpho menelaus* and *M. helenor*, were also spectacularly abundant: they sat in dozens along the portage road, rather than flitting, alone, high overhead in dark forest.

Curiously, Wallace decided to go no further north into Venezuela, but to return instead to the Rio Negro. He left Javita after forty days with much regret. Everything went remarkably smoothly on his return journey. Indians were sent ahead to bring his canoe up to the southern end of the portage, and the comisario arranged others to carry his belongings back to Pimichin. With the rains, the rivers had risen spectacularly so that he sped downstream, his men paddling by day and floating by night. The journey to Marabitanas in Brazil took a third of the time it had taken upstream. Wallace indulged in some tourist shopping. At one village he traded fish hooks and calico for Indian baskets and blow-guns with quivers of curare-tipped darts. At Tomo he bought one of the 'patriarch' Antonio Dias's beautiful featherwork hammock fringes, for the princely sum of £3 in silver dollars. He paid his Indian paddlers with calico and cotton cloth (here worth fifteen times its cost in England), soap, beads, knives and axes.

At the decrepit little Brazilian frontier fort Marabitanas, the commandant, Lieutenant Filisberto Correia de Araújo, treated the young Englishman with the greatest kindness and hospitality. (This Lieutenant Filisberto later proved to be less admirable, as we shall see. Thirteen years previously, the great explorer Robert Schomburgk was similarly charmed by Pedro Ayres, commandant of another northern fort, unaware that he was an Indian-hunter and brother of Bararoá, the psychopathic scourge of rebels in the Cabanagem revolt.) The people living at Marabitanas were famous for celebrating every notable saint's day. They seemed to spend half their time on these fiestas, and the other half getting ready for the next. Their parties lasted several days, with prodigious drinking of raw spirit distilled from sugar cane and manioc.

Moving south for about 100 kilometres (60 miles) to Guia, Wallace spent the month of May in that little village awaiting the arrival of his friend Lima. The rains were too heavy for most collecting, so he concentrated on fish. Wallace was one of the first naturalists to take an interest in the amazing ichthyology of Amazonian rivers. He wrote: 'I have now figured [drawn] and described 160 species from the Rio Negro alone.' These were usually the brightly coloured little fish that were later collected for private aquaria. He reckoned that the number of species in those waters must be immense. He was right: in the twenty-first century, some three *thousand* species of freshwater fish have been described in the Amazon – about 30 per cent of the world's total – and more are still thought to be undiscovered.

Six months previously, on his way upriver, Wallace had already spent many 'dull days and weary evenings' in Guia. At that time he had to listen to the 'oft-told tales of Senhor Lima and the hackneyed conversation on buying and selling calico, on digging salsa [sarsaparilla], and cutting piaçaba.' Guia's Indians were now awaiting the arrival of Friar José dos Santos Inocentes, one of only two priests functioning north of the Amazon in Brazil. The devout Baniwa felt that their children must be christened by a real consecrated priest. When the Friar finally came and was carried up from the river in a hammock, he was 'a tall, thin, prematurely old man, thoroughly worn out by every kind of debauchery, his hands crippled, and his body ulcerated.' But he made a fine ritual of the baptisms of about twenty children of all ages, with water, holy oil, rubbing eyes, signs of the cross, kneeling and prayers – all similar to 'the complicated operations of their own "pagés" (conjurors) to make them think they had got something very good, in return for the shilling they paid for the ceremony.' The Friar then performed marriage ceremonies, with homilies about the spiritual rewards of marital life. (The two other white men present, Lima and Lieutenant Filisberto, commandant of Marabitanas, were both unmarried but with large illegitimate families. Friar José told his parishioners that they should not risk going to Purgatory like those sinners, at which the white men laughed heartily and 'the poor Indians looked much astonished'.)

Wallace, with his strict upbringing, was shocked but delighted by the worldly Friar, famous as the most amusing and original raconteur in the region, with an inexhaustible fund of anecdotes. 'He seemed to know everybody and everything in the Province, and had always something humorous to say about them.' But the 'innocent saints' in his spiritual name was a misnomer. 'His stories were, most of them, disgustingly coarse; but so cleverly told, in such quaint and expressive language, and with such amusing imitations of voice and manner, that they were irresistibly ludicrous.' He told Chaucerian tales of his convent life. 'Don Juan was an innocent compared with Frei José; but he told us he had great respect for his cloth, and never did anything disreputable – *during the day!*'*

In June 1851 Lima and Wallace embarked on a journey up the Uaupés river (spelled Vaupés in Colombia). This great torrent rises in the Andes of Colombia and joins the Negro from the west not far above São Gabriel da Cachoeira: it is actually longer than the Negro and could be said technically to be the main river. Lima was going to trade for sarsaparilla – and, as it transpired, illegal human cargo. Wallace went for adventure, and because it was virgin territory for natural history collectors. There was the usual problem of manpower: even the experienced regatão Lima could recruit only two Indian boatmen. Thus under-manned, progress against the rainy-season current was almost impossible, even when they moved through the flooded forest. Their men had to heave the boat forward or push with long poles, often for miles on end. In the process they got the boat overrun 'and ourselves covered with stinging and biting ants of fifty different species, each producing its own peculiar effect, from a gentle tickle to an acute sting; and which, getting entangled in our hair and beards, and creeping over all parts of our bodies under our

* The Friar was indeed a turbulent priest. As an army chaplain he had been sent to an isolated fort on the Guaporé river facing what is now Bolivia; he was recalled from this because he mocked local Portuguese settlers. He was in Belém in the troubled times before the Cabanagem rebellion, but his political manoeuvres got him exiled to missionary work on the Rio Branco. On that northern river he tried to convert Makuxi on the frontier with British Guiana. This led to clashes in 1839–1842 with a Protestant missionary and the explorer Robert Schomburgk, who were doing similar missionary work from the British side. At one point Friar José tried to sell Schomburgk some cattle and Indian headdresses and artifacts – trading activity that the German-British explorer condoned because the priest had not been paid for ten years. The Friar also begged the British to get his music-box mended, and was almost in tears when they failed; but a present of a couple of bottles of wine cheered him up completely. There were tense moments, with small squads of soldiers on both sides and diplomatic fluttering in the two capitals, London and Rio de Janeiro. But the clerics avoided armed conflict. The Makuxi village was called Pirara, so this episode was known as the Pirara Incident. The two governments decided to put the dispute on hold; so the frontier remained unresolved for sixty years, until arbitration in the early twentieth century.

clothes, were not the most agreeable companions.' An even nastier hazard was hornets, whose nests were concealed among leaves but which attacked in furious swarms if disturbed. Known as *vespão* or 'big wasps', hornets remain a dreaded hazard to anyone cutting a picada: the trail has to curve around their nest and everyone has to move gently and silently. Another discomfort, for all travellers moving through flooded forest, was that it was rarely possible to find a dry rock or hillock on which to light a cooking fire. So their meagre food was cold.

Wallace's reward for travelling up the Uaupés was to see unspoiled indigenous peoples living according to their magnificent traditions. He was among these Indians for ten weeks in June to September 1851 and then for twelve weeks in February to April 1852. Wallace was thrilled by his first sight, after three years in Brazil, of 'truly uncivilized Indians... in places where they retain all their native customs and peculiarities.' He knew that very few travellers had met such people, because they could be found only by going far beyond white men's dwellings or trade routes. They surpassed all his expectations, and were 'something new and startling, as if I had been... transported to a distant and unknown country'.

Throughout the seventeenth and early eighteenth centuries Portuguese slavers had denuded the main Amazon and the accessible parts of all its great tributaries. The Manau people of the middle Negro blocked that river until they were defeated in the 1730s. Then, in 1750, enslavement of Indians was formally forbidden. So although human traffickers started to prey on the tribes of the upper Negro and Branco, they had not yet destroyed peoples of these rivers' tributaries, such as the Uaupés. As we have seen, missionaries of the monastic orders had converted many living on the main Rio Negro into devout Christians; but with the expulsion of the Jesuits and then the other orders, there were no missionary priests to proselytise among the remoter tribes.

Wallace came to know many tribes of the Uaupés and its tributaries. He eventually named fourteen peoples on the main river, including Tukano, Wanana, Desana, Tariana, Tuyuka, Carapanã, and Cubeo. All these Indians speak variations of Tukano (or Tukanoan), apart from the Tariana who speak Arawak.*

These peoples generally lived in harmony with one another and had come to be almost identical in dress, ornament and every aspect of society. The men were fairly tall,

* These are modern spellings, although some anthropologists now use 'k' rather than 'c' (Kubeo/Kobewa, etc.) and spell the language Aruak rather than Arawak.

and with both sexes 'their figures are generally superb... living illustrations of the beauty of the human form'. Their hair was jet-black, smooth and thick, their skins coppery brown with no body or facial hair (the little that grew was plucked), and regular facial features often differing only in colour from good-looking Europeans. Each man had a prized possession hanging from a necklace: a long cylinder of white quartz. These pendants were very valuable, because they had come from the distant Andes and because the hole for suspending them took literally years to drill, using only the shoot from a wild banana leaf twirled in fine sand and water. Men's only clothing was a small pouch of inner bark tied between their legs; a few also had prized jaguar-teeth belts. Both sexes had tight 'garters' bound around their upper calves, apparently just to make their legs bulge attractively. Women were normally entirely naked. Like most other observers, Wallace said that there was nothing remotely erotic about this natural nudity – it was less suggestive than the diaphanous dresses of some European dancers. 'Paint with these people seems to be looked upon as sufficient clothing; they are never without it on some part of their bodies.'

During dances, men were adorned with beautiful feather ornaments. Desana men wore their long hair parted in the middle, but hanging far down their backs in a tail firmly bound with monkey-hair cord and perhaps adorned with a cascade of white heron feathers [PL. 49]. On the top of his head each wore a wooden comb or coronet decorated with red or yellow toucan feathers. Together with their lack of bodily hair and plucked eyebrows, all this gave them an almost feminine appearance. (Wallace wondered, very plausibly, whether the 'Amazons' that Orellana's first Spaniards saw in 1542 might have been similar female-looking warriors.) When dancing, women wore only 15-centimetre (6-inch) square *tanga* aprons, of beads arranged in elegant diagonal designs. Their naked skins glowed sumptuously with diagonal or diamond-shaped patterns of black (genipapo), red (annatto/urucum) or yellow (mineral) body paint, and their faces were adorned in bold red stripes or spots.

One notable common trait was magnificent communal huts – called *malocas* in lingua geral. Wallace saw his first maloca belonging to the Desana four days up the Uaupés. This hut had a rectangular plan, roughly 30 metres (100 feet) long by 12 metres (40 feet) wide. Wallace admired how a central beam ran the length of the house, and noted that 'the roof is supported on fine cylindrical columns, formed of the trunks of trees, and beautifully straight and smooth.... The main supporters, beams, rafters, and other parts are straight, well proportioned to the strength required, and bound together with split creepers, in a manner that a sailor would admire.' These tree-trunk pillars were 9 metres (30 feet) or more high. From the roof-beam sloped great palm-thatched gables, right down to the ground on either side. These buildings thus resembled early Christian basilicas. The thatch of the roof was laid with great compactness and regularity, and was totally watertight; and the thick walls could resist arrows or even gunshot. Even when it rained torrentially the maloca was bone dry. The triangular facade was covered in bark-cloth decorated with bright geometric designs. This wall contained a large door, the main source of light for the interior [PL. 44].

Inside, malocas had the comforting gloom of a church. Each housed about a hundred people from a dozen families, but three times this number came in for dances. On either side of the central aisle, each family had 'a private apartment like a box in a London theatre' formed by a palm-thatch partition; the chief and his family resided at the curved far end. There was a central space for children's play and dances, and the huts were spotlessly clean and tidy. People of course slept in hammocks, so a family's only other furniture was low stools and storage baskets. Personal weapons and adornments might hang from rafters. Larger ovens, pots and other apparatus for communal food preparation were in the hut's central aisles.

These great houses were the cement that united the peoples of the Uaupés. The Owenite Wallace was impressed: 'I could not help admiring the degree of sociality and comfort in numerous families thus living together in patriarchal harmony.' Tragically, decades later, the malocas were deliberately demolished by bigoted Salesian missionaries. In the 1920s, the anthropologist Curt Nimuendajú watched in horror as they destroyed of one of the last of these great houses. He knew that these communal halls were 'the symbol, the veritable bulwark' of the societies on which Christianity was being imposed. 'The Indians' own culture is condensed in the malocas: everything in them breathes tradition and independence. This is why they have to fall.'

Six days up the Uaupés, Wallace and Lima reached a second maloca, of the Wanana people. A ceremony was just ending, with some of the 200 Wanana men and women still dancing. Wallace was dazzled by the beauty of 'these handsome, naked, painted Indians, with their curious ornaments and weapons', dancing to rhythmic music of rattles and songs, in the great smoke-filled gloomy house. They, in turn, admired and watched him – 'principally on account of my spectacles, which they saw for the first time and could not at all understand.'

Lima knew these Wanana well, so they chatted, over prodigious quantities of caxiri. But these indigenous people lived in fear of other Brazilians. When a stranger appeared, men would disappear into the forest to escape being pressed into work as boatmen. Wallace knew that some of the worst characters of the Rio Negro came to trade on the Uaupés.

They would force Indians into their boats at gunpoint and if they resisted shoot them. The rule of law, tenuous on the Rio Negro, was non-existent on this great tributary.

On 12 June they reached São Jerónimo (now Ipanoré), a village above which thundered this river's most dangerous waterfall, Urubuquara ('Vulture's Nest'). This rapid could be attempted only by a small and empty dugout canoe, with its passengers and cargo carried around on a portage path. The river was about three times as wide as the Thames in London and, in this rainy season, very deep. Here it was constricted into a narrow gorge. 'Although the actual fall of the water is trifling, its violence is inconceivable... with boiling foam and whirling eddies... [and] immense whirlpools which engulf large boats. The waters roll like ocean waves, and leap up at intervals forty or fifty feet into the air.'

Three more days brought them to Iauaretê ('Jaguar Falls'), a beautiful little place below another fine waterfall. This is the boundary between Brazil and Colombia. Iauaretê was, and still is, inhabited mostly by Tariana people. There was another splendid maloca here, flanked by about twenty huts along the brow of a steep bank of reddish soil. 'The number of pupunha palms [*Bactris gasipaes*; peach palms] standing in clusters on the hillside and among the houses gives a very pretty appearance to the village.' (The peach palm is a most useful tree: the trunk makes building-beams; the fronds yield bows and arrowheads; its reddish fruit tastes like a chestnut, is full of nourishment, and can be fermented into a drink; and its core is edible *palmito* palm-heart.) 'Stretching away from [the great maloca] towards the forest is a very broad sandy path... which is kept constantly cleared and levelled for the dances of the Dabucuri' – a lingua geral word for a ceremonial gathering at which tribes meet for dancing, drinking and ritual exchange of goods. The broad avenue was flanked by other useful plants: gigantic umarí trees (*Poraqueiba sericea*), which yield a rich, oily fruit the size of a potato; and between these 'there twines the indispensable caapi (a powerful intoxicant)'. Wallace liked Iauaretê's chief, Callistro (called Uiáca in his own language), because of his 'benevolent countenance and dignified manner'; but Lima called him an untrustworthy rogue, largely because he stood up to white traders.

Lima asked Callistro to organize a dance for Wallace's benefit, and he readily obliged. It all started with brewing great quantities of caxiri. The girls then painted one another with 'a neat pattern of black and red all over their bodies, some circles and curved lines occurring on their hips and breasts.' Women wore small square tanga aprons, but only while dancing, and occasional copper earrings. Once their own body paint was complete, the women and girls painted their men, discussing how the lines and patterns would look most attractive. 'Necklaces, bracelets and feathers were entirely monopolized by the men.' Wallace described their luxuriant plumage, notably a coronet of regular rows of red and yellow feathers and, attached to a comb on top of the head, a broad plume of white feathers, either from egrets or rare harpy eagles [PL. 50].

The dance itself was as elaborate as the ornaments. There was rhythmic movement by a circle of men, with a regular sideways step, stamping feet, gyrating body, and a deep, martial chant. At times the young women joined in, each clasping a man in either arm and bending forward as she moved. Outside the maloca, the younger men staged a serpent dance. This consisted of two undulating lines, each holding a 12-metre- (40-foot-) long 'snake' made of bushes, with a head of bright-red cecropia leaves – 'altogether a very formidable-looking reptile'. Inside the great house, some 200 people of all ages danced and celebrated throughout the night. There was continual music from fifty little fifes and flutes, and murmuring conversation. When night fell, a great fire lit up the beautiful and exotic scene. Wallace was enchanted. He went to thank Callistro, and delighted the chief's wife by drinking an entire calabash of the mildly alcoholic drink. He found the caxiri 'exceedingly good', even though he had watched 'a parcel of old women' start its fermentation by chewing manioc cake and spitting into the brew.

Seeing this festival was a rare privilege. Wallace was the first non-Brazilian foreigner to describe such a ceremony by unacculturated Indians. Portugal had jealously excluded other Europeans from Brazil throughout the colonial period. The few foreigners admitted after the Napoleonic wars and then Brazilian independence travelled in the more open parts of eastern and southern Brazil – where indigenous peoples had long been subjected to colonial oppression and missionary conversion. The first to be allowed on the Amazon included the Bavarians Spix and Martius, the Austrian botanist Johann Pohl and, as we have seen, the zoologist and ethnographer Johann Natterer. Of these, Natterer was the only one who visited the peoples of the Uaupés and could have seen their ceremonies; but since his papers were lost we shall never know whether he did.

Wallace noted various customs peculiar to these peoples of the Uaupés river. One was that a mother usually gives birth in the great house, and while she is in labour everything is removed from the hut 'even the pots and pans, and bows and arrows' until the following day. She takes her baby to the river and washes herself and it, and she then rests in the house for four or five days. Before puberty, young children eat mostly manioc cakes and fruit – no meat of any kind and only a few small, bony fish. At puberty, girls have far tougher ordeals than boys. Each is secluded in the maloca for a month, with only a little bread and water. Then she is brought out, naked of course, into the midst of her relatives and friends, and 'each person present gives her five or six severe blows with a length of sipó (an elastic climber) across the back and breast, till she falls senseless and, it sometimes happens, dead. If she recovers, it is repeated four times, at intervals of six hours....' The girl was then considered a woman. She might eat all foods, and was marriageable. Boys underwent a similar, but less severe, initiation.

Another 'most singular superstition' found only among these peoples concerned the music of *botuto* sacred trumpets, played at some festivals. Each maloca had several pairs of these trumpets or horns. For Wallace, each pair 'gives a distinct note, and they produce a rather agreeable concert, something resembling clarinets and bassoons'; but Spruce (who was there in 1852–1853) said that they boomed lugubriously. The trumpets were kept hidden some distance from the maloca and brought out only for special ceremonial. When their sound was heard approaching, all women and children ran and hid in the forest or some shed, and remained invisible until the ceremony was over. This was because the botuto horns were so mysterious 'that no woman must ever see them, on pain of death.' Should a woman catch sight of one, even by accident, 'she is invariably executed, generally by poison, and a father will not hesitate to sacrifice his daughter or a husband his wife...'.

Richard Spruce acquired two of these dangerous trumpets and sent them to William Hooker for his Museum at Kew. They were 'of inconvenient bulk' for packing, and 'innocent as they may look, [they] have been the cause of not a few scourgings and poisonings in their native country.' Because of their magical powers, 'I had no small difficulty in carrying them off.' 'I wrapped them in mats and put them on board [my boat] myself at dead of night, stowing them under the cabin floor, out of sight of my Indian mariners, not one of [whom would] have embarked with me had they known such articles were in the boat.' The precious artifacts are still in the Economic Botany collection of the Royal Botanic Gardens at Kew. Each botuto is about a metre (3 feet) long, with a thick black mouthpiece at one end, and a flaring soundbox like a huge megaphone or cornucopia. The soundbox is a spiral coil of wide strips of dark-brown bark from the yévaro (*Eperua grandiflora*, a tall riverbank tree common on the Uaupés) held in place by long canes and cord. It is an ugly instrument, endowed with sinister magic.*

Each maloca had a *tuxaua* (hereditary chief) whose family lived in the curved apse at the far end of the rectangular great house. But Wallace learned that chiefs were just first among equals, with little governing power over other members of the tribe.† There were

* Thirty years after Wallace and Spruce were there, Italian Capuchins attempted to convert the Tukano-speaking peoples. A crass priest, Friar Giuseppe Coppi, sought to ridicule tribal beliefs by suddenly exposing a botuto to a congregation of both sexes. There was pandemonium: women desperately tried to hide and not to look, and the fanatical missionaries had to flee for their lives.

† The French philosopher Michel de Montaigne had grasped this 'primus inter pares' in his essay *Des cannibals* over two centuries earlier, but Wallace could not have known this.)

31

32

31 The village of Nazaré, where Wallace and Bates stayed when they arrived in Belém in 1848. Wallace's painting shows the everyday white clothes of people in Pará.

32 When Bates returned to Belém in 1859 he found it being transformed into the city it became in the later nineteenth century. The inner harbour is flanked by a market and warehouses.

33

34

33 Spruce's drawing of Santa Isabel, the hamlet created by the dynamic runaway slave Custodio in the forests of southern Venezuela, far up the Pacimoni river and below Mount Neblina.

34 Santarém was in many ways the most agreeable town on the Amazon. Bates and Spruce each chose to spend years and months collecting around Santarém, midway between Belém and Manaus.

35 Spruce's drawing of Tarapoto in northern Peru, where he spent twenty-one months of happy botanizing in 1855–1857. His house and garden are on the right.

36 In January to March rains in the Andes cause the Amazon river to flood vast areas of forest known as várzea. It was easy to move upstream in those floods, and travellers marvelled at the ability of boatmen to find their way through the labyrinth of trees. But there was no dry land on which to camp or light a fire.

37 Movement on the upper reaches of Amazonian rivers was (and still is) punctuated by constant rapids. Wallace and Bates marvelled at their Indian boatmen's skill and courage in surmounting these obstacles, which change throughout the year with a river's rise and fall.

38 A typical riverside campsite. A European scientist slings his hammock, surrounded by boxes and alongside his boat with a thatched tolda cabin.

37

38

39

40

41

39 After 1853 a subsidized steamship service transformed movement and communications on the main Amazon, for the elite. A steamer went monthly from Belém to Manaus, and quarterly from Manaus to Nauta in Peru. Half the time of the voyage was taken in loading wood, and the logs occupied half the deck space.

40 A Tukanoan shaman teaches a boy about the powerful hallucinogenic vine caapi. Spruce brilliantly established that this was the same hallucinogen that other indigenous peoples called yagé, cadana and ayahuasca. He gave it the botanical name *Banisteria caapi.*

41 The great tropical botanist Professor Richard Evans Schultes of Harvard revived Spruce's reputation as the foremost Amazonian botanist of the nineteenth century. Here he is inhaling tobacco snuff through a Y-shaped tube.

43

42, 43 Spruce and Wallace were pioneers in studying indigenous rock art, which they recorded near the lower Amazon, in the middle Rio Negro, on the Casiquiare canal and on the Uaupés. This study is now an important branch of Amazonian archaeology. These photographs are of petroglyphs on tributaries of the upper Uaupés.

also plenty of shamans, whose healing powers were highly prized. And, as with some other Amazonian indigenous peoples, there was a belief that every death of a youth or young adult was caused either by poisoning or the evil sorcery of an enemy. Both imagined causes of death cried out for revenge, and this could lead to a spiral of retaliation.

Spruce wrote a moving account of the burial of a Tariana woman in the great house of Urubuquara (Ipanoré). Lamentations started immediately after she died, and continued without interruption as her grave was dug inside the maloca and her body placed in it. 'By sunset the whole population had assembled on the spot... seated themselves around the grave... and beat the earth hard down over the corpse [to prevent its removal by sorcery].... When night closed in, a large fire was lighted on the grave.... Into the fire was thrown everything that had belonged to the deceased – her hammock, saya [skirt], baskets, tinder-box, etc. The fire was kept up and the people sang and wept around it through the night.' In the case of this woman's death, the female shaman who was suspected of having caused it was the aunt of Ipanoré's chief, Bernardo – which saved her from revenge execution. Before long, some houses had over a hundred graves in them; but when a maloca was full, burials continued outside in a form of churchyard. Wallace observed among the Tariana and Tukano that corpses were dug up after a month and the decomposing body was cooked in a great pan. When 'all the volatile parts are driven off with a most horrible odour' the resulting 'carbonaceous mass' is pounded into powder, mixed in caxiri, and all drunk by the assembled company. 'They believe that thus the virtues of the deceased will be transmitted to the drinkers.'

The most distant people that Wallace met were the Tukano-speaking Cubeo. He saw them on his second voyage up the Uaupés at a village called Uarucapurí, twelve days and countless rapids northwest of Iauaretê, and then at Mucura, the farthest west that he travelled. He spent a fortnight at Mucura, hoping to do fruitful collecting. The Indian men 'came in great force to see the "Branco", and make an attack on my fish-hooks and beads, bringing me fish, pacovas [a medicinal plant, *Renealmia exaltata*], farinha, and manioc-cake, for all of which one of these two articles was asked in exchange.' The Cubeo were 'handsome men, all clean-limbed and well painted, with armlets and bracelets of white beads', but they differed from tribes downriver in at least two respects. One was that they wore ear-plugs, the size of bottle corks and capped with a piece of white porcelain. The other was that they were true cannibals. They ate enemies whom they had killed in battle, and even made raids in order to get human meat for food; and they smoked and stored surplus human flesh for later consumption. This was in addition to eating symbolic fragments of dead relatives soaked in caxiri, similar to the funerary ritual practised by all

Tukanoan peoples. There was no belief in an afterlife, but reincarnation as an animal seemed likely. For this reason these peoples shunned eating large animals such as peccary, tapirs or brocket deer. One said to Spruce: 'How should we kill the stag, [if] he is our grandfather?' So they were essentially fish-eating. The Indians of the Uaupés had all the skills of fishing and hunting of other indigenous peoples. Wallace described their different fishing nets, basketry traps and elaborate weirs. With fish trapped by a weir, they took special care in extracting electric 'eels' and biting piranha.

Wallace noted that most of the women's working day involved harvesting, processing and cooking food. The staple was manioc (*Manihot esculenta*), the super-crop of Amazonia. Its tubers are rich in carbohydrates and calories (although deficient in protein) and when roasted it is perfect explorer's food – easy to carry, impervious to most moulds and very durable. The myrtle-like manioc bush grows easily, and can be harvested three or four times a year. But women have to dig up the tubers; carry bundles of them back to their huts in large baskets called *aturá*; peel the tubers; grate their white flesh (or, in other parts of Amazonia, pound it in hollow tree-trunk mortars); and pulp it with water. Manioc is also the world's only major food crop that requires the removal of a lethal poison – cyanide or prussic acid – before it can be eaten. Many centuries ago, indigenous people made an invention to remove this poison. This was a long tube of diagonally woven basketry called *tipiti*. The tube expands when filled with manioc pulp. Suspended vertically, with a weight at the bottom (a rock, or log, or person), the poisonous liquid is squeezed out of the food [PL. 54]. The cyanide-free manioc is then roasted on broad circular pans to form delicious farinha ('flour') granules. It can also be eaten as *beiju* tapioca pancakes, mingau broth, or fermented into mildly alcoholic caxiri.

Three of the utensils in this manioc process – aturá baskets, graters and roasting pans – were peculiar to the Uaupés and were prized items of trade with other tribes. Spruce sent a manioc grater to Kew for Hooker's nascent museum (now in the Economic Botany collection). It is a beautiful piece, the size of a European washboard, gently concave, studded with diagonal rows of hundreds of quartz teeth, and decorated with Grecian-looking geometric designs. The round roasting pans were 1.2 to 1.8 metres (4 to 6 feet) in diameter with raised rims, made of clay mixed with caraipé (*Licania*) bark before firing. Another valuable trade item was a low stool carved from a single piece of wood, with four legs on sled-like runners and the concave top decorated with geometric incisions.

Wallace drew nice sketches of these and other artifacts, as well as many of rock engravings. He made an ethnographically valuable list of sixty-five articles manufactured by these tribes for domestic and hunting use. He gave far more information than is summarized here. In the few months that he spent on the Uaupés, and with so many other preoccupations of travel, rapids, subsistence and collecting, he could not make a deep anthropological study of these peoples' societies and beliefs. Anthropology scarcely existed as an academic discipline when Wallace left England, so he had no training, or even examples of it to study. But it was amazing how much he did learn and how well he reported it. His anthropological notes earned him nothing in his livelihood of selling natural history specimens. They were done solely to satisfy his intellectual curiosity and demonstrate his delight in meeting unspoiled Indians.

Wallace also compiled vocabularies of the peoples of the lower Uaupés river. For someone with no linguistic training, and spending only a few weeks in tribal villages, he also did a remarkably good job of this. Since he was the first, there were no earlier studies for him to build on (Natterer's having been lost). He made a chart of eleven languages, with comparable words for ninety-eight parts of the body, family members, utensils, foods, colours, some animals, numbers and even a few common phrases. He did this for three Tukanoan-speakers (Tukano and Cubeo (of the Uaupés) and Cueretu (of the Caquetá/Japurá)), as well as for six Arawak-speakers (Tariana, Baré, Wainambeu and three groups of Baniwa, from the Uaupés, Içana and upper Rio Negro), and for the linguistically isolated Juri of the Solimões. Alongside these he included English and the Tupi-based lingua geral, the 'simple and euphonious language' that was spoken by all ordinary people of the main rivers, was the preferred tongue of some Europeans, and the main means of communication between regatão traders and Indians. Wallace did his best to spell native words phonetically. Thus, although later vocabularies by anthropologists were of course more thorough and accurate, the young naturalist did better for these tongues than the great Bavarian Carl von Martius in his two-volume study of the ethnography and languages of Brazilian Indians, published in 1863.

The Tariana of Iauaretê were beginning to feel the destructive impact of Brazilian society. Their remoteness had spared them the murderous slave raids of earlier centuries. But in the 1850s and later decades, the threat was from unscrupulous regatão traders and the military garrisons that represented the government on these distant, empty rivers. Conversion to Christianity was yet to come – in the 1880s by the bumbling Italian zealots who exposed botuto flutes; then in the twentieth century by the efficient but inflexible Salesian missionaries who demolished the great malocas.

Wallace's friend Lima now revealed his ugly side. He engaged an Indian called Bernardo from the lower Uaupés to 'procure' Indian boys and girls for him. 'The procuring consists of making an attack on some maloca of another nation, and taking prisoner all that do not escape or are not killed.' Lima gave this slaver Bernardo powder and shot for his gun and other trade goods. He was to go to seize Indians, particularly a girl each for the Chief of Police and another man in Manaus. This slaving and human trafficking was of course illegal. But Wallace saw that the scant authorities on the upper Rio Negro condoned it – or engaged in it themselves. Lima tried to justify it to his young companion by the specious and time-worn argument that kidnappers were saving their victims from being killed in inter-tribal fighting. On this occasion, the raider hired by Lima returned empty-handed a few weeks later: he said that he tried to use a circuitous route, but his intended victims learned of his approach and decamped. He promised to do better next time.

Ever adventurous, Wallace became excited by the thought of pressing far upriver to the Jurupari ('Devil') Cataract. Traders and knowledgeable Indians told him that those upper districts were full of new species of birds and animals. 'But what above all attracted me was the information that a white species of the celebrated umbrella-chatterer was to be found there.' He doubted whether a white or albino variety of this cotinga (*Cephalopterus ornatus*) actually existed – he had collected the normal black version of it on the lower Rio Negro. But, for an enthusiast like Wallace, 'I could not rest satisfied without one more trial'.

Wallace, however, was a professional collector and he had to send the product of a year's collecting back for sale in England – the specimens he had with him on the Uaupés or had lodged at Guia and Manaus. If he left these on the Negro during the rainy season they might be destroyed by damp and insects, and only he himself could properly organize their packing and dispatch to England. So there was no alternative but to go all the way down to Manaus and then back up the Negro – a journey of a staggering 2,400 kilometres (1,500 miles) – 'which was very disagreeable'. His plan was to spend the rainy months of September to November 1851 on this round trip to Manaus; the dry, collecting months at the start of the following year going far up the Uaupés; and then return to England in July 1852 personally accompanying 'a numerous and valuable collection of live animals'. The thought of eventually returning to beautiful countryside, intelligent conversation and wholesome food made him desperately homesick.

When he started back down the Uaupés at the end of July 1851, Wallace collected insects and found masses of orchids – one clearing near São Jerónimo 'was a complete natural orchid-house.' He did not know the flowers' names – many were probably new to science – but he rapidly collected specimens from over thirty species. 'Some, minute plants scarcely larger than mosses, and one large semi-terrestrial species which grew in clumps eight or ten feet [2.5 to 3 metres] high.... One day I was much delighted to come suddenly upon a magnificent flower growing out of a rotten stem of a tree... a bunch of five or six blossoms which were three inches [8 centimetres] in diameter, nearly round, and varying from a pale delicate straw-colour to a rich deep yellow on the basal part of the labellum. How exquisitely beautiful did it appear, in that wild, sandy, barren spot!' On another day he collected a handsome orchid that opened for only a single day. He packed them in empty farinha baskets, shielded from the sun by layers of plantain leaves. He also collected many new species of tropical fish.

A collector's routine was never easy. São Jerónimo was plagued by 'countless myriads' of biting pium black-flies. In the equatorial heat Wallace had long since ceased to wear stockings, which would have protected his ankles. So 'the torments I suffered when skinning a bird or drawing a fish can scarcely be imagined.... My feet were so thickly covered with the little blood-spots produced by their bites, as to be of a dark purplish-red colour and much swelled and inflamed.' All that he could do was to try to wrap his hands and feet in blankets while he worked.

Back at Guia, Wallace met Manoel Joaquim, one of the worst villains on the Rio Negro. This former soldier had been banished to that remote river because he had been involved in uprisings and was thought to have murdered his wife. He was notorious for threatening Indians at gunpoint, to seize their wives and daughters; he beat his own Indian woman so badly that she fled to the forest; and the people of Guia accused him of murdering two Indian girls and committing other horrible crimes, including setting fire to Lima's house and shooting an old mulatto soldier. The Sub-Chief of Police took statements against this man, from Senhor Lima and others, and wanted to arrest and send him under armed escort to Manaus. But Lieutenant Filisberto, the commandant of Marabitanas, refused to provide soldiers for this because Joaquim was his compadre. The villain therefore went to Manaus in his own boat, persuaded its Chief of Police (the man who wanted Lima to procure an Indian girl for him) that he had no case to answer, and returned upriver 'in great triumph – firing salutes and sending up rockets in every village he passed through.... So ended the attempt to punish a man who, if half the crimes imputed to him were true, ought... to have been hanged or imprisoned for life.' Wallace commented how easy it was for anyone with friends or money to defeat justice on that frontier. The Indians of Guia were terrified by this monster's return, and fled to their sitios or into the depths of the forest. The result for Wallace was that he could find no Indians to man the igaraté boat that Lieutenant Filisberto had generously lent him for his journey. He and Lima went to São Joaquim, but were equally unsuccessful there. So it was not until 1 September that Wallace finally set off, with two Indian paddlers.

The voyage down the almost-empty river showed the agreeable side of Amazonian travel. At the great rapids of São Gabriel, Wallace paid a handsome sum to an expert pilot and, the river being at its fullest, shot down in a mere two hours. He was frightened and bewildered by 'the conflicting motions of the waters. Whirling and boiling eddies, which burst up from the bottom at intervals, as if from some subaqueous explosion, with short cross-waves, and smooth intervening patches, almost make one giddy.' It all depended on the skill of the pilot, and the crew obeying his every order. After that excitement, it was a pleasant journey in alternating sunshine and rain. It was often impossible to find dry land for a cooking fire, so Wallace had to make do with cold manioc farinha and water. But he was now a seasoned explorer so did not mind what, a year before, would have been a great hardship. Wallace stopped at the few sitios and plantations in the otherwise unbroken forested banks. Such was the camaraderie of this sparsely populated river, that everyone who had met the tall young Englishman in the previous year welcomed him as an old friend. They also took 'a friend's privilege' – asking him to buy things for them in Manaus. One wanted a pot of turtle-egg oil, another a large flagon of wine, one police chief wanted a couple of cats, and his clerk two ivory combs, another gimlets, and another a guitar. None gave money for these purchases, but they promised to reward Wallace with the equivalent value in coffee or tobacco or another local commodity on his return.

The voyage was enlivened for Wallace by the never-failing amusement he got from two pet parrots. One was a beautiful anacá (*Deroptyus accipitrinus*), now known as a red-fan parrot, with green wings and tail and pretty red and blue-grey feathers on its head and breast. Wallace described this bird as 'of a rather solemn, morose and irritable disposition', but it was great friends with his other bird, a black-headed parrot (*Pionites melanocephala*) that Wallace called a marianna and his Indians macaí. This pretty parrot had a black cap, bright orange-yellow feathers around its neck and thighs, a white breast and green wings. It was 'a lively little creature, inquisitive as a monkey and playful as a kitten.' The bird was constantly running about the boat, climbing everywhere, and was an omnivorous eater who also loved tasting Wallace's coffee. The Indian boatmen teased it by imitating its clear whistle, to which it would reply and search in vain for a companion.

The journey down the lower Rio Negro took only two weeks. Wallace was pleased by the white houses of Manaus after so many months in the 'mud-walled, forest-buried villages of the Rio Negro'. But he was even more delighted to find Richard Spruce there. The two stayed in a house that was 'made classic to the naturalist' because it had been occupied two decades previously by Johann Natterer. During his days Wallace was very busy, buying trade goods for his Uaupés journey, selling produce from upriver, and arranging and packing his precious collections. The only carpenter in Manaus had quit to become a river-trader, so Wallace had to make his own insect boxes and packing cases. But in the evenings he and Spruce 'luxuriated in the enjoyments of rational conversation – to me at least the greatest and, here, the rarest of pleasures.' They spent a day and a night collecting together on the far side of the Rio Negro. However, sadly for both of them, they could not travel up the river together. Wallace had two boatmen who wanted to return to their homes near Guia, and the boat his friend Lima had lent him was too small for both collectors and their baggage. Spruce was, as usual, waiting helplessly for paddlers – none could be obtained in Manaus for love or money, not even by the authorities. So Wallace set off alone at the end of September 1851; and, as we have seen, Spruce followed two weeks later – in the boat he had bought for himself and with boatmen whom their friend Henrique Antonij had summoned down from São Gabriel.

Wallace's second voyage up the Rio Negro was slow, because the river was at its highest, and because he stopped frequently to visit friends on the banks. At one sitio he was repairing his boat, in the sun, but was 'quite knocked back with headache, pains in the back and limbs, and violent fever'. This was the first time that he had suffered malaria, which was rare on black-water rivers: Wallace had probably caught it from a mosquito bite in Manaus, and the disease had taken the usual fortnight to manifest itself. He dosed himself with the malaria palliative quinine, and two other curious remedies: plenty of cream-of-tartar water (from the African baobab tree) and some purgative medicine. He was so weak and apathetic that he scarcely had the energy to prepare these nostrums. 'During two days and nights I hardly cared if we sank or swam.' The quinine then brought his fever down.

By the end of October Wallace had reached the sitio of a friend called João Cordeiro (the mulatto Brazilian who had traded piaçaba and sarsaparilla with Lima in 1850). This was opposite the decayed village Santa Isabel and at the mouth of the Urubaxi river. Wallace spent a week here, partly because nearby lakes abounded in manatees (*Trichechus inunguis*). When he was at Manaquiri on the Solimões, he had eaten this aquatic mammal and found its flesh to be very good, 'something between beef and pork... and an agreeable change from our fish diet'. Bates also ate a manatee and felt that its meat tasted like very coarse pork, but its thick layers of greenish fat had a 'disagreeable, fishy flavour'. [PL. 70]

Wallace's manatee was a female, a big animal about 1.8 metres (6 feet) long and 1.5 metres (5 feet) in circumference at its widest; Bates's was even larger. Its head was fairly small with a large mouth and fleshy lips like a cow's – hence its local name peixe-boi or 'cow-fish'. 'Behind the head are two powerful oval fins, and just beneath them are the breasts, from which, on pressure being applied, flows a stream of beautiful white milk.' Although Bates's Pasé Indians were very familiar with this animal, they were baffled by 'its suckling its young at the breast, although [it was] an aquatic animal resembling a fish'. One theory is that the mermaid legend came from a vision of fish-tailed manatees clasping their young while they breast-fed them. The manatee has only fore-limbs, but the bones of these are just like those of a human arm, even down to the five jointed fingers. Its fat was boiled down, and one animal yielded from 20 to 110 litres (5 to 20 gallons) of oil – used for cooking or lighting lamps. Tragically but unsurprisingly, this harmless mammal, with its tasty meat and quantity of useful oil, was hunted almost to extinction. It is now endangered and largely bred in captivity.

Now, on the Negro, Wallace wanted to collect a specimen manatee. The Indians caught a 2-metre (7-foot) male – no easy task because these animals had remarkably acute sight and hearing, and could swim very quickly with their tail and flippers. Hunters would approach very cautiously, and either harpoon the manatee or catch it in a strong net across a stream. Preserving the catch proved to be a horrible task, an object lesson in the trials of taxidermy in the wild. Three men skinned it, while Wallace himself worked on the paddle-like flippers and head – a delicate task because he had to see that no bones were carried away. They then scraped off the layer of fat beneath the skin, followed by all the meat from the ribs, head and flippers, leaving the skeleton bare. Wallace then separated the skeleton into sections, cleaned out the bone marrow, and put it into a barrel of brine. But they found that wood-boring beetles had riddled their barrel with tiny holes, so Wallace and his men spent hours pegging scores of these. It still leaked, and another day was spent plugging more holes under the hoops, losing much brine. They then had to dry the barrel and cover it with pitch, then cloth, then more pitch. Wallace commented that he spent 'two most disagreeable days' work with the peixe-boi'. He also skinned a small turtle as well as one of the strangest Amazonian reptiles: a matamatá or bearded tortoise (*Chelus fimbriatus*). This primeval-looking creature has a flat shell that Wallace described as 'deeply-keeled and tubercled' and from which its huge flat neck and head protrude. Little tufts grow above and below its head – hence the 'bearded' name – and its nostrils are on tubes so that it can breathe when walking below water.

Wallace was as usual in awe of his Indian boatmen. Moving up the broad Rio Negro they were often able to sail. But that river was (and still is) famous for its sudden, violent storms. 'Many of [these] were complete hurricanes, the wind shifting round suddenly through every point of the compass.' One of the many storms hit them at night, with great waves, driving rain and bitter cold. They had difficulty getting their sail in, under this fierce wind. Wallace saw that his montaria dinghy had swamped and was being towed waterlogged and damaged. He ordered his men to cut it loose and abandon it. But one Indian disagreed, dived into the turbulent water, and saved the boat by swimming it to shore. Next morning, Wallace rewarded him and the others with a cup of cachaça each. The men were, as always, brilliant at manoeuvring the boat up the São Gabriel falls. But then things went wrong: two men went down with malaria. Wallace found two more Indians and paid them in advance, but they absconded; and when he reached São Joaquim on the lower Uaupés, two other Indians stole the rum he was keeping for preserving fish – they doubtless felt that this was a waste of good spirit.

Now, in December 1851, Wallace himself succumbed to a terrible attack of malaria. Unlike the black-water Rio Negro, the Andes-fed Uaupés was a 'white' river in which mosquitoes could breed; and there was just enough traffic along the Uaupés for men carrying malaria to transmit it by mosquito bites. All that Wallace could do was describe his symptoms. He was completely devastated. The fever then abated, but on every other day he experienced a great depression and lethargy that was always followed by a feverish night in which he could not sleep. He would recover somewhat for two days, but then 'the weakness and fever increased, till I was again confined to my rede [hammock] – could eat nothing, and was so torpid and helpless that Senhor Lima, who attended me, did not expect me to live. I could not speak intelligibly, and had not strength to write, or even to turn over in my hammock.' He took plenty of quinine, but it was a fortnight before 'the fits ceased and I only suffered from extreme emaciation and weakness.' The disease continued to return regularly, for two months until February 1852. Raging fever was followed by drenching perspiration lasting half a day. When the fevers appeared to have abated Wallace could eat heartily, but he gained so little strength 'that I could with difficulty stand alone or walk across the room with the assistance of two sticks.'

We now know that there are four types of malaria in South America, but the most common are *Plasmodium falciparum* and *P. vivax*. Wallace probably had the worst of these, falciparum, in which the attacks usually recur every day, unlike vivax where the interval is two days. But he precisely described the symptoms of both strains: the 'cold-dry' phase of chill and shivering; the 'hot-dry' phase of raging temperature, headache and possibly delirium; and the 'hot-wet' phase of profuse sweating and lethargy. These three phases correspond to: the parasites bursting out of the liver, where they had quietly but profusely replicated for two weeks after the infected mosquito's bite; then the body's white blood corpuscles reacting and causing fever; and then the malarial parasites continuing to multiply in their thousands. The disease leaves its victim badly anaemic, which explains Wallace's extreme weakness.

In late December Wallace wrote a letter to Spruce, which someone took down the Negro in the hope of meeting the other Englishman on his way up the river. The courier found Spruce when he was staying at the Uanauacá farm of Manuel Jacinto de Souza below the São Gabriel rapids. (Wallace had stayed there, a month earlier, and described it as by far the prettiest and most efficient plantation on the river.) On 28 December, Spruce wrote from there to John Smith of Kew Gardens that he had just heard from 'my friend Wallace... that he is almost at the point of death from a malignant fever which has reduced him to such a state of weakness that he cannot rise from his hammock or even feed himself.' For some days he had taken nothing but orange and cashew juice. Once Spruce had surmounted the great rapids, he hurried to visit Wallace at São Joaquim and was appalled to see how devastated he was. And when Wallace had recovered somewhat, he went down to São Gabriel to spend a few days with his botanist friend.

8

INTO THE UNKNOWN

Richard Spruce's journey up the Rio Negro in November 1851 to January 1852 was a delight. As he wrote to his friend John Teasdale, because he owned his boat he was 'master of my movements – I could stop when I liked and go when I liked'. He had had the boat's tolda made long enough for his hammock; he also had a soft sofa-bed of thick layers of bark of the Brazil-nut tree; his boxes along the sides served as a table and seats; and he suspended his gun and other essentials from the roof. Above all, the crew of six Indians (four Baré, one Tariana from the Uaupés, and one Manau) were superb. In addition to smooth handling of the boat, they were also excellent fishermen so that 'we rarely passed a day without fresh fish.' They would glide up to a fish in his montaria and unerringly shoot it with a bow and arrow. This uncanny skill has to be seen to be believed. The archer stands motionless in a canoe, sees a fish below the surface of the black water, instantly calculates the angle of refraction and speed of swimming, releases his arrow just hard enough to hit the fish, and the arrow (attached to a cord) then bobs about in its thrashing victim [PL. 58]. In half an hour one morning, two of Spruce's men killed twenty large fish in a side stream with their arrows. The Manau was also an eagle-eyed hunter who would take Spruce's gun into the forest before daybreak 'and surprise the birds still asleep in the trees, when I could no more discern them than I could the fish in the waters'. This man kept the team supplied with the most delicious game birds: mutum curassows, partridge-like inambu (tinamous; *Tinamus tao*), and water birds. Spruce persuaded the Manau to stay with him after the others had returned to their homes. For six months he kept his boss fed, in a country where apart from manioc everyone had to catch their own food. He was also invaluable on Spruce's collecting forays, paddling the canoe, climbing and felling trees, and expert on forest lore. But, tragically he had to retire because of a back injury – acquired when he had earlier been press-ganged and taken to Belém for forced labour as a stevedore.

Years later, Wallace edited Spruce's papers and described his friend's collecting successes on this voyage: 'Everywhere he finds [plants] in flower on the very margins of the river.... [It was] ever-changing vegetation, the various trees and shrubs and palms which his experienced eye detected as novelties, and the many beautiful flowers he was able to gather which were not only new species but were so peculiar in structure as to constitute new genera – all this rendered the journey a continuous intellectual enjoyment to so enthusiastic a botanist.' Spruce himself wrote to John Smith at Kew, that after they passed Barcellos on the middle Rio Negro everything was new. In such a cornucopia he had confined himself 'to those which presented the greatest novelty of structure. Nothing like this has ever happened to me before.' For instance, 'I counted no fewer than fourteen species of Lecythis [a tree related to the Brazil-nut] in flower, and all but one new to me! ...But the glory of the Rio Negro is a Bignoniaceous tree (apparently an undescribed genus) with whorled leaves and a profusion of pink flowers the size of those of a foxglove. It grows 90 feet [27 metres] high!' It was this new genus that Spruce asked Bentham to name *Henriquezia*, after his Italian friend in Manaus. The Indians shared his enthusiasm. They would peer into the riverbank trees and call out (in lingua geral, which Spruce had wisely learned) 'Master! Here's a pretty flower' – and it did often prove to be something new. They generally travelled by night so that they could occasionally go ashore in daytime. The men would then rest, but Spruce worked – spreading out his paper in the sun, drying plants and exploring the forest for blossoms. When he found something, he called a man over to climb the tree for him.

Although the vegetation was luxuriant, the river was depopulated – as noted by Wallace in the previous year and by many Brazilian observers. Spruce wrote to a friend that 'The Rio Negro might be called the Dead River – I never saw such a deserted region. In Sta. Isabel and Castanheiro there was not a soul as I came up, and three towns, marked on the most modern map I have, have altogether disappeared from the face of the earth.'

Spruce rested at Manuel Jacinto's lovely farm Uanauacá, the place that Wallace had described as the finest on the Rio Negro (not that there was much competition). Then, on 9 January 1852, he started to tackle the rapids below the big fall. He himself rashly took the helm, while some men pulled at a rope on shore to keep the boat straight, and the pilot and others jumped into the water and shoved with their shoulders. But the boat hit a rock, almost capsized, spun round and jerked the rope from the men on the bank. Spruce was alone on his boat. It almost capsized to the other side, spun again, and then shot downstream like an arrow. By good luck, he managed to get its head into a small bay. It was all over in an instant, but Spruce vowed never again to try being helmsman. (This author has twice slipped in a rapid and been swept down the river at terrifying speed until I, too, was saved by being swept into calmer water.) At another rapid, they had a brisk following wind to help the heaving men but, even so, advanced only by inches against the raging current. It was not until 13 January that they finally reached the main falls. Spruce recruited workers from a nearby farm, so he had eleven men through most of the day. They progressed by the Indians carrying a five-inch-thick rope cable across the rushing water and over granite rocks – 'a very laborious and perilous task' – then fastening this to a protruding rock upstream. This cable was tied round the boat's mast, some men heaved while bracing their legs against a cross bar, and others pulled on a rope from the bow to keep the boat inshore. One great danger was that if the boat caught on a sunken rock this became a pivot that made the current irresistible. The men on the shorter rope would then be dragged under and 'dashed to pieces' if they did not let go. Those on board would leap into the water to prevent the boat capsizing catastrophically. This happened several times. The men performed heroic manoeuvres, once to save a heavy Welsh-made cooking cauldron which pitched overboard. Another time the montaria raced down and slid under the batalão, but the man in it leapt clean across the larger boat, dived in on the far side, caught the capsized dinghy, sat astride it, and rode it to still waters.

The following day was spent climbing more rapids, often with wearisome unloading of cargo and carrying it overland. Then, on 15 January, there were more rapids, followed by the toughest of all, beneath the hill on which the town of São Gabriel was built. By now Spruce had fifteen men yoked to his boat's ropes, and others helping him to carry his heavy cargo up and down winding paths in the heat. During these days of toil 'my Uaupé Indians did not hesitate to swim down the most furious of the falls; they even seemed to delight in doing it,' swimming with their legs and holding their arms in front to fend off from sunken rocks. He paid them off that night, with mirrors and machetes, and for the leader 'a gay handkerchief'. 'They all seemed highly contented, and went on their way rejoicing. They were really a set of fine fellows, always in good humour, and almost competing to do anything that "the patron" [Spruce] requested.' Thanks to them, he had not lost his boat nor a single article of cargo.

Spruce's boatmen's only weakness was alcohol. When they were briefly in Manaus, they all got terribly drunk on cachaça. The oldest of them sold all his worldly goods apart from his trousers – his hammock, shirt, knife and tinder-box – and 'got so gloriously drunk on the proceeds as to be in a state of utter helplessness for a couple of days.' But, when sober, this man was the best of that admirable crew. Spruce had to keep his men loyal by occasionally giving them money for 'a skinful of cachaça'. Alcohol was a threat to unwary Indians in another way. The shortage of boatmen was acute, with far too few for all the boats plying Amazon waters. So traders stole one another's Indians by getting them 'dead drunk, then tumbling them into the boat like so many logs and setting sail immediately. When the Indian wakes up from his drunken sleep he finds himself far from port and embarked on a voyage he dreamt not of undertaking.' For official business, the authorities did not even bother to get paddlers drunk: they just sent soldiers to seize men at gunpoint. Spruce had brought a Uaupés-river Indian called Ignacio up from Uanauacá, and this man was an excellent hunter and collector. But one day Spruce's man was in a friend's sitio helping women grind sugar cane, when he was press-ganged to row on the postal packet to Manaus. Whenever a 'courier' boat was due to sail, 'a detachment of soldiers is sent by night to enter the sitios and seize as many men as are wanted, who are forthwith clapped into prison and there kept until the day of sailing – in irons if they make any resistance. The voyage averages fifty days, and these poor fellows receive neither pay nor even food for the whole of this time.' The boatmen have to beg food from friends in riverbank farms. As Spruce commented, 'such treatment is a great disgrace to the Government' in a country where slavery of Indians was theoretically illegal.

As soon as he had surmounted the cataract, Spruce hurried to visit the desperately ill Wallace, some 30 kilometres (19 miles) upriver at São Joaquim. Back in São Gabriel, Spruce decided to stay, probably because he wanted to collect plants that grew among the great rapids and in nearby hills. In the event, he stayed for seven months, from January to August 1852. Located right on the equator, the village had a curious weather pattern, so that it rained during much of Spruce's time there.* The town, the largest on the Rio Negro above Manaus, was a miserable little place. Because Spruce was constantly going up and down the falls, he chose to stay in a pilot's hut below one of them. The thatch was alive with rats, scorpions and cockroaches, and the earth floor was riddled with leaf-cutter saúba ants. Spruce had 'terrible contests' with these stubborn insects. 'In one night they carried off as much farinha as I could eat in a month; then they found out my dried plants and began to cut them up and carry them off. I have burned them, smoked them, drowned them, trod on them and, in short, retaliated in every possible way.' For a while he seemed to have won that struggle against the leafcutters. Another plague was termites, who had their tunnels all over the old hut's beams. They ate up a towel and bored into a wooden

* Broadly, the rainy season in the southern hemisphere is from November to March, the exact opposite north of the equator, but irregular on the line itself.

packing case. São Gabriel was also infested with vampire bats. Spruce's house had large blood stains on its floor from them; but he himself suffered less than most people because in his hammock at night he wore stockings and wrapped himself in a blanket. These small bats attack in the dark, so the best defence was to keep a lamp burning; but lamp-oil was scarce and expensive. Another problem was wasps. On a collecting excursion, Spruce swiped a low branch with his machete 'not noticing that a wasps' nest was suspended from it.... A cloud of the vile insects buzzed out... and attacked me tooth and nail.' He ran back, beating away the wasps, but many 'got into my hair and stung me all over my head and neck.' When he finally got rid of them, he was 'dizzy and stupefied, and it seemed as if my head were bursting, for I suppose that I had not fewer than twenty stings in the head and face alone.' Pondering on this ferocious attack some years later, Spruce wrote that 'I have been stung by wasps I suppose hundreds of times.... Yet I have always admired their beauty, ingenuity, and heroic ferocity.' There spoke a true naturalist!

Worse than all these tribulations, 'the greatest nuisance at São Gabriel [was that] its sole [non-Indian] inhabitants are the soldiers of the garrison' and these fourteen men were all criminals whose punishment was army service in this godforsaken posting. Every soldier had committed a serious crime, and half of them were murderers. They twice burgled Spruce's hut, stealing two gallons of spirits, a quantity of molasses and vinegar, and other things.

There was absolutely nothing to be bought in the village beside the fort, 'not even an egg or banana'. When Spruce's expert hunter-fisherman had to retire because of his back problem and the other was press-ganged into paddling the postal courier, the botanist had nothing to eat. He wrote to Bentham that the situation became acute during July, when all the Indians spent a month drinking and dancing to celebrate the local saint, Gabriel, and no one fished or hunted. 'Never was I so near dying of hunger. I was reduced to take the gun on my shoulder and go out early in the morning into the capoeiras [secondary-growth scrubland] in quest of parrots and jacús [guan game birds].' If the rain was not too heavy, he usually shot something. But 'I once passed three days solely on xibé (farinha mixed with water)... which causes great flatulence and does not allay hunger.' These shooting forays took up many mornings, so that Spruce could collect only briefly in the afternoons. Apart from some aquatic plants in the rapids collecting was not good, with few trees in blossom during those rainy months.

Spruce tried an excursion to a hill called Serra do Gama, but this was largely washed out by torrential storms that were very cold and with pelting rain. His men rigged up a palm shelter for the hammocks, but Spruce still spent a miserable sleepless night. 'We were serenaded by the lugubrious croaking of frogs..., to which the raindrops pattering on the leaves and plashing in the stream formed an appropriate accompaniment.' Climbing the hill was difficult and slippery. On the summit clouds obscured the view, and there was nothing of botanical interest.

Spruce wrote sadly to George Bentham that he had had no post or newspapers from England for about a year. 'I seem to have taken my last leave of civilisation....' Despite this, on 21 August 1852 he plunged further into the wilds, finally leaving São Gabriel and taking his boat up to the Uaupés river.

In January 1852 the malaria-stricken Wallace had his twenty-ninth birthday, but he was probably too ill to appreciate this. Spruce briefly visited his stricken friend. Then in early February, as soon as Wallace could walk with a stick, he hobbled to the landing place and was taken down to São Gabriel. Both men were delighted to converse with a compatriot who was also an environmental enthusiast – although they had each spent so many months speaking Portuguese that they found it hard to remember English. One topic of conversation was nothing less than evolution. Years later, after Darwin had published *On the Origin of Species by Means of Natural Selection* and Wallace had written a paper with the same ground-breaking idea, Spruce reminded him that they had discussed this at São Gabriel. He recalled that he had then said: 'I have never believed in the existence of any permanent limits – generic or specific – in the groups of organic beings.' Spruce also felt that it was during those travels on the Rio Negro that he came to the conclusion that the laws of nature were eternal, that 'the evolution of organic forms is continuous, without a break' so that the variation of living beings is an incessant progression.

The weakened Wallace bought some biscuit and wine from the commandant of São Gabriel, and was carried back up to São Joaquim. After three weeks his friend João Antonio de Lima appeared with seven Indians from the Uaupés, to take him up that river. So on 16 February, with superhuman determination, Wallace set off on his quest for the white umbrella bird, 'painted turtle' and other rarities thought to exist near its headwaters. 'I was still so weak that I had great difficulty in getting in and out of the boat.' But he reasoned that he might just as well be prostrate in the boat as confined in the house he had occupied for three agonizing months. He longed to get back to England, and therefore wanted to complete this voyage 'and get a few living birds and animals to take with me.'

After six days he reached São Jerónimo (Ipanoré), just below the big Urubuquara falls. An affable trader called Agostinho gave the Englishman a cordial welcome, paid off his Indians, arranged for his boat to be carried up past the rapids and recruited new men for the next stage. In late February the river was low. At Urubuquara it plunged through two narrow channels and was so dangerous that 'even Indians who fall in here mostly perish'. So the boat and all its cargo had to be carried up and down a steep half-mile trail through the forest.

Wallace had two 'half-civilized' Indians, known as *guardas* or headmen, who hunted for him, interpreted with the Tukano-speaking local people, looked after his baggage, and ate with him. They were too important to row with the ordinary boatmen. He was disappointed when these guardas could not resist breaking into his flagon of cachaça (partly intended to preserve specimens) and getting drunk. But he pretended not to notice, because they had been paid in advance and he depended on them to continue the journey. Wallace's pilot Bernardo found him ten paddlers, whom he paid with axes, mirrors, knives, and beads.

They reached Iauaretê at the end of February. Getting up the waterfall that spectacularly overlooks the village proved tough. It involved long portages of cargo and back-breaking hauling the boat up the rocks. Wallace's malaria was still bad. And there was the usual trouble with boatmen. First the pilot Bernardo 'coolly informed me' that he was leaving; then his brother departed, followed by two other Indians who deserted even though they had been paid in advance. Above the falls the river was low, meandering and studded with rocks and rapids. For over 100 kilometres (60 miles) it forms the boundary between Brazil to the northeast and Colombia to the southwest – not that either nation had any official presence in those virgin forests in the 1850s.

On the Negro and lower Uaupés Wallace had been accustomed to days of placid river interrupted by a few major rapids. Now the cachoeiras came thick and fast. On one day they had to climb ten, five of which were 'exceedingly bad' and could be tackled only in the middle of the river 'where the water rushes furiously'. A dozen Indians, standing chest-deep in the current, had great difficulty heaving the loaded boat against the foaming waters. At other rapids the rocks were bare, and it took almost a day to pull the unloaded but heavy boat around the edges of the barrier. At a rapid called Caruru the wide river rushed among huge rocks down a drop of 4 to 6 metres (15 to 20 feet). They unloaded the boat, cut many poles and branches to protect its bottom, and sent to a nearby Indian village for help. But even with twenty-five men pushing in the water or pulling on liana ropes, the boat moved only 'by steps and with great difficulty'. Wallace decided that his craft was too heavy to proceed. So he bargained with the chief of a village of Tukano-speaking Wanana to build him a big dugout. He paid an axe, a shirt and trousers, two machetes and some beads. It took five days to find and fell a suitable tree and shape and hollow it out. They then pressed on, struggling up fearful waterfalls throughout the month of March, including two weeks at Mucura, a village of Cubeo people where the Cuduiari river joins the Uaupés.

Wallace then decided to turn back, and not press on for a further week just to see the great Jurupari Cataract, deep inside Colombia. It was a wise decision. He had lost three months from malaria, from which he was still weak, and the weather was now changing against him. However, he was justifiably proud to have penetrated so far into Amazonian forests 'that no European traveller had ever before visited'. He had gone without accident 'up a river perhaps unsurpassed for the difficulties and dangers of its navigation. We had passed fifty cachoeiras, great and small; some mere rapids, others furious cataracts, and some nearly perpendicular falls.' He itemized these. About twenty were rapids up which the boat (or, later, the dugout canoe) was pulled by liana rope; eighteen were 'bad and dangerous', necessitating partial unloading and heaving by all the Indians; but twelve were 'so high and furious' that the craft had to be fully unloaded and pulled over 'dry and often very precipitous rocks.' Little has changed in present times. Rivers are still the main arteries in a forested land with very few airstrips, and the rapids are as formidable and fearful as ever. The only change is the advent of outboard motors and lighter metal boats. But indigenous people are still by far the best at navigating the dangerous rocks and currents, which of course change constantly with the annual rise and fall of each river.

From a collecting point of view, Wallace's brave journey yielded only modest returns. He failed to find his white umbrella bird or 'painted turtle'. But he did get a haul of birds and insects, including an anambé de caatinga (a rare species of cotinga), a beautiful topaz-throated hummingbird and a new species of *Asterope* butterfly. He also started gathering a living menagerie that he hoped to take to England. On the Uaupés he had four monkeys, a dozen parrots and eight or ten small birds, and he continued to buy more during his descent of the Rio Negro: two parrots at one sitio, five more at another, and then 'a blue macaw, a monkey, a toucan, and a pigeon'.

Wallace's return journey started with delays, some caused by his crossing swords with a villainous army lieutenant, others from the usual lack of Indian boatmen. He finally left São Jerónimo on 23 April 1852, São Joaquim on the 27th, and reached São Gabriel on the 29th. He paid his respects to its commandant 'and then enjoyed a little conversation with my friend Mr. Spruce' who was still there. Descending the middle and lower Negro, Wallace visited various friends on the banks, some of whom 'embraced me with great affection at parting, wishing me every happiness'. The voyage was fast, and marred only by three tribulations: recurring attacks of malaria, constant rain and customs duties. Also, 'To attend to my numerous birds and animals was a great annoyance, owing to the crowded state of the boat, and the impossibility of properly cleaning them during the rain.' Some died almost every day. Then, at Barcellos and again on arrival at Manaus, Wallace had to declare the contents of his boat and 'pay duties on every article, even on my bird-skins, insects, stuffed alligators [caiman], etc.' This was because the newly created province of Amazonas was desperate to get all the income it could. Despite these irritations, Wallace reached Manaus on 17 May and was 'kindly received by my friend Henrique Antonij'.

The most important legacy of Wallace's two journeys to the Uaupés (June to August 1851 and February to April 1852), and the seven months that Spruce spent there (August 1852 to March 1853), was their descriptions of the indigenous peoples. Such reporting could never have come from the handful of Brazilians who were there. Those so-called *civilizados* were usually traders intent on bartering for indigenous produce. They were uninterested in the Indians' customs or beliefs, but were anything but disinterested about abuse. During those decades, there were one or two assessments by officials or missionary priests; but these were dry reports for the provincial government – which was intent on trying to turn indigenous peoples into docile caboclos. There were no journals or travel-writing by Brazilians. Thus, this first glimpse by intelligent young Englishmen was valuable.

The main trade item that attracted Lima and the other regatão traders to the Uaupés was sarsaparilla. They wanted the long roots of these *Smilax* shrubs, that were difficult to locate and laborious to dig up (an operation that killed the plant). As Wallace explained, 'it is not generally found on the great rivers, but far in the interior, on the banks of small streams, and on dry rocky ground. It is principally dug up by the Indians, often by the most uncivilized tribes, and is the means of carrying on considerable trade with them.' The Cubeo were skilled at collecting sarsaparilla, which was why Wallace met two young Brazilian traders when he visited Mucura, at his farthest west.

As we have seen, traders were not interested solely in sarsaparilla. Some also trafficked in people. In June 1851 Wallace had seen his friend Lima arming an Indian to catch Indian women and children. Nine months later, when he was among the Cubeo at Mucura, he found that there were almost no men in the village. All had been taken up the Cuduiari tributary of the Uaupés by Brazilians called Jesuino and Chagas, to help them attack the Carapanã tribe (another Tukano-speaking people, mostly in Colombia). 'They hoped to get a lot of women, boys, and children to take as presents to Barra [Manaus].' A few days later, further down the Uaupés at the Caruru rapids, Wallace met them 'with a whole fleet of canoes, and upwards of twenty prisoners, all but one women and children. Seven men and one woman had been killed; the rest of the men escaped; but only one of the attacking party was killed. The man was kept bound, and the women and children well guarded, and every morning and evening they were all taken down to the river to bathe.' The captives were taken either to the little forts at Marabitanas and São Gabriel or all the way down to Manaus. In the provincial capital, the children were placed in 'orphanages', which allocated them to citizens for 'civilising instruction' – but in reality for unpaid domestic service. The claim that these kidnapped children were orphans was a fraud for circumventing the prohibition on enslaving indigenous people.

The slaver inciting the Cubeo was Lieutenant Jesuino Cordeiro, whom Wallace described as 'an ignorant half-breed'. He had been sent by the new provincial government in Manaus to be Director of the Indians of the Upper Rio Negro.* The first president of the new province of Amazonas, João Batista Tenreiro Aranha, declared in 1851 that he hoped to tame some of the indigenous peoples in his vast province by appointing Directors of Indians and vigorous missionaries. The missionary sent to the region was a Capuchin Franciscan, Friar Gregório José Maria de Bene. In the year 1853, the Friar baptized hundreds of people on these rivers, married fifty couples, and set up chapels in twenty-four villages – he would have done more, but the Indians of the Içana had fled to escape having to 'pay their debts' to the Director of Indians. Between them, the Friar and Cordeiro forced several tribes to congregate in new settlements. But the two men then fell out. The Friar accused Cordeiro of filling his house with Indian captives and enriching himself by trading. Director Cordeiro then antagonized many traders by trying to prevent them from entering 'his' Indian villages, and from denouncing his excesses to the authorities. In the event, the depredations of this pair were short-lived. A later provincial president, Albuquerque Lacerda, reported that the region was more abandoned than ever: the only visitors to indigenous villages were regatão traders 'who exploit, deprave and dishonour them under the pretext of commerce'.

Jesuino Cordeiro was back on the Cuduiari river in the following year. His report to the provincial president was flagrantly mendacious. He alleged that he was 'peacefully'

* The position Director of Indians, hated in colonial times, had been revived in 1845 in Brazil's important legislation on indigenous policy. This job title may have implied championing and protecting indigenous peoples. It was nothing of the kind. The Director of Indians was supposed to 'civilize' his charges and turn them into detribalized and docile workers.

trying to found a settlement for the Cubeo, when some canoes of Carapanã and other tribes appeared at his camp early one morning. He allowed them to land, thinking that they had come to talk to him. But they were armed with bows and arrows, clubs and poisoned darts for their blow-guns (as any hunters would be), and appeared 'ill-intentioned'. So 'I ordered a volley to be fired to intimidate them, but because of this volley two of the wretches were wounded – they died a day later. I immediately asked for help from the commandant of the Fort of São Gabriel, to send fifteen soldiers for my defence.' These men (Spruce's convicted criminals) arrived two weeks later and Cordeiro took them up the river and into the forest in pursuit of the 'wicked heathen'. They of course found everything tranquil. Hardly surprisingly, Cordeiro did not mention his slaving raid against the Carapanã in the previous year.

Wallace had a brush with Jesuino Cordeiro in April 1852. He was at Caruru, about to descend the mass of rapids down the Uaupés. He had two boats: the one lent by Lima and the dugout he had paid to have built. His Indian boatmen were happy because Wallace had hit a swimming deer on the head and this provided plenty of food. A Cubeo chief's son had undertaken to guide him through the ferocious rapids in return for knives, beads and a mirror. Wallace himself was still too weak from malaria to go into the forest to collect. The trader Chagas then appeared with a request from Lieutenant Jesuino that Wallace sell him the big dugout canoe: he claimed that he needed it to transport chiefs from the Uaupés and Içana rivers to Manaus to receive letters-patent and presents from the provincial President. (This was probably genuine, since it was Brazilian policy to win over chiefs by giving them certificates and uniforms.) Wallace refused to part with his boat. Jesuino was furious. So he ordered the Indians to desert the Englishman as he descended the river – which most did, for fear of the Director of Indians. Wallace was left with only two men in each boat, whereas six or eight good paddlers were needed to shoot the rapids safely. But Wallace took the risk, 'drifting down stream after Senhor Jesuino, who no doubt rejoiced in the idea that I should probably lose my canoes, if not my life, in the cachoeiras.' Wallace managed to recruit some Indians in a village. But the wicked Director of Indians almost got his revenge. At the last major rapid above Iauaretê, Wallace's boat was being lowered by rope, with him alone in it, when it was twisted broadside to the current, was forced up against the fall, and was poised there for a considerable time, on one side and about to capsize and drown him. Wallace finally got free, and he reached Iauaretê 'much to the surprise of Senhor Jesuino'.

When Richard Spruce arrived at São Jerónimo almost six months after Wallace, he found that the village was 'very lively'. The traders Chagas (Cordeiro's accomplice) and Agostinho (with whom Wallace had lodged) were there, employing almost all the Tariana men in building two large boats. Spruce wrote to Wallace that he enjoyed the company of these rascals and another white Brazilian called Amansio. 'We generally all supped together, and passed the evening very agreeably, laughing and amusing ourselves.' The traders told yarns about naughty friars and girls, and of men who could turn themselves into river dolphins or large snakes. Spruce said that Chagas was known locally as 'a very useful man'; but he was a great scoundrel and 'with a face exactly like the back of a Surinam toad. He rendered me much assistance in my excursions.' Chagas was still trying to capture Indian children. Spruce lent a gun to Bernardo, Tariana chief above the Urubuquara falls, unaware that he and Chagas were about to raid up the Papuri river to steal Indian boys and girls. Spruce thus became an 'unwitting accomplice' in this sordid kidnapping. He wrote to Wallace, in the following year, that 'for this, and for other of his good deeds, our friend Chagas is now in prison in [Manaus]'. Wallace had of course seen 'this man's former evil deeds and his escape from punishment' when Chagas had been slave-raiding with Jesuino Cordeiro in the previous March. But both Englishmen could not help but like the friendly villain; and neither knew for which crime Chagas was finally convicted.

Spruce spent most of his time collecting near São Jerónimo (Ipanoré) on the lower Uaupés. He was so busy botanizing that he even stopped writing a journal, apart from during an excursion to Iauaretê and then on the lower Papuri river (which joins the Uaupês from the southwest at that village). He found the scenery on the Papuri 'really beautiful', with occasional Indian sitios on the banks. But 'in its lower part [the Papuri] is an uninterrupted and dangerous rapid', one of which had three narrow channels with perpendicular cascades some 4.5 metres (15 feet) high. The granite of these rapids contained 'the clearest and best picture-writing I have met with'. He drew some rock art at Iauaretê and asked Indians to interpret it for him. Some designs appeared to be trees with buttress roots; others 'merely fanciful geometrical patterns'; one showed a river dolphin, another the three-toed footprint of a tapir ungulate; some appeared to depict domestic utensils such as manioc ovens. Spruce felt that these had been engraved with chips of hard rock crystal, and had been made by ancestors of the peoples he met – not by some mysterious 'higher civilisation' using metal. Like Wallace, he was a pioneer in the study of rock art. It is therefore a pity that he failed to write more about these splendid indigenous peoples. This was partly because he knew that Wallace had studied them a few months earlier, but particularly because he was overcome by the wealth of botanical marvels that surrounded him. He did, however, make a list of the nine fruit trees and eight edible plant roots that Indians planted in their roça gardens.

Spruce made pencil portraits of some individual Indians at Iauaretê. Wallace later wrote (rather unkindly) that 'though my friend was no artist he was a very painstaking and accurate draughtsman' so that these sketches 'give a faithful idea' of these very handsome people. Spruce drew Callistro, the paramount chief of all the Tariana [PL. 46]. He described Callistro as a fine old man of about sixty with a powerful head 'which would not have disgraced a European'. During all their centuries of artistic body-paint, no indigenous peoples ever attempted human portraits, and nor had any visitor. The sketch so delighted his people that every one of them came to look at it; then they wanted more. So Spruce depicted Callistro's son, daughter-in-law and young granddaughter. He later drew Bernardo, the chief above the Urubuquara falls (and the kidnapper to whom he had lent his gun) and his daughter, as well as a forty-year-old Pira-Tapuya woman and a young man of the Carapanã, both peoples who lived on the Papuri river. These and other drawings are now in the archive of The Royal Society in London.

A few months later, when Spruce was at Marabitanas fort on the Rio Negro, he sketched two beautiful Maku girls aged about sixteen and nine [PL. 48]. These children, 'taken in a marauding expedition at the head of the Içana [river] had been purchased by the Commandant of Marabitanas.... The poor creatures were downcast, as might be expected of captives, and could converse with none of those around them....' The Maku are a remarkable people, who survive to this day as one of the world's few totally nomadic forest peoples. (Maku is actually a pejorative Arawak name for these people: they call themselves Hupda.) They never build a hut or village, but are constantly on the move between bivouac sites deep in the forests of this part of Brazil and Colombia. Despite their roaming existence they enjoy a very healthy diet, because they are superb hunters and fishermen, and they plant their favourite trees at their temporary forest camping sites. The wretched little girls drawn by Spruce were understandably bewildered, having been snatched from their families and the nomadic world under the comforting embrace of the forest canopy. They found themselves thrust among ugly, clothed aliens who spoke incomprehensible languages. They had no idea what fate awaited them, as enslaved captives.

Spruce was at Iauaretê for only a fortnight, during which it rained heavily. When he sailed back down the river, the weather brightened and masses of trees sprang into blossom. The frustrated botanist wrote to George Bentham: 'I well recollect how the banks of the river had become clad with flowers, as it were by some sudden magic, and how I said to myself, as I scanned the lofty trees with wistful and disappointed eyes, "there goes a new Dipteryx – there goes a new Qualea – there goes 'the Lord knows what'!" until I could no longer bear the sight and covering up my face with my hands, I resigned myself to the sorrowful reflection that I must leave all these fine things "to waste their sweetness on the desert air".' But once back at Ipanoré, Spruce was wonderfully happy botanizing. He told Bentham: 'I have occupied no station so rich in respect of plants, and not at all to be complained of in respect of eatables.' So he spent most of his seven months on the Uaupés at that village halfway up the lower river.

One thing that helped to make Ipanoré so delightful was a tame grey-winged trumpeter bird (agami; *Psophia crepitans*) that 'so attached itself to me that it would follow me about like a dog.' This dignified bird, looking slightly like a guinea-fowl, was extraordinarily immune to snake bites and 'it never failed to kill any snake that came in our way'. Once when Spruce was preoccupied collecting tiny liverworts, the agami, which had already brought him three or four snakes, brought another live one and 'apparently to attract my attention... because I had not praised its prowess as I usually did', laid its catch on his naked feet. The reptile promptly twined itself up Spruce's leg and he had to snatch it off and hurl it into the brush.

Spruce wrote that 'by means of fish-hooks, Jew's harps, and beads I was able to enlist a troop of little Indians in the search for plants, and they were a great help to me, especially when I could go into the forest with them and point out what I wanted. They were also expert at hunting out fungi....' He himself was, of course, also collecting every day. He wrote to Sir William Hooker of Kew that he had made himself ill 'by working too hard both in and out of doors in the heat of the day' so that he suffered from distressing attacks of vertigo (possibly sunstroke). He commented that the mechanical labour of drying plants there was so great that it left him little time for other studies. However, he summed up his collection from the Uaupés as containing 'a greater number than any preceding one of the tallest forest trees, among which are several undescribed Vochysiaceae and Caesalpinieae.' (The second of these families includes brazilwood (*Caesalpinia echinata*), whose purple dye first attracted Europeans to Brazil and gave the country its name.) In his journal, Spruce mentioned many other trees, including masses of jauari palms (*Astrocaryum jauari*) on seasonally flooded islands, and various members of the bean family including *Campsiandra laurifolia*, which was abundant on sandy beaches. Spruce watched Indian boatmen play 'ducks and drakes' with large flat kidney beans from this tree. As always, lianas festooned trees at the river's edge, notably broad clusters of *Strychnos rondeletioides* from tree-tops right down to the water, whose 'small cream-coloured flowers... perfumed the whole igapó... with delicious odour.' But Spruce's passion, since boyhood, had been for tiny mosses and liverworts. He wrote to Hooker that 'There are also a great many new things among the minutest tribes of flowering plants, such as Podostemeae, Triurideae, Burmanniaceae, and the leafless Gentianeae (Voyrieae).' He commented that from his seven months on the Uaupés: 'I suppose that of the whole collection, numbering some 500 species, about four-fifths are entirely undescribed.' Spruce was the first serious botanist to have collected in that region, so he was like the proverbial child in a sweet shop. Even so, to have collected some 400 species of plants new to science was an amazing achievement.

Spruce would have liked to continue working on the Uaupés, but he feared that the Indians – who lived communally and were not obsessed by private property – would help themselves to his belongings when he was out in the forest. So when the other Brazilians (whose wives guarded his possessions) left, in March 1853, he felt that he must also depart.

Wallace had enthused about the upper Rio Negro and its watershed with the Orinoco as a perfect location for collectors. Spruce therefore hurried down the Uaupés and up the Negro. It took nine days to reach Marabitanas, where he did some collecting and drew the portraits of the two pathetic Maku girl captives. Then he was paddled on past the great granite outcrop Cucui and into Venezuela. On 11 April, after eleven more days of plain sailing, Spruce's boat reached San Carlos de Río Negro, the largest town in this part of South America.

Ten weeks after arriving at San Carlos, Spruce was caught up in a curious episode in that normally tranquil little place. Most of the population were Baré or Baniwa Indians. There were a few Spanish-origin whites, and these resented other foreigners who were more enterprising than them in dominating local trade. It was feared that these Spanish-speaking Venezuelans had been inciting the Indians, whom they bullied, to expel or even massacre Portuguese-speakers. The feast of San Juan, 24 June, was traditionally the day on which old scores could be settled, so the town was buzzing with a rumour that the Indians would rise up against foreigners and other whites. The comisario – the aged Baré who had been kind to Wallace two years previously – and his deputy both left San Carlos, perhaps to be absent should there be a disturbance. The only remaining foreigners were Spruce and two young Portuguese who had settled there with their families. The housekeeper of one of these Portuguese told him that she had heard her Indian relatives planning a massacre of whites – which meant the two Portuguese families, and Spruce because he was white and had a little stock of merchandise for them to loot. Local Indian women brought their men great quantities of cachaça rum that they had distilled, and the celebration started with firing of muskets and blowing of bamboo horns. The Portuguese were frightened by the rumoured uprisings. Their families and Spruce therefore fortified a house and assembled a small arsenal: Spruce's three guns, four other firearms, two swords, plenty of machetes and a good supply of ammunition. On the first day there was nothing more than drunken Indians coming to demand more rum. That night the foreigners waited in 'a state of anxiety' with their arms at the ready; drunken groups occasionally approached, shouting, beating drums and firing muskets, but 'it passed over without our being attacked'. On the evening of the following day there was eerie quiet and they 'were filled with apprehension that this unwonted silence was the prelude to an attack'. During the second night, pairs of Indian scouts patrolled outside the house, but they never 'screwed up their courage... to attack us.' Spruce had decided not to let himself be taken alive had the 150 native people stormed the three men in the house. But nothing happened. Spruce was convinced that this was only because the Indians of San Carlos reckoned that they would suffer too many casualties in an onslaught. He and the Portuguese undoubtedly exaggerated the danger. But their fear was real.

Spruce spent seven months collecting around San Carlos de Río Negro. His main excursion was to climb the granite rock of Cucui [PL. 9]. Spruce climbed only halfway up the sugarloaf, but he was thrilled by the view towards its summit. His Indians went higher and brought down a haul of bromeliads and orchids. He was excited to think that Alexander von Humboldt had been there, that Joseph Natterer had climbed it, and that Robert Schomburgk had also described it.

Spruce's gratification at being in 'Humboldt territory' was considerably lessened by the tribulations of life in San Carlos. A major problem was getting food – 'indeed it occupies nearly all a person's time, and we think ourselves well off at San Carlos when we can eat once a day.' 'I might easily have died of hunger, for the forests near San Carlos are quite exhausted of game'. He survived because he had arranged for a salted ox to be sent up from the Orinoco, and he still had some rice and manioc from Brazil. Spruce heard that in colonial times there had been missions on all these rivers at which travellers could be fed. But after independence officials and ecclesiastics were expelled or had left. 'A country without priests, lawyers, doctors, police, and soldiers is not quite so happy as Rousseau dreamt it ought to be.'

Collecting was very poor. It was the wet season, with few trees in blossom. The riverbank vegetation was very similar to that on the Rio Negro and Uaupés, so there were few novelties. The only consolation was that he collected some new mosses, and many new species of liverworts; and, as he wrote to Hooker, 'my predilection for these tribes, Mosses and Hepaticae, is known to many'. The humidity was appalling: a piece of paper dropped onto the ground would be too wet to write on. Specimens that were dried and stored in a box would be covered in mould in a month's time; but if left out on a table mould would get them in one night. So Spruce wrote to Bentham that he was sending only one box downriver after months of collecting.

Insects were a nuisance, with mosquitoes, pium black-flies, tiny sand-flies, and mutuca horse-flies. One day Spruce was botanizing in capoeira (*rastrojo*; brushwood on destroyed forest) close to San Carlos. He cut a patch of moss from a decayed tree stump, turned to put the moss into his vasculum collecting case, but failed to notice 'that a string of angry tucandéras [tocandira ants; *Paraponera clavata*] poured out of the opening I had made.' He felt a prick on his thigh and thought it was a snake bite, but then saw that his feet and legs were being covered by the dreaded big black tocandiras. He ran through entangled undergrowth, 'and finally succeeded in beating off the ants, but not before I had been dreadfully stung about the feet' since he was wearing only sandals one of which came off 'in the struggle'. He hurried back across a strip of burning sand then waded in a shallow lagoon, but 'both these increased the torture'. Once home, his Indian-woman cook bound a ligature tightly above each ankle and rubbed in hartshorn (ammonia in water) and oil; but this did not lessen the pain. For three hours, 'my sufferings were indescribable – I can only liken the pain to that of a hundred thousand nettle-stings. My feet and sometimes my hands trembled as though I had the palsy, and for some time the perspiration ran down my face from the pain.' He resisted 'a strong inclination to vomit'. A dose of laudanum helped to dull the pain, but it continued through the night and next day, particularly whenever he trod on his bitten feet. Spruce then returned to recover his vasculum, 'which is to me a priceless article', and the sandal. He found them, and cautiously drew them towards him with a stick without disturbing a single ant.

Hardly surprisingly, Spruce wanted to leave San Carlos and continue his journey northwards. He had a *piragoa* boat built for himself and by the end of July thought that it was almost finished and he could pack for the voyage. But there were endless delays. Since the comisario had left San Carlos, its Indians went off to their manioc clearings and there was no one to organize their labour. Although Spruce offered double the usual pay, he could get none of the town's five or six caulkers to finish his boat. For six weeks she lay 'baking in the sun, which opened all the seams and split some of the timbers.' Then termites attacked some ribs that had wrongly been made of soft- rather than hardwood. Spruce found this only after his boat was launched, 'and it has cost me infinite trouble to kill them, with boiling water, for they were in thousands on thousands.' Finally a new comisario was appointed. It took him several weeks to get some Indians back, and each brought a supply of cachaça rum that he had brewed on his cane-patch. Two caulkers spent several more weeks working on Spruce's boat. They were constantly intoxicated and the caulking was so badly done 'that the night after the boat was launched it went to the bottom'. It had cost Spruce 16 litres (3½ gallons) of rum to get the boat launched, and it would cost him another 2 litres (½ gallon) to get it afloat again.

Spruce's piragoa had a single tree trunk as foundation keel, to which side-boards were added. She was just over 9 metres (30 feet) long, over 2.5 metres (8 feet) broad at the widest, and about 75 centimetres (30 inches) deep. The cabin occupied almost half the length, with a board deck and sides, a palm-thatch roof high enough for Spruce to sit comfortably inside on a low stool, and small windows in the sides and stern. There were benches for rowers in the forward part of the boat, with stores below. The crew was seven men and a little boy. All had to be paid in advance for the planned three-month journey, since Indian carpenters were all in nominal debt to some white man and anyone hiring one had first to pay his debt and then 'a further advance of goods'. It says much for the honesty of people on those rivers that, two years after leaving Manaus, Spruce still had plenty of trade goods to meet these costs. He finally embarked at the end of November 1853 – four months later than he had originally planned.

Spruce's long-delayed journey in his comfortable new boat started badly. A few hours north of San Carlos there is a rapid, before the Casiquiare Canal joins the Negro-Guainía from the east. Spruce's men toiled against the current for a couple of hours. But just as the boat was in the middle of the fall, a cable broke – even though it was new, 10 centimetres (4 inches) in diameter, and made of piaçaba fibre. 'She whirled round three or four times, and barely escaped being dashed in pieces against a projecting point of rock.' This made a hole under the boat, through which they could hear water hissing. So they moored to a bank and the men spent the night baling. Spruce 'did not venture to sleep a moment, and roused the men in turn to their necessary task.' He encouraged them with plenty of cachaça – his pilot, Carlos, 'was especially thirsty, and got so large a share that he became tolerably "well drunk".' In the morning, they found the hole and caulked it with mud. But when Spruce examined his piaçaba cable, he found that it had been cut half-through with a knife. Weeks later, his men told him that the sabotage had been done by his 'thirsty' pilot, 'a merry, lazy scamp' who reckoned that a shipwreck 'would have saved him the toil of a voyage for which he had already received pay.' The Indian boatmen thought nothing of being thrown into a rapid; but Spruce's boat might have been smashed and he and his belongings drowned.

The botanist had some large glass flagons and a demijohn of powerful cachaça (*bureche* in Venezuela). He planned to use it for preserving specimens and occasional drinking. But the boatmen, after consuming a quantity of it during the night of baling, were constantly begging for more. So Spruce made a bold move. At midnight on the fourth night after leaving San Carlos, like a Federal agent in the Prohibition era, he surreptitiously poured the entire supply into the river. He was wasting a very valuable commodity and trade good. His Indians were aghast. But he rightly reasoned that it was more important to sober up his men and stop their importuning him.

The Casiquiare 'Canal' is really a river, a branch of the upper Orinoco that bifurcates and never rejoins the main stream. Instead it meanders sluggishly southwestwards, for some 300 kilometres (almost 200 miles) across flat forests, to join the Rio Negro and flow down into the main Amazon. It is the world's only fluvial link between two vast river basins. This connection had been known to indigenous people for thousands of years. The first European to describe it was a Spanish Jesuit, Padre Manuel Roman, in 1744. After that the Casiquiare was used by missionaries (mostly Franciscans), traders and perhaps slavers until the end of the colonial era. Twice a year a boat took wages, salt and other supplies through it to the small Spanish garrison at San Carlos. In 1800 Humboldt and Bonpland had ascended the Orinoco and Atabapo, crossed the portage to the Guainía that Wallace used, and – after they were refused entry to Brazil – returned northwards through the Casiquiare. Confirmation of the link was one of Humboldt's main objectives, so he took a star-fix to determine its latitude and made a remarkably accurate traverse survey as they sailed up it. The only other non-Iberian who had navigated the Casiquiare was Robert Schomburgk, in 1839. Spruce knew about these earlier explorations, from Humboldt's famous book translated into English, and Schomburgk's reports in the *Journal of the Royal Geographical Society*. He was proud to be moving in their footsteps. Nevertheless, it was bold exploration.

It took Spruce's men twenty-four days to paddle and sail up the Casiquiare, and a further two to ascend the main Orinoco to a village called Esmeralda. This was a white-water river, so there was a plague of sand-flies. A typical entry in Spruce's journal was that these tiny insects were 'terrible today, especially when we stopped to cook supper.... Looking into the cabin afterwards, it was like a beehive.' There was also the deep roaring sound of perhaps a million bats flying out at dusk to catch the insects. The boatmen sometimes had to heave on riverbank trees to progress against the current. One day a hook caught a branch containing a large wasp's nest. 'The wasps sallied out in thousands, and the men... leaped into the river.' Spruce was working in the cabin 'when the fierce little animals came buzzing in and settled on me in numbers'. He lay motionless and was not harmed. But on a different occasion he was standing on the tolda's roof trying to gather flowers from a tree, when he hooked another wasps' nest. He kicked this into the river and continued to collect despite 'battling all the while with the wasps and getting severely stung'. This was because he saw that the tree was new – a *Hirtella* that he named *casiquiarensis* – and no amount of painful stings could prevent him gathering it.

As a white river, the Casiquiare had splendid fish like the tambaqui, which does not swim in black waters. Botanically there was little of interest, just a great many paxiúba palms, varieties of inga, cecropias, swartzias and piaçaba. This was a reminder of another problem confronting collectors plunging into the unknown: they might not find novelties that would appeal to European collectors. Spruce was, however, sufficiently expert to spot new variations of what was otherwise familiar: four species of *Vismia* nutmegs 'which seemed new' and unfamiliar varieties of other trees. He also noticed a subtle shift from the vegetation of the Rio Negro to that of the Orinoco.

Climbing a steep granite outcrop, Spruce surveyed the carpet of vegetation on this very flat land, stretching unbroken to the horizon in all directions. His 'practiced eye' could distinguish the main types of forest or landscape. There was the mass of what he called virgin or great forest (*caaguaçu* in Portuguese, *monte alto* in Spanish). Such forest had 'sombre foliage of densely-packed, lofty trees, out of which stand [emergents]… overtopping even the tallest palms… [all] interwoven with stout, gaily-flowering lianas'. This is now known as *terra firme* ('mainland' or 'upland') forest, and it dominates the greater part of Amazonia. Such 'dense forest is the formation with the greatest biomass, with a clear understory, and occurs where environmental conditions are optimal and there are no limiting factors such as a scarcity or an excess of water.' Then Spruce saw patches of lower 'white forest': caatinga (*monte bajo* in Spanish) with 'humbler but surpassingly interesting vegetation… of low, neat-growing and thinly-set trees and bushes', plenty of ferns but few palms or creepers. Spruce could not know, but caatinga vegetation is caused by nutrient-poor sandy soil and the stress of occasional flooding or severe drought. On seasonally flooded lowlands bordering rivers grew igapó (várzea in other parts of Brazil, *rebalsa* in Spanish), where bushes and young trees 'must have the curious property of being able to survive complete and prolonged submersion, constituting for them a species of hibernation.' From his rocky vantage point, Spruce distinguished igapó by its varied tints of foliage, many 'herbaceous lianas which drape the trees and often form a curtain-like frontage', and abundant palms, particularly fan-shaped trees that 'often stretch in long avenues along… the shore.' Two other types of vegetation were not visible from the Casiquiare outcrop: capoeira secondary forest, full of weeds and shrubs that spring up where man has deforested; and campo savannah, of which there are great plains further east near Mount Roraima and that has a distinct and precious vegetation – currently becoming endangered because this is good terrain for ranching or agriculture.

Spruce reached Esmeralda on Christmas Eve 1853. The traveller wrote to his friend John Teasdale that the village 'stands on the most magnificent site I have seen in South America'. It was surrounded by wide grassy savannahs, studded only with a few fan-like moriche palms (buriti or miriti, *Mauritia flexuosa*). Beyond rose the 'abrupt and frowning mass' of Cerro Duida, 2,400 metres (7,850 feet) high, and between the village and river a ridge of 'fantastically-piled granite blocks'. In the setting sun, with deep ravines furrowing the forested mountain, and patches of mica schist glittering like silver, it was 'a scene which has few equals' [PL. 75]. The attractive name was because early travellers thought that they had seen emeralds in its quartzite rocks.

Sadly, the first beautiful glimpse was an illusion. 'To the sight Esmeralda is a Paradise – in reality it is an Inferno, scarcely habitable by man.' When Spruce entered the village's small square surrounded by six miserable huts, a warm wind whipped up some dust, but there was no living creature to be seen – no people, animal, bird or even butterfly. The problem was gnat-like sand-flies (*Culicoides*). Spruce wrote: 'If I passed my hand across my face I brought it away covered with blood and with the crushed bodies of gorged [sand-flies].' Throughout the day, the air was alive with the midges. 'I constantly returned from my walks with my hands, feet, neck and face covered with blood, and I found I could nowhere escape these pests.' It was this plague of insects that kept Esmeralda's few inhabitants sealed inside their huts as best they could during the day. They drowsed there, 'bat-like... and only steal forth in the grey of morn and evening to seek a scanty subsistence.' It was ever thus. In colonial times, Esmeralda marked the end of the known world for Venezuelan Spaniards, with its sky black with insects and almost nothing to eat, dreaded even by dedicated missionaries who were required to go there. When Humboldt arrived in 1800, there had also been prodigious swarms of sand-flies. The village's handful of inhabitants were then governed by an old officer who taught a few children their rosaries and rang the church bells occasionally to amuse himself. When Schomburgk came almost forty years later he found Esmeralda abandoned apart from three huts occupied by a single family of a patriarch and his offspring. The great explorer suffered from 'the number of sand flies, which proved from first dawn to the setting in of night an unceasing torment, [and] surpassed whatever I had seen in that respect.' He could not understand why anyone chose to live in that horrible place, because the locals suffered just as much as he did from 'these bloodsuckers'.

Hardly surprisingly, Spruce spent only four days in gnat-infested Esmeralda; but he collected assiduously. The scarp of tumbled rocks near the Orinoco yielded fascinating small plants; a stunted form of *Humirium floribundum* shrub; an unusually short remija; under large stones 'the most delicate little fern I have ever gathered'; an asclepiad with narrow leaves and minute white flowers; and a completely new shrub about 1.2 metres (4 feet) high 'with long pinnate branches, minute rigid leaves ending in an arista, and solitary axillary fruits the size and colour of haws.'

While Spruce was at Esmeralda he met Ramon Tussarí, chief of a group of Maquiritare, who invited him to visit his village on a nearby river. This was the Cunucunuma, which flows into the Orinoco from the north a few miles below the Casiquiare outlet. Spruce was just as adventurous as Wallace, so he immediately set off for this new challenge. The Cunucunuma rises in the Quinigua mountains, behind and higher than Cerro Duida, so it was full of rapids. Spruce's boat entered the tributary river on 1 January 1854 and passed the first rapid. On the following day seven Maquiritare were sent down to help them surmount further rapids. These men were tall, remarkably pale-skinned and with long noses – but in Spruce's view not quite as handsome as the Tukanoan peoples of the Uaupés. The Carib-speaking Maquiritare are, confusingly, also known as Maiongong in Brazil and Yekuana in Guyana. They have always been – and still are – tremendous travellers and traders, masters of the rapid-infested rivers of southern Venezuela and Roraima in Brazil. They had shown Schomburgk how to use the Uraricoera river to get from the Rio Branco in Brazil over a watershed to the upper Orinoco. One of the Indians sent down to help Spruce had been one of Schomburgk's guides fifteen years earlier. Each man wore a long rectangular apron suspended from a string belt, in front to below his knees and behind thrown up over his shoulder. Schomburgk wrote that, thus adorned 'he stalks about as if the world was his own'. Like peoples on the Uaupés, they admired bulging calves and upper arms, so they wore tight garters (woven from their own hair) below their knees and below their shoulders. They also had thick necklaces of mostly blue beads and belts of white ones.

As experienced boatmen, the Maquiritare were noisily curious about Spruce's piragoa. But it was too big for that river. It was now the end of the dry season and the water level was falling daily. All the men struggled for hours to haul the craft up the second rapids, but it grounded frequently and risked swamping. So 'with sorrowful heart, I gave the word to return' to the base of the fall. Spruce grabbed some goods as presents, and continued up the Cunucunuma in his small *curiara* bark canoe. After a day he reached Tussarí's village at the foot of the third fall. The chief himself had a whitewashed house, but Spruce was more interested in other traditional huts of buçu palm thatch. These were great dome-like hemispheres, but tapering at the top to a point that reminded the traveller of a Turkish helmet [PL. 45].

Spruce found Tussarí a remarkable man – and his wife even more impressive because she was so skilled at bartering their produce. Both were much travelled. They had lived for years on the upper Rio Branco in Brazil, where Tussarí raised some cattle and traded with the British in Guiana. When the paramount chief of the Maquiritare died, Tussarí was summoned back to Venezuela to take his place. He was famous for making dugout canoes, from a heavy laurel wood; but his wife and daughters were even more industrious. Spruce was struck by the quantity and quality of their output: 'several mapires [baskets] of manioc, masses of circular shallow baskets, and a sort of reticule... for carrying a tinder-box, tobacco, and other indispensables.' There were blow-pipes, gourds, clubs and pottery. Spruce made a pencil portrait of the handsome fifty-year-old chief [PL. 47].

Hurrying back down the river, with the water falling fast, Spruce recovered his big piragoa but almost lost it when attempting to shoot the final rapid. It leapt down one fall, but stuck on a rock in the next. It was in acute danger of swamping or being dashed on the rocks. All the men jumped into the raging water, but they shoved in vain at the heavy boat. So they used a small canoe to unload the cargo little by little and, after two hours of labour, freed the boat – which was surprisingly undamaged. This must have been a terrifying ordeal for Spruce. That boat and its cargo were his entire worldly goods, and he was on an obscure river off the upper Orinoco, with no one for hundreds of kilometres having any inkling of his whereabouts. However, he was in the safe hands of honourable and tirelessly helpful Maquiritare people.

Thwarted by the rapids and low water of the Cunucunuma, Spruce lost no time in starting back down the Casiquiare. He was of course collecting as often as he could, including a four-day excursion up a small, unexplored and uninhabited eastern tributary called Vasiva (now Paciba, and still empty of people). The rainy season had suddenly broken out. So 'the first three days and nights were dreadfully rainy, and... our position was gloomy and lonely in the extreme.' There was some good collecting in this dripping virgin forest, but Spruce returned to the Casiquiare because he had a more ambitious plan. He wanted to be the first scientist to explore a long eastern tributary called the Pacimoni or Baria. This enters the Casiquiare from the southeast at a place where that river-canal turns westwards towards the Rio Negro. It was mentioned by both Humboldt and Schomburgk as they sailed along the Casiquiare, along with many other tributaries, but neither said anything noteworthy about it. Spruce probably hoped that it would lead him to the botanically unexplored hills in which it rose.

Spruce found that the Pacimoni river was 'wide, black, and still, and so continues for a long way up.' At first the forest was all low, seasonally flooded igapó. On one day the flooding was so bad that, although they started well before dawn, they could not find a dry place to cook breakfast until after midday. After five days, they reached the mouth of a southern headwater called Baria.* Spruce chose to continue up another, eastern, headwater called Yatua. As the land rose slightly it ceased to be flooded and there was some high forest, but dry, low caatinga predominated.

Surprisingly, there were three small settlements far up these otherwise empty rivers. An escaped Brazilian slave called Custodio had created them. One was called San Custodio after him, with some sixty people; and another Santa Isabel, a day's journey upriver, with perhaps 140 inhabitants in fourteen huts [PL. 33]. Custodio filled his villages with Arawak-speaking Mandauacá and Cunipusana peoples as well as his Yabahana wife's family from the Marauiá tributary of the middle Rio Negro.

Spruce liked and admired Custodio, a tall, well-built, almost-black mulatto of about forty-five. His extraordinary story gives a glimpse into the world of the relatively few slaves in Brazilian Amazonia. Custodio was born as a slave in Bararoá (now Tomar) on the middle Rio Negro. His childless master treated him well, almost as a son, and took him on trading voyages up the Uaupés and other rivers. But a false rumour that Custodio was sleeping with a motherly housekeeper turned the master violently against his slave: he tried to shoot the young man, but the gun misfired and Custodio fled for his life in a canoe. Halfway up the Negro some Indians attempted to catch him for a reward, but Custodio fought his way out and went on the run again. He fled into the forest with only a knife, made himself a canoe and paddle, and survived on wild figs and fruits and any game he could catch – he rubbed sticks to make cooking fires. He was saved by reaching a Yabahana village, far up the Marauiá river, where he lived for two years, married and had two children. Crossing through the Neblina sandstone hills, Custodio and his young family descended the Pacimoni to San Carlos in Venezuela – where escaped Brazilian slaves were supposed to be free. But in about 1838 a Portuguese trader treacherously seized Custodio and took him in irons all the way down to Manaus. He was put onto a prison sloop with Cabanagem captives. Their jailer allowed them ashore for Christmas celebrations and, on the third day of these, Custodio escaped again. He managed to steal a montaria from an old woman and paddled day and night up the Rio Negro, avoiding all

* Spruce was told, correctly, that from the upper Baria there is a short portage to the Cauaburi, which enters the Negro below the São Gabriel rapids. Because of this, modern Brazilian cartographers refer to the huge flat forested area enclosed by these rivers and the Rio Negro as Ilha Pedro II (Pedro II Island).

humans and sleeping only in the middle of the day hidden up creeks. After five weeks, eating nothing but forest fruits and 'the thin pulp of the Miriti palm, which is barely sufficient to sustain life', he managed to reach his indigenous friends up the Marauiá. Descending the Pacimoni again, he returned to his family in San Carlos and worked for three years as a blacksmith. He became such a respected townsman that the Venezuelan authorities sent him to manage the Pacimoni river, since he knew its Indians. As Spruce wrote to Hooker: 'Custodio – the mulatto – the slave – the captive – now figures as "Chief of the River Pacimoni and founder of the villages of Santa Maria and San Custodio"', as well as reviver of the village of Santa Isabel.

On the lower Pacimoni, the vegetation was almost identical to that of other tributaries of the Casiquiare. A mimosa, *Parkia americana*, 'was exceedingly frequent, always overhanging the water's edge, and very ornamental from its large pendulous crimson tassels.' But beyond the caatinga, with better soil, the vegetation improved. The forest was loftier, with the first nutmegs on this river, and palms – bacaba, inajá, açaí and buriti – in such dense clumps that the narrowed river wound interminably through them. Spruce had to abandon his big boat at one village and continue in a dugout canoe. He did not know that strong squalls were frequent, and that these blocked the narrow stream with fallen trees, so he had failed to bring an axe. At times they had to drag the heavy canoe up and over logs that were far too large and petrified to be cut with a machete. The stream became a tunnel flanked with dense brush, with sunlight filtering through overhanging trees and lianas. (This author has been on the Pacimoni, and it has not changed at all since Spruce's day. The only tiny settlements in this part of Venezuela are still San Custodio and Santa Isabel.)

When the expedition moved from the boat to the canoe, Spruce took his loaded gun but forgot his bag of shot, and his Indians forgot their fishing lines. On the first evening his hunter used the gun to bag two guans, which fed them for two meals, but after that there was nothing for three days. Spruce wrote in his journal: 'I was near famishing.... When I have nothing to eat I find it impossible to work.' He had room in the canoe for only two small bundles of drying paper, which he was saving for small plants in the hills; so he had to abandon some interesting specimens. Such were the tribulations of a collector on his own in unexplored country.

Spruce finally managed to buy some fowls from their owner at Santa Isabel and, thus fed, he set off for the hills with a young guide and two of his Indians. The Yatua and Baria headwaters of the Pacimoni both rise in the Serranía de la Neblina, the southwesternmost *tepui* (table mountain) of the Guiana Shield. These sandstone mountains are

Precambrian, the most ancient geological era, and they therefore antedate the plate tectonics that separated South America from Africa. Being a watershed, they form the boundary between Brazil and its northern neighbours. Mount Neblina ('Mist') at the southwestern end of this great massif is, at 2,994 metres (9,823 feet), technically the highest point in Brazil – 184 metres higher than Mount Roraima.

From a low hill behind San Custodio hamlet, Spruce had a fine view of this range of flat-topped mountains: Aracamuni to the east, then sweeping round to Tibiali topped by a cone called Avispa, and to the south the Imeri group that contains Neblina, all rising out of 'the forest plain, like an immense heath'. In a letter to Hooker, he pondered these 'magnificent mountains, uninhabited and all but inaccessible, and scarcely known to geographers even by name.'

Spruce had no hope of walking up these gigantic cliffs, which proved to be far farther away than he had imagined. There were no trails through the thick undergrowth. During the first morning they 'crossed streamlets forty-three times' as well as expanses of pools in the flooded forest. They started up the lower slopes, but not high enough to reach 'good plants'. It rained, and Spruce feared a thunderstorm. So he sadly had to turn back, 'having reaped nothing but a wetting'. They had no supper and Spruce passed a miserable night 'in a tiny hammock of very open texture, shivering with cold and tormented by zancudos [mosquitoes]'. Next day he had a painful walk back, for 'I had torn my naked feet on the previous day and I contrived… to deepen the wound by treading on a sharp stump, so that, what with bleeding feet and an empty stomach, I found the journey sufficiently toilsome. But this did not prevent me gathering… plants in flower.'* Spruce later wrote to Hooker that there were many novelties in his Pacimoni collection, including a new genus of strangling *Clusia* trees that had white cones thickly scattered among their dark-green foliage. There were eight species of orchid, including three type specimens, and *Trisetella triglochin*, which was not collected there again until the 1960s.

The explorer took first the canoe and then his boat back down the Yatua and Pacimoni to the Casiquiare, arriving at San Carlos on 28 February 1854. He had been away for three months. He wished that he had collected in disciplines other than just botany, but he excused himself to Hooker because 'one person cannot do everything, that the preserving of plants in this climate involves great mechanical labour, and that the daily cares and *contretemps* of a voyage where one's only workmen are Indians, and where food

* The Imeri table mountain group that contains the conical Neblina was in fact not climbed until a century later, and then by a substantial American-led expedition. This author was lifted onto the tepui by helicopter as part of a Venezuelan research project that found a wealth of new botanical and entomological species on the dense vegetation along the summit.

must be sought from day to day in the rivers and forests, consume no little time.' Despite these challenges, Spruce had accomplished a great deal. He had explored and made the first maps of the Cunucunuma and Pacimoni rivers; drawn a few sketches, including many of picture-writing engravings; compiled 'more or less complete' vocabularies of six indigenous languages – including the first of Yanomami (then known as Guaharibo); written something about Maquiritare ethnography; and of course collected a wealth of plants. He modestly wrote that someone with more health and strength, more manpower, and more financial resources 'would have done much more'.

9

CALAMITIES

While the adventurous Wallace, and then Spruce, were exploring far up the Rio Negro and its remote tributaries, Henry Walter Bates had remained quietly on the main Amazon. We saw how lonely and depressed he had become after a year in Ega.

He reckoned that in twenty months on the Amazon he had collected 7,553 insect specimens. His frugal expenses for living and travel during that period had been £67. In his gloomy state, he calculated that insects would sell for only 4 pence each, so that after deduction of Stevens's 20 per cent commission and other costs, all that work during almost two years on the world's greatest river had earned him only £27. But his income in fact proved to be better than this.

Bates left that remote town on the Solimões on 21 March 1851. He made the 2,250-kilometre (1,400-mile) journey to Belém in a schooner heavily laden with jars of turtle oil, brazil-nuts and a great pile of sarsaparilla. With generally good winds and the river in full flow, the descent took only four weeks. He brought with him a small nocturnal monkey (*Aotus trivirgatus*) known in the upper Amazon as *ei-á*. This little creature's body was 25 centimetres (10 inches) long, with a straight tail of 30 centimetres (12 inches). Its fur was soft and thick like a rabbit's, and generally brownish-grey with three darker stripes on top of its head. These monkeys sleep during the day, often in hollow trees. A passer-by would be 'startled to see, aroused from their sleep, a number of little striped owl-eyed faces crowding a hole in the tree-trunk'. During the day, Bates put his pet into an old chest, because it hated daylight, and it hid in a glass jar. Approached gently it could be caressed, but 'when handled roughly it takes alarm and bites severely, striking with its fore-hands and making a hissing noise like a cat.' It loved to eat papaya and bananas, but also spiders and cockroaches – it could totally clear a house of these vermin. People in Belém were curious about this strange monkey, and Bates liked it so much that he wrote a piece about it for the *Zoologist* journal.

Arriving in Belém, Bates was shocked to find 'the once cheerful and healthy' town decimated by two terrible epidemics. Yellow fever had broken out during the previous year. Three-quarters of the population caught the violent disease and 5 per cent died of it. Then, on the heels of yellow fever came smallpox. This hideous disease devastated people with indigenous blood, since they had less inherited immunity to it. Out of a population of perhaps 15,000 Indians, about 800 died.

In mid-May 1851, a few weeks after Bates's arrival, Wallace's 22-year-old younger brother Herbert arrived from Manaus – where he had been staying since Alfred left for the Rio Negro in August 1850. He was in 'robust good health' after almost two years in the tropics. He booked himself a passage to England, leaving on 6 June. Bates liked Herbert Wallace, who was only four years younger than him, and saw him frequently. On 2 June the two went shopping and then had tea with the shipping agent Daniel Miller. But that evening young Wallace was struck by shivering, fever and vomiting – the dreaded symptoms of yellow fever. Bates got him into a house in the centre of town, where a Doctor Camillo treated him as best he could.* For a while Herbert seemed to recover. For four nights Bates slept by his side. But then, as Bates wrote to the young man's mother Mrs Mary Wallace, 'the fever struck inwards and black vomit declared itself… until he died… after suffering fearfully', on 9 June 1851.

Bates himself also caught yellow fever. Soon after tending Herbert Wallace, he was 'seized with shivering and vomit', perspired 'famously… and felt very weak and aching in every bone of my body'. Bates, who was always concerned about his health, took weird remedies: a decoction of elder blossoms to make himself sweat, then a small dose of Epsom salts and 'manna' as a purgative. He recovered after a week. Their friend and helper Daniel Miller was not so fortunate. Later that year, he went out to inspect the wreck of an English ship near the mouth of the Amazon, caught 'a severe chill', and died of 'brain-fever'.

Despite the tragedies from disease, Bates's professional work flourished. When he reached Belém he found a letter from his agent Stevens with 'better news than ever'. The three

* No one at that time knew anything about the spread of yellow fever, which is a mosquito-transmitted virus. There is still no treatment, although there is now an effective vaccine to prevent it.

collections he had sent – one from Manaus, then his first Ega collection and another sent on the ship *George Glen* – had all sold far better than he had hoped, and a new Ega collection had arrived safely. Stevens also conveyed the wonderful news that William Hewitson, a great collector and authority on butterflies, had named a new species *Callithea batesii* (now *Asterope batesii*). Bates wrote to Stevens: 'I feel very much [underlined] the compliment that Mr. Hewitson has paid me'. In that same letter, Bates revealed that he had shipped another consignment from Ega on 2 January 1851 and that this had been forwarded from Belém in March. It could then take weeks or months to reach London, in those days of sail and little traffic between northern Brazil and England. But it was remarkable that all the collections *did* reach Stevens safely and in excellent condition; and the payments, by letters of credit and then currency, found their way back to the isolated collector. So Bates cheered up and, as his friend Edward Clodd later wrote, 'happily for biological science' decided not to return home as he had planned, but to resume his collecting along the Amazon.

In June Bates shipped a magnificent consignment to Stevens, the product of months of work at Ega. There were of course 'new and fine things' among beetles, since Bates had always been passionate about Coleoptera. But there were always problems. There was a new species of the genus *Gorgus* that was covered in a short silky 'pile' but, despite many experiments, Bates could not retain this pile on dead beetle specimens. Another problem was that when approached these beetles 'feign death, and fall to the ground, where half the specimens one sees are lost among fallen leaves.' There were several new species of butterfly. Among birds there 'appear to be some rare things, as the blue Piosoca [parrot], a pair of species of Aracari [related to toucans] I have not before seen, a male Ouramimeu, &c.' Among animals there was a caiman skull, but this was now imperfect owing to another unforeseen hazard: it was missing teeth which 'I think have been stolen by people in the house to make *charms* of.' Writing four months later, in October 1851, Bates mentioned other threats. He had lost a consignment when the little ship *Mischief* foundered. (By coincidence, this was the barque in which he and Wallace had originally reached Brazil, three-and-a-half years earlier.) He could not recall which specimens of his private collection had been lost, so he begged Stevens to set aside for him one good example of everything he was now sending, and to sell the rest.

Stevens had repeatedly advised Bates to send perfect material. He tried to do this, 'but it is not so easy a matter to get good specimens' – the ones he caught might have a defect, or 'even when taken fine, flutter so much... that they require the greatest care and long practice to secure fine.' Then there was the danger of ants, mites, termites and other voracious pests. Bates had himself lost six weeks from yellow fever and two bouts of diarrhoea. But when he recovered in July, for ten weeks 'I have applied myself very closely to collecting, and think I have got together a very superior lot of insects, as to variety and the quality of the specimens.' In return for these collections, he asked Stevens to send him a batch of reference books: simple catalogues of natural history museums, some issues of the *Zoologist* (particularly those that contained his letters or that mentioned the butterfly named after him), any cheap manual of South American birds or animals (in any language), and two of the latest maps of the Amazon.

When he had descended the great river, Bates's boat had stopped at Santarém. He was reminded what a charming place it was, and he determined to return there and also to collect up the Tapajós river, which joins the Amazon at that town. He wrote to Stevens that he was sure that Wallace was doing a good job on the Rio Negro; so the southern Tapajós tributary should prove a fruitful new hunting-ground for him. The drawback was that he had heard that this clear-water river had been badly hit by malignant malaria – in just one settlement, 400 people had died of it – and the survivors were so short of food that the government had to send emergency provisions to prevent them starving.

Bates reached Santarém at the end of November 1851, and it took him just one day to find a house and servant. He was delighted with both. The house was solidly built, with three rooms and a little verandah that happened to overlook a neighbour's 'beautiful flower-garden, a great rarity in this country'. It was at the edge of the town and 'pleasantly situated' near the beach. Its rent was only £16 a year, with no further taxes to pay. For a servant, 'I had the good fortune to meet with a free mulatto, an industrious and trustworthy young fellow named José... [who] assisted me in collecting [and] proved of the greatest service in the different excursions we subsequently made.' The people of José's family cooked for Bates and him, so that they were free to collect. The young naturalist knew how lucky he had been to find José, because free people in Santarém were generally too proud to be servants, and 'slaves too few and valuable to their masters to be let out to others.'

During the first month at Santarém, Bates and José spent most of their time trying to collect the beautiful *Asterope sapphira* butterfly, because Stevens had particularly asked for it. When he sent his first consignment, Bates explained that they had worked 'six days a week... both of us with net-poles 12 feet [3.5 metres] long.' The males in particular were hard to catch since they flew high, were difficult to see 'in showery gloomy weather, in certain parts of the woods', and three-quarters of those he netted were imperfect specimens. So he told his agent 'not to think it is an abundant species because I now send you so many', and to charge a high price for fine specimens. The collector then thought that he could *breed* these and erycinidae butterflies, 'I having hit upon a few broods of larvae and chrysalides.' By May he was writing that he was rearing many Lepidoptera, and had discovered that the butterflies known as mysceliae were in fact females of the epicalia: to prove this he had some 'bagged *in copula*, leaving no mistake about it.'

Bates was as meticulous about dispatching his collections as he was about catching and mounting them. In January 1852 he sent a box with 1,095 insects and 152 shells, which he valued at £30 and which he hoped would catch a ship called *Windsor* sailing on 25 February. He was apologetic about the next collection, sent in April, because it was 'the work of the wet season, which has been very severe here.' In early May he sent another collection, with 'copious letters and notes', which included material collected in a glorious forest 8 kilometres (5 miles) from Santarém that teemed with new and rare things. A final consignment went in June. Bates boasted justifiably that he now found that from his own memory he could add a great deal of information to that in his reference books. And as a collector, 'I have now taken, with this pair of hands, 1000 species of butterflies! I have now about 50 specimens of small reptiles, about 30 of woods and medicines, 50 shells, &c.' – not to mention flowers, molluscs, and other exotics. He was thrilled by the number of lovely butterflies near Santarém, and became lyrical about 'a new Thecla, far surpassing in silky brilliancy the beautiful blue ones I sent from Pará [Belém]; there is a band of deeper blue traversing the wings, forming a startling and unexpected style of Thecla beauty.'

Bates planned to leave the main Amazon, to collect on its southern tributary, the Tapajós. He wrote to Stevens that this 'glorious river stretches away southward from Santarém, with its high banks, blue hills, and white sandy shores seeming to invite its exploration. Every one from there speaks in raptures of the scenery, and the novelty of the birds and animals.' He rented for almost £2 a month a two-masted cuberta of 6 tons burden. This was strongly built of itaúba, harder than teak and the best for boat building. Bates organized the square cabin as his sleeping and working apartment, with his chests arranged on each side. These were 'filled with store-boxes and trays for specimens... and above them were shelves and pegs to hold my little stock of useful books, guns, and game bags, boards and materials for skinning and preserving animals, botanical press and papers, drying cages for insects and birds, and so forth.' He would have loved to take a thermometer, barometer, quadrant and other surveying instruments, but these were 'all quite beyond the means of a poor man like myself.' In the bow of the boat was an arched covering where the crew slept. This held Bates's stock of salt and groceries, as well as trade goods for riverbank caboclos of the interior – cachaça, powder and shot, pieces of coarse checked-cotton prints, fish hooks, axes, machetes, harpoons, metal arrow-heads, mirrors, beads, and other trinkets. Bates and José spent many days organizing all this, including salting meat and grinding coffee for themselves (all three Englishmen adored coffee as one of their few luxuries). They stored the perishables in insect- and damp-proof tin boxes – as explorers in Amazonia still do.

As always, finding boatmen was the greatest problem. Santarém had fewer paddlers than any other town on the Amazon, and it was impossible to find any Indian or mulatto who was not technically in debt to some trader or headman. So on 8 June 1852 Bates finally set off with a crew of only two: José and an Indian from Minas Gerais who claimed to know the river. This man was 'a coarse specimen… [who] proved a great annoyance, …[and] understood less navigation than myself; but he was insolent, and would have his own way.' Bates dismissed him at the earliest opportunity.

The Tapajós is a clear-water river of great beauty, very wide (Bates reckoned 13 to 16 kilometres or 8 to 10 miles), and flanked by low hills in the distance. 'Sometimes a narrow margin of alluvial land skirts the banks, with beaches of white sand, forming lovely bays and harbours; but a great part consists of rock-bound coasts and precipices, at the foot of which the swell breaks with threatening roar.' Sailing up this river was very different from the quiet of the main Amazon or the Rio Negro. 'In fact, the voyage… turned out a serious affair, full of peril and anxiety. …Squalls, with thunder and lightning, occurred daily; at times, violent gusts came suddenly from the hills before sudden showers.' Hardly surprisingly, with his incompetent skeleton crew, Bates had sailing adventures – losing his montaria dinghy, trying to tack to rescue it but having ropes snap, having sails fly into rags, the ship heel over frightfully, then drag anchor on a sandy bed, drift broadside-on to a rocky beach, and clear a rock point 'at a close shave with our jib-sail'. However, after a few days' collecting at the charmingly situated but scruffy village Alter do Chão ('Earth Altar', after a rectangular hillock), they made it up the river for about 180 kilometres (110 miles) to another village called Aveiro.

Bates stayed at Aveiro for six weeks and accomplished some excellent collecting. 'The time was spent in the quiet, regular pursuit of natural history: every morning I had my long ramble in the forest… and the afternoons were occupied in preserving and studying the objects collected.' He loved 'the soft pale light' of moonlit nights. 'After the day's work was done, I used to go down to the shores of the bay, and lie at full length on the cool sand for two or three hours before bed-time.'

Aveiro was a sleepy little place, happily free of pium black-flies, mutuca horse-flies and mosquitoes. But it had three discomforts. One was the small shiny-red fire-ant

(*Solenopsis saevissima*). Bates called these ants the scourge of the Tapajós, with Aveiro as their headquarters; and they were a greater plague than all the other pests put together. The soil of the entire village was undermined by fire-ants, the ground perforated by entrances to their galleries, and the houses inundated by them. 'They dispute every fragment of food with the inhabitants, and destroy clothing for the sake of the starch.... They seem to attack persons out of sheer malice: if we stood for a few moments in the street... we were sure to be overrun and severely punished, for the moment an ant touched the flesh, he secured himself with his jaws, doubled in his tail, and stung with all his might.' The sting caused great pain and irritation. The only consolation was that at the end of each rainy season, male and female fire-ants leave their nests in a vast flying swarm. These were often swept out to the river and drowned in droves. Ascending the Tapajós, Bates saw a line of tens of thousands of dead or half-dead winged ants, piled several centimetres in height and breadth, and stretching for miles on end along the edge of the river.

The second problem at Aveiro was malaria, but the third, milder, annoyance was its lively old priest. His sole topic of conversation was homoeopathy. He had been 'smitten by the mania... [had a Dictionary about it] and a little leather case containing glass tubes filled with globules, with which he was doctoring the whole village.'

The best collecting at Aveiro was insects. Within half an hour's walk of the village, Bates 'enumerated fully 300 species of butterflies... [which] is a greater number than is found in the whole of Europe.' Many species were 'old friends' from Belém, the Tocantins or Óbidos; two were ones seen in Ega, including *Callithea batesii*, named after him; and 'the new species are in Leptalis, Heliconia, Eresia, Heterochroa, &c., and in Eurygona, Calospilus, and other groups of Erycinidae.' There was equally great diversity in other orders, such as Coleoptera (beetles) and Hymenoptera (bees, wasps and other insects with four transparent wings) – with an even greater number of species new to the entomologist than among butterflies.

Bates acquired another pet. Out on the Tapajós a pretty little parrot fell into the water, apparently stunned in an aerial fight among its flock. It was green with a patch of scarlet feathers under its wings: the Indians called it maracanã (a name they also gave to a species of macaw) and Bates thought that it was a *Conurus guianensis*. It hated captivity – for a week 'it refused food, bit everyone who went near it, and damaged its plumage in its exertions to free itself.' So Bates lodged the little parrot with an old Indian woman who was famed as a bird-tamer. After two days she brought it back 'almost as tame as the familiar love-birds of our aviaries'. It learned to talk quite well and remained a devoted companion of the Englishman for two years. Bates reckoned that the local people's secret was to treat birds with uniform gentleness and allow them to roam wherever they wished in a hut. His parrot went out on collecting rambles, often sitting on the head of one of his boy helpers.

In August, Bates decided to go up the Cupari river, which joins the Tapajós from the east some 13 kilometres (8 miles) upriver, south from Aveiro. The Cupari was 'hemmed in by two walls of forest, rising to the height of at least 100 feet [30 metres], and the outlines of the trees being concealed throughout by a dense curtain of leafy creepers. The impression of vegetable profusion and overwhelming luxuriance increases at every step.' There were four or five settlers on this river. Bates spent each night in a different sitio, and was invariably entertained with amiable hospitality – as was the norm throughout Amazonia. The last 'civilised settler' on the river was 'a wiry active fellow and capital hunter' called João Aracú. Bates spent three weeks with this pioneer. He and young José scoured the woods and streams, adding twenty new species of fish and many small reptiles to his collection, while 'a great number of the most conspicuous insects... were new to me, and turned out to be species peculiar to this part of the Amazons valley.' The dry season was now nearing its height. The middle of the day was like a furnace, and Bates wore nothing but a pair of loose thin cotton trousers and a light straw hat. But he never wavered in his daily collecting.

He was delighted to catch a white-whiskered coatá or spider monkey (*Ateles marginatus*). This variety had a white patch on the top of its head. These coatás are easily domesticated and make jovial and comic pets, but they have the misfortune of being known as the tastiest game in all Amazonia. Bates ate several and thought it the best meat he had ever had – like beef, but richer and sweeter. He was feeling weak from a diet of fish, so he smoked his coatá meat and lived off it for a fortnight. His final piece of it was an arm with a clenched fist – only great hunger drove him 'so near to cannibalism as this'. Food was always a problem and the Indians often went off to hunt or fish and brought back unusual catches such as tortoise, anteater or iguana eggs.

Meanwhile his two Indian paddlers spent a week building a canoe to replace the lost one. Senhor Aracú expertly supervised this operation. They first found an itaúba tree, hauled it through the forest, hollowed it with axes and adzes, then widened it over a slow fire and added plank sides, to make a handsome dugout montaria. It was launched with great merriment and rejoicing, decked in coloured cloths as flags.

Bates had three particular objectives in venturing up the Cupari. One was to collect hyacinthine macaws, another was to see a waterfall near its source, and the third was to visit a village of indigenous people – the Mundurucu. Aracú agreed to accompany Bates, with an Indian to hunt and fish for them. It took only a couple of days to reach the first Mundurucu village. Bates learned that these indigenous people formed the largest and most warlike tribe in that part of Brazil. The Munduruku had once been feared enemies of the Portuguese settlers; but at the end of the eighteenth century they had abruptly changed, and ever since remained 'firm friends of the whites'. They gave a warm welcome to any white visitor, and had greatly helped the authorities defeat Cabanagem rebels, particularly against their traditional enemies the Mura (the people whom Bates had disliked, as scruffy mendicants). The Munduruku on the Cupari were far more acculturated than the peoples of the Uaupés, whom Wallace was encountering at that time and Spruce was to visit in the following year. Munduruku women were naked but hurried to put on petticoats when they saw visitors. The warriors had just returned from trying to destroy a nomadic tribe and were all relaxing in their hammocks. And the chief was sophisticated: this hereditary tuxaua was a tall, broad-shouldered young man 'with handsome regular features… and a quiet good-humoured expression'. He wore a shirt and trousers of blue-checked cotton cloth, and spoke very good formal Portuguese – acquired on visits to Santarém and Belém. 'There was not the slightest trace of the savage in his appearance or demeanour.' The Munduruku warriors, however, were still covered in geometric tattoos of black genipapo.

Bates entertained a crowd of men and women by showing them his copy of *Knight's Pictorial Museum of Animated Nature*. These consummate hunters, who knew every creature in their forests, were of course astonished by pictures of unfamiliar animals. In return, they gave him an outline of Munduruku ethnography: how they made superb featherwork including capes of dazzling blue and red macaw feathers, 'sceptres' covered in white and yellow toucan feathers (two of which Bates bought), and of course gorgeous headdresses; how they had ceased to decorate their huts with mummified enemies' trophy heads; and how each group had a shaman who cured the sick and predicted propitious times for attacking an enemy. Bates and Aracú asked a shaman to cure a sick child, which he did by incantations and blowing smoke over the patient before extracting a pernicious 'worm'. They then tricked him into revealing that the worm was in fact a length of a plant's air-root. Bates dismissed the pajé as a 'conscious impostor' practising 'a shallow secret'. But his faith-healing was little worse than some of the nostrums that Bates himself took from western doctors baffled by tropical diseases.

Bates was impressed by this village of Munduruku and another he visited upriver. He admired their 'regular mode of life, agricultural habits, loyalty to their chiefs, fidelity to treaties, and gentleness of demeanour'. But he noted that, like most indigenous peoples, they 'showed no aptitude' for settling in or near towns and thus 'seem incapable for further advance in culture'. Actually, he could see that they gained by 'not being corrupted by too close contact with the inferior whites and half-breeds of the civilised settlements….

I could not help contrasting their well-fed conditions and the signs of orderly, industrious habits, with the poverty and laziness of the semi-civilised people of Alter do Chão [the village he visited near Santarém].' But Bates, like many observers, was dismayed by these Indians' apparent lack of ambition, abstract thought, curiosity about the causes of natural phenomena, and religious belief.

The Munduruců had large plantations of manioc, and sold a substantial surplus of this to traders. They also gathered quantities of sarsaparilla, tonka beans and India-rubber. In 1852 trade in rubber was in its infancy, but the great Amazon rubber boom was about to take off. Within a few decades the Munduruců were to become vigorous and efficient rubber-tappers, more successful at this than any other indigenous people; and they continue to tap wild rubber along the Tapajós to the present day.*

Besides visiting a genuine indigenous village, Bates achieved his other goals on this river. He took his cubertá to the Cupari's first waterfall, the limit of navigation, some 110 kilometres (70 miles) from the Tapajós. Setting out in a dawn mist, 'the lofty wall of forest, with the beautiful crowns of açaí palms standing out from it on their slender arching stems, looked dim and strange through the misty curtain. The sudden change a little after sunrise had quite a magical effect; for the mist rose up like a [theatrical] gauze veil... and showed the glorious foliage in the bright glow of morning, glittering with dew drops.' The boat stayed for four days in 'a little rocky haven' at the foot of the falls. Bates remained with it, collecting, while two men went above the falls in the montaria and returned each evening with game they had hunted. The collecting included a big guariba howler monkey (*Alouatta belzebul*) and Bates's third objective: six sought-after hyacinthine macaws. These were feeding on hard nuts of the tucumã palm (*Astrocaryum vulgare*), so Bates found that their beaks were filled with the sour paste to which these fruits had been reduced. 'Each bird took me three hours to skin, and I was occupied with these and my other specimens until midnight, after my own laborious day's hunt; working on the roof of my [boat's] cabin by the light of a lamp.'

The water in the pool was totally transparent, so that Bates could watch the fish that swarmed in it. There was a shoal of handsome black-banded acará (*Mesonauta insignis*) 'gliding through at a slow pace, forming a very pretty sight'; needle fish (*Hemaramphus*) with long slender jaws scattering hosts of smaller fry; and a strangely shaped sarapó

* In 1875 the Englishman Henry Wickham removed *Hevea brasiliensis* rubber seeds from this part of the river, for eventual planting in Malaya. Then, eighty years after Bates was there, the American automobile magnate Henry Ford embarked on a huge plantation of rubber trees behind Aveiro. His employees built a town called Fordlândia where the Cupari river joins the Tapajós. But Ford's millions of rubber trees were annihilated by leaf blight, and this plantation enterprise was the greatest failure of his career.

(*Carapus*), 'wriggling along one by one with a slow movement'. There were masses of piranha. 'When nothing was being given to them, a few only were seen scattered about, their heads all turned one way in an attitude of expectation; but as soon as any offal fell from the boat, the water was blackened with the shoals that rushed instantaneously to the spot.' In their frenzy they wrested bits of meat from one another. But when a larger fish appeared, the piranha took alarm and flashed out of sight. Bates wrote to Stevens that he collected many small fish, most of which were 'very curious and new to me'. But he could collect no larger examples because these all went into the cooking pot: it would have risked 'mutiny to abstract a specimen for my pickle-jars from my hungry people.'

On 21 September 1852 Bates's boat returned to the broad waters of the Tapajós, after seven weeks on the Cupari. For all its beauty he was glad to leave the 'confined and stifling gully... where heat, mosquitoes, insufficient and bad food, hard work and anxiety, had brought me to a very low state of health.' But, like both Wallace and Spruce, he could now claim an Amazonian river that he was the first educated outsider to have explored.

Bates was usually as uncomplaining as the others, but he had to tell Stevens that the journey down the Tapajós was surprisingly difficult. This was the end of the dry season, so the river was a massive 9 metres (30 feet) lower than it had been in June. Its banks and bed were studded with rocky points and shoals. 'There is no current... to aid in descending the river and, to complete the difficulty, the trade-winds blow furiously upward from the Amazons all day and often at night. The slightest wind is sufficient to obstruct all progress by oar, but the wind generally blew a gale, and sometimes a regular hurricane. Three times we were near being wrecked.' 'Near being wrecked' was an understatement for hours of battling in raging winds to save the boat from smashing on rocks. They tended to travel at night when the contrary winds were milder, and they often poled along the very broad but shallow river.

Back in Santarém in October 1852, Bates sent the fruits of his Tapajós journey to his agent in London. It consisted of 'three boxes of insects, a case of medicinal and economic botany, a barrel of fishes and reptiles in spirits, a few birds, eggs, nests, shells, mammals, &c.' He was nervous about sending these precious specimens downriver in a heavily laden boat but had to risk it, because another opportunity would not arise for two months. He was, however, cheered up by two things. One was that his own private collection was growing splendidly, well preserved in Downie's boxes. It already contained 2,200 specimens, 'nearly all Lepidoptera and Coleoptera, and of course has a glorious amount of new things.' The other was that Stevens had told him that his paper on 'Megacephali' (large-headed) beetles had been accepted for publication in the *Transactions of the Entomological Society*.

Bates had only a rough idea of what he would earn from his latest consignment, but he knew that the four-month expedition up the Tapajós had cost £50. Most of this was for the charter of the boat, plus £6 on rum for his Indians. He had lived and worked as frugally as possible, denying himself any luxury. But he sadly concluded that he could never again afford to have his own craft and 'a lot of lazy natives'. In future he would have to travel 'as a favour' in someone else's boat and must stay only in the larger towns. He proceeded to do just that.

Alfred Russel Wallace reached Manaus in mid-May 1852, after a year and eight months up the Rio Negro and Uaupés (apart from a fortnight in Manaus in the previous September). He found the atmosphere in the town changed for the worse, largely because it had become the capital of the new province of Amazonas. The place was full of provincial politicians and bureaucrats. These included barely literate, fashionably dressed young civil servants – with polished leather boots and gold watch-chains, but no idea how to perform their functions. Housing was at a premium. Even with help from Henrique Antonij, Wallace had great difficulty finding somewhere to rent and had to settle for 'a small mud-floored, leaky-roofed room'. It was five months since a boat had arrived from Belém. With the river very high, its current powerful, and adverse winds, vessels had to warp upriver and could take two to three months over the voyage. As a result Manaus had no flour, bread, sugar, biscuit, wine or cheese, and even cachaça was rationed. 'Everybody was reduced to farinha and fish, with beef twice a week and turtle about as often.... Joined to the total absence of amusement or society' this destitution made Wallace's stay in Manaus disagreeable. His misery was heightened by being lame, from an inflamed toe caused by jigger (*Tunga penetrans*) eggs under its nail. He was also very weak 'and quite unable to make any exertion' from recurring malaria. So he was content to stay in his tiny room, tending to what was left of his menagerie of animals and birds. This included twenty parrots, but the best of these 'would persist in wandering about into the street' and getting lost.

Finally, on 10 June Wallace got a boat to take him and his mass of luggage down to Belém. He had a great many cases and boxes. Two-thirds of his hundred live animals had died or escaped, but he still had five monkeys, two macaws, twenty parrots and parakeets, five smaller birds, a pheasant and a toucan. The boat stopped at Santarém, where he found old Captain Hislop. But by bad luck 'Mr. Bates, whom I most wished to see, had left a week before on an excursion up the Tapajós'. So he pressed on to Belém. Unlike Manaus, this town was much improved since he was last there almost three years before. It had new avenues and drives, some of them lined with almond trees, and new buildings. But in other ways it was as cheerful and scruffy as ever. 'The dirty, straggling, uncovered market, the carts of hacked beef, the loud chanting of the negro porters, and the good-humoured smiling faces of the Indian and negro girls selling their fruit and "doces" [sweetmeats] greeted me as of old.' But Wallace had a sad task: to visit the cemetery to see the grave of his beloved younger brother Herbert. His cross was among a multitude belonging to other victims of yellow fever.

After only ten days in Belém, Wallace sailed for England on the 235-ton brig *Helen*. Besides the naturalist and his boxes, she had cargo of 120 tons of rubber, some cocoa, red annatto dye, piaçaba fibres and aromatic balsam.

What should have been a smooth voyage turned into tragedy. Wallace himself was seasick, and so feverish that he feared that he had contracted the yellow fever that had killed his young brother and so many others. Then, after three weeks at sea the captain, John Turner, who owned the vessel, came to him and said 'I'm afraid the ship's on fire; come and see what you think of it.' They saw dense smoke coming out of the after hatchway, opened it, and had the men start to throw out cargo and then pour in seawater. The thick yellow smoke became suffocating, and was soon billowing all over the ship. All efforts to break into the hold and extinguish the fire were in vain. The aromatic balsam was the cause. They could hear it bubbling like a great boiling cauldron, knowing that it would soon burst into flame.* The captain of course made a great mistake opening the hatches to pour in seawater: he thus 'admitted an abundance of air, [which] changed a smouldering heat into actual fire'.

Captain Turner ordered the longboat and gig launched, and the crew and passenger into them. In the bustle and confusion, Wallace recalled that he did not think rationally. 'I went down to the cabin, now suffocatingly hot and full of smoke, to see what was worth saving.' He grabbed a small tin box containing some shirts and, as he wrote to his friend Spruce, 'put in it my drawings of fishes and palms, which were luckily at hand; also my watch and a purse with a few sovereigns.' He scrambled up on deck with these few possessions, but did not dare go down again to save precious journals, a large portfolio of drawings, hundreds of specimens, or more clothing. Meanwhile, the captain put some nautical instruments into the two boats, and the crew loaded them with food and water.

* Wallace later learned that this balsam, called copaiba or copaiva, was 'liable to spontaneous combustion by the constant motion on a voyage', so that more experienced captains carried it in small kegs embedded in sand at the bottom of the hold. It came from the copaiba tree (*Copaifera martii*), which is now sometimes known as the 'diesel tree' because its oil burns so well.

Wallace clambered off the burning brig's stern, rubbing his fingers raw as he slid into the gig that was heaving on the ocean swell. Both boats leaked badly and men had to bale constantly. They watched in horror as 'the flames caught the shrouds and sails, making a most magnificent conflagration up to the very peak, for the royals were set at the time.' One by one, the two masts crashed overboard. After an hour 'the decks were now a mass of fire and the bulwarks partly burnt away.' Wallace watched his monkeys, parrots and other animals burn; some escaped to the bowsprit, but would not jump to safety in the boats; so only one parrot who fell into the water was rescued. The men stayed near the burning vessel throughout the night, hoping that some passing ship might see the light of its conflagration. They baled ceaselessly, occasionally dozed off, but woke in the red glare of their burning vessel. 'It was now a magnificent spectacle, for the decks had completely burned away, and as it heaved and rolled with the swell of the sea, presented its interior towards us filled with liquid flame – a fiery furnace tossing restlessly upon the ocean.'

Taking stock, they reckoned that they were 1,125 kilometres (700 miles) from the nearest land, the island of Bermuda. With sails set and a good breeze they hoped to reach it in a week. In the event, they were at sea for eight days, with some storms, veering winds, calms, sunburn and thirsts, before they were miraculously spotted and rescued by a passing ship. They were still 320 kilometres (200 miles) from Bermuda, and their water and food were almost exhausted. Their saviour was a cargo vessel, the *Jordeson*, bound for London from Cuba. But their travails were not quite over. After a few days they were hit by a very heavy gale. Some sails were blown to pieces, the ship rolled about fearfully, and water poured over it, drenching men in the cabin. 'The ship creaked and shook, and plunged so madly that I feared something would give way, and we should go to the bottom after all; all night, too, the pumps were kept going, for she leaked tremendously....' There were further delays and storms. Even in the English Channel there was a violent storm that sank several sturdier ships. Captain Turner thought that they might drown. But finally on 1 October 1852, eighty days after leaving Brazil, Wallace was 'glad once more to tread on English ground.'

Saved from shipwreck, Wallace could ponder the magnitude of his loss. His sole income came from selling specimens, but the last consignment to Stevens had left Manaus in September 1851. Everything collected since then, as well as earlier material he had stored with Senhor Henrique, had been with him on the *Helen*. Apart from the living birds and animals, these included ten different species of river turtles, a hundred species of fish from the Rio Negro, the skin and skeleton of the manatee whose preservation had given so much trouble, and stuffed monkeys, an anteater and other mammals. But Wallace minded more about the loss of his personal collection, and the memories that went with it. 'With what pleasure had I looked upon every rare and curious insect I had added to my collection! How many times, when almost overcome by the ague [malaria] had I crawled into the forest and been rewarded by some unknown and beautiful species! How many places where no European foot but my own had trodden would have been recalled to my memory by the rare birds and insects they had furnished to my collection! How many weary days and weeks had I passed, upheld only by the fond hope of bringing home many new and beautiful forms from those wild regions: every one of which would be endeared to me by the recollections they would call up.... And now everything was gone, and I had not one specimen to illustrate the unknown lands I had trod, or to call back the recollection of the wild scenes I had beheld!' Although he had previously dispatched duplicate specimens for sale, he had lost 'all my private collection of insects and birds [caught in the three years] since I left Pará... hundreds of new and beautiful species, which would have rendered (I had fondly hoped) my cabinet... one of the finest in Europe' for American items. Before the age of photography, travellers' journals and sketches were all-important, but Wallace now had neither. Quite apart from losing the income from sales, he was about to try to write up his Amazonian travels largely from memory. But he realised that he must be philosophical and stoical, to brood as little as possible about what was gone, and to live for the present and future.

When Bates heard about the shipwreck, he wrote to Stevens: 'I am really sorry for Mr. Wallace's loss. Had it been my case, I think I should have gone desperate, because, so far as regards the unique specimens, the journal &c., such a loss is irreparable.' Wallace himself described the tragedy in a letter to Spruce, which he hoped would reach his botanist friend far up the Rio Negro. He asked him to give 'kind remembrances to everybody, everywhere, and particularly to the respectable Senhor João de Lima.' Spruce did indeed tell Lima about the shipwreck, and he later gave Wallace a reply from the kind-hearted (but not altogether 'respectable') trader. Wallace translated this in his autobiography: 'I was much grieved at the misfortunes which befell our good friend Alfredo! My dear Senhor Spruce, what labours he performed for mankind, and what trouble to lose all his work of four years; but yet his life is saved and that is the most precious for a man! ...When you write to Senhor Alfredo, give my kind remembrances... from the bottom of my heart.'

We can appreciate the magnitude of the loss from the few drawings and notes that survived – either because Wallace had sent them home with earlier correspondence or because he had grabbed them from his smoke-filled cabin. Some, of freshwater fish, are now in the archive of the Linnean Society of London, and were reproduced in Wallace's autobiography *My Life*, published over half a century after the shipwreck. They are beautiful pictures, meticulously accurate and drawn with a very sure hand [PL. 28–30]. The notes on them, and also about palms and insects, are in admirably tidy handwriting.

What had Wallace achieved during his four years in Brazil? He covered a lot of ground, in the lower Amazon around Belém, and then up the rivers Amazon and Negro to the Orinoco watershed. But he was of course most proud of getting his Indians to take him far up the Cubaté and then the Uaupés – beyond where any European (other than a handful of rascally regatão traders) had ventured. He made remarkably accurate traverses of both rivers, which the Royal Geographical Society drew into a fine map. This was achieved despite terrible handicaps – alone, prostrated by malaria, almost devoid of surveying instruments, battling appalling rapids and trying to collect for a living.*

Endearing traits of Wallace were his enthusiasm, curiosity, courage and stamina, and his hubris. His enthusiasm took him up the Tocantins (with Bates) to seek hyacinthine macaws and marvel at the Tucuruí falls, to Mexiana island for caiman, up the Guamá to watch its pororoca tidal bore, hacking beyond Monte Alegre to see rock art, on the lower Rio Negro for umbrella birds, then up the Cubaté for cocks-of-the-rocks, and far up the rapid-infested Uaupés for white umbrella birds and painted tortoises. His amiability – and command of Portuguese and lingua geral – won him friends wherever he went, among planters, traders and Indians. He was one of the first travellers to be curious about rock art, unspoiled indigenous peoples and their languages, and of course the geography, geology, fauna (particularly fishes and birds) and flora of the Amazon. Wallace showed his courage throughout his travels, surviving terrifying rapids, near-shipwrecks, tough explorations, spartan living, insect pests and almost fatal malaria. His hubris was that he immediately plunged in and wrote about all these topics – even though he was self-taught after primary schooling and had no formal training in any of the sciences. All this was on top of his job and sole livelihood: natural history collecting. Like Bates and Spruce, Wallace spent hours every day collecting, mounting and stuffing, then packing and dispatching his thousands of specimens. He learned as he worked, and his experience in Amazonia was to prove essential when he moved on to Southeast Asia and the Malay archipelago.

* Over sixty years later, that Society's secretary wrote to Wallace's son that 'your father's work still holds good' as the only good map of the Uaupés.

After his exploration of the Pacimoni river, Richard Spruce returned to San Carlos de Río Negro at the end of February 1854. Despite the scare in the previous June about a possible uprising, and the drunkenness of its Indians, Spruce liked Venezuela's southernmost town. In three visits, he spent a total of almost twelve months living there. He studied the region and its people, and also learned Spanish as well as the local Baré version of the Arawak language – a skill that was later to save his life. He spent the month of March sorting, mounting, packing and dispatching his latest collections. One feature of this consignment was oil from palm trees. In a covering letter to Sir William Hooker, Spruce gave much information about these oils. The palm tree that yielded most oil was caiaué (*Elaeis oleifera*). The next in order of productivity was the familiar bacaba (*Oenocarpus bacaba*), whose oil was colourless and slightly sweet, excellent for both lamps and for cooking. Spruce confirmed: 'I can bear testimony that, for frying fish, oil of Bacaba is equal either to olive-oil or butter'. He was also 'passionately fond' of the oil from patauá, another species of *Oenocarpus* palm that was super-abundant near the headwaters of the Rio Negro. But of all the vegetable oils, Spruce supposed that andiroba oil from *Carapa guianensis* 'holds the first place'. This had the advantage of being so bitter that ants and other insects would not touch it; and it gave out excellent light in lamps. Andiroba oil had been one of Amazonia's exports ever since colonial times. The expert botanist also identified two *new* oils: one called cunuri from a eurphorbia related to the rubber tree; the other from uaçu, a leguminous tree 'with pretty pink flowers of very curious structure'.

As so often, Spruce reminded his mentor Hooker about the difficulties of collecting. The population of San Carlos was scanty, and it was 'exceedingly difficult' to get its people, with 'their listless, lazy habits' to gather oil for him. He apologized for the small quantities he was sending. Another great problem was that Indians had to be paid in advance and were often in debt when they failed to deliver sufficient produce. Traders had to keep returning to recover these 'debts'; but for a visiting foreigner like Spruce this was impossible.

After another eight weeks' collecting, Spruce left San Carlos towards the end of May 1854. It was Wallace who had told his friend about rich pickings near the watershed between the Rio Negro and Orinoco river basins. So Spruce followed Wallace's route, taking his beloved boat for nine days up the Guainía to the village of Tomo. He had to moor it and proceed in a smaller boat, which involved four days of drying and packing his plants during heavy rains and with nothing to eat apart from two toucans. Two days on a small tributary brought him, on 11 June, to Pimichin and across the portage road to Javita on a headwater of the Atabapo. This was the furthest that Wallace had reached three years earlier, but Spruce kept going northwards into Venezuela. He found that Javita and the next village, Balthasar, were 'the neatest in the whole district, and their inhabitants the least demoralised' thanks to a zambo (the Spanish for an Indian-black mestizo) called Padre Arnaoud. This preacher had gained great influence through his talent for singing masses and litanies and his insistence on strict religious observance of his own semi-Christian cult.*

Proceeding northwards from Padre Arnaoud's pious dominion, conditions deteriorated alarmingly. A trader lent Spruce a boat for the three-day descent of the Atabapo to San Fernando de Atabapo, where that tributary joins the Orinoco. This once-thriving Franciscan mission had looked after Humboldt and Bonpland in 1800; but it was now a large scruffy village, with its church and convent in ruins. Its inhabitants were mostly mestizos, and included many fugitives from justice who seemed to be 'the scum of Venezuela'. It was in the midst of low, flooded forests and was considered very unhealthy between June and August – which was when Spruce was there. The botanist did not linger. He sailed down the main Orinoco in the boat of a trader called Lauriano. The big river was almost as deserted as the Negro. There were only two tiny Indian villages during the 250 kilometres (155 miles) of voyage to his destination: the mighty rapids at Maipurés. One night was spent at a place that had a sugar-cane plantation and distillery. The refuse-dump of cane attracted millions of biting fire-ants that swarmed everywhere, ate all the food, and made it 'impossible to walk anywhere without being overrun and bitten' (this was the scourge that had made Aveiro on the Tapajós intolerable for Bates).

When Spruce reached Maipurés it had just six families of permanent residents, all of mixed blood, but swarms of mosquitoes at night. Its only attractions were proximity to the thundering rapids, and occasional visits by indigenous peoples: Guahibo from the Colombian llanos to the west or Piaroa from Venezuelan forested hills to the east. Spruce lodged with Macapo, a Piaroa who was the leading pilot to get boats down the waterfalls. The Englishman congratulated Macapo on his skill and dexterity in passing the fearful rapids so often. But the Indian said that it was entirely due to a tattered picture of San José to which he prayed before each descent. This picture was the only relic that Spruce could see from the 'devoted missionaries' who had brought Christianity to this

* Arnaoud was in fact the founder of a messianic movement that, under his pupil Venancio, came to inspire and dominate the indigenous peoples of the Upper Rio Negro during the rest of that decade – much to the alarm of the religious, civil and military authorities who did their utmost to suppress the cult. Venancio called himself Christo and said that God had told him how to alleviate Baniwa and Tariana sufferings, bring them plentiful food and reverse their subjugation by whites. One of the elders of his cult was called God the Father, others Saint Lawrence and Saint Mary. Part of its success was reaction against slaving activities of Wallace's old foe Lieutenant Jesuino Cordeiro. At its height, hundreds of Arawak-speakers on the Içana, Xié, Uaupés and Negro were involved. It was finally suppressed at the end of the 1850s. For more about Arnaoud and Venancio, see John Hemming, *Amazon Frontier* (1995).

region – before they were destroyed by 'sacreligious iconoclasts' at the time of Bolívar's 'revolutionary troubles'. It is interesting that Wallace deplored the demoralizing effect of Christianity on detribalized Indians; whereas the more conservative Spruce regretted their priests' expulsion. Both naturalists had been enchanted by the unspoiled peoples of the Uaupés.

The main reason for Spruce's descent to Maipurés was to obtain an ox. After a few days' delay for celebration of the Feast of San Juan, four horsemen rode out to fetch cattle from the great herds that roamed the Colombian llanos. They returned the same evening with perhaps a hundred animals, from which Spruce was allowed to select the best for slaughter. It took him several days to salt and dry his steer into jerked beef – by himself because the Indians disliked the smell. He was interrupted by frequent showers, and then 'the thousands of maggots [that] bred in the flesh as it hung to dry demanded my constant vigilance' in picking them out.

Spruce climbed a 300-metre (1,000-foot) hill above the town for magnificent views in all directions: to the west, flat plains to the horizon; east, the lofty and picturesque Sipapo range; and north, the Orinoco flowing on towards its other mighty rapids, Atures. He also crept out on a slippery and dangerous jumble of rocks for a close look at the Maipurés falls, with spray on his face and a roar in his ears. His reward was to gather thrilling orchids and mosses on the drenched rocks and the fallen trees lodged in them. Collecting on these rivers was otherwise slightly disappointing, because the Atabapo seemed to him a counterpart of the Pacimoni (the tributary of the Casiquiare that he had explored earlier that year) and at its source the vegetation was the same as on the nearby upper Guainía-Negro, with plenty of palms. The Orinoco reminded Spruce of the Solimões above Manaus, apart from having no willows. But Maipurés was rich in plants, perhaps because it had the waterfall and the Colombian llanos nearby. Spruce spent four days there, collecting 'a few specimens of everything in flower or fruit'. When Wallace edited Spruce's papers after his death, he found his friend's list of plants collected at Maipurés, which contained no less than 102 species, many of them gathered 'On rocks at falls'.

As soon as the ox was salted and dried, Spruce started the return journey. It was now the rainy season, so travel up the Orinoco was possible only in a small boat that hugged the banks. Spruce's stock of pressed plants and his dried beef occupied all the boat's tolda; so he himself had to 'half-sit, half-lie in a very uneasy posture at its entrance'. It was now Spruce's turn to suffer malaria. The disease takes a couple of weeks to manifest itself, so he must have caught it in the pestilential San Fernando de Atabapo. He thought that exposure to sun and rain had combined 'to sow the seeds of fever in me'. The symptoms worsened during his four-day ascent of the Orinoco back to San Fernando. On the final two days he was 'nearly helpless with continued fever, and to have proceeded farther… would have been almost certain death.' He was thus condemned to spend what proved to be a horrible thirty-eight days in that miserable village. His Indians remained with him, but they were 'sorry nurses… and I might as well have been alone'. He had violent attacks of fever at night, with short respites in the middle of the day. On the second night he asked his men to help him during a violent fit of vomiting. But they were 'all so completely stupefied with rum' that none was able to help – they had drunk a bottle of rum he gave them, but then traded some of his jerked beef for more.

The malaria Spruce had caught was evidently the most violent form, *Plasmodium falciparum*. His attacks soon fell into a punishing rhythm of thirty-six hours of fever followed by twelve hours of exhaustion. Coldness of his extremities speedily passed into fever that continued throughout a night and day. 'Then the fever increased in violence – pulse so rapid that it could not be counted – thirst unquenchable – breathing laborious and with great fatigue of chest – mouth constantly filled with viscid saliva, and towards morning the attack would begin to pass off and leave me with forces completely prostrated.'

Spruce felt that he must get better help. So he made a deal with a friend: to lend him his Indians as boatmen and, in return, get a woman to nurse him. He moved into the house of his nurse's family. 'This woman – Carmen Reja by name – I shall not easily forget.' She was a zamba, the class of Indian-black mestizos whom Spruce abhorred – he had picked up the ludicrous rumour that zambos committed 90 per cent of the heinous crimes in Venezuela. (Interestingly, Spruce liked pure black people, often freed slaves, more than any of the other races in Amazonia – as did his companions Bates and Wallace.) When Carmen Reja was in a bad mood, which she usually was, 'her face put on a scowl which was almost demoniacal.' If Spruce needed something from a shop, Carmen's little granddaughter was sent with a note from him; but the paranoid nurse, who could not read, was convinced that the notes were accusations against her. Spruce could hear her talking to her daughters 'always in a high pitch of indignation, and muttering a few curses against the *foreigner*.' She loved rum, so Spruce kept her well supplied; but its mollifying effect on her was short-lived. His own condition deteriorated alarmingly. He had some quinine, which indigenous people had long known to be a malaria palliative (this 'fever-bark' would later play an important role in Spruce's life). But he also took the 'fever root' ipecacuanha (*Carapichea ipecacuanha*), an emetic to induce vomiting that was thought to help. Nurse Reja also prescribed local pills, one of which was a violent purgative that she gave him frequently. Not surprisingly, this 'produced no good effects'. Instead, the fever worsened in intensity and duration. For days and nights on end, Spruce got no sleep and – being drained by the emetic – could take no food beyond watery gruel of the starchy plant arrowroot (*Maranta arundinacea*). These nostrums and the malaria reduced him to extreme weakness and exhaustion. He had unquenchable thirst, difficulty breathing and violent bouts of sweating.

Spruce himself and those around him 'were nightly and almost hourly expecting his death... which he awaited in a state of almost complete apathy.' The demonic nurse often left the house for hours on end, evidently expecting and hoping that he would be dead when she returned. At night she would light his lamp and leave water by his bed, and then fill the house with her friends. He could hear them discussing or abusing him with all the colourful Spanish terms of abuse. 'Among other things she would call out: "Die, you English dog, that we may have a merry watch-night with your dollars!"' They scolded him for giving them the burden of disposing of his belongings, planned how to divide them, and even considered finishing him off with poison. Spruce himself tried to arrange with the village's comisario how to dispose of his worldly goods in the event of his death.

Finally, on the fifteenth night of fever, Spruce realized that the purges were worse than useless. He decided to stop them and instead take more of his precious stock of quinine – it is a mystery why he had not done this earlier. He started with two or three grains (130 to 195 mg); the malaria improved, so he increased the dose to six grains (390 mg) four times a day. The effect was marvellous: he slept better, ate a little and even drank some red wine. There was one last attack on 4 August. After that, fortified by coffee, quinine, rum and tapir meat (locally considered to be a malaria cure), the worst of his malaria was over.

On 13 August, he heard that the Portuguese trader Antonio Dias was returning to the Rio Negro; so he seized the chance to escape from the hell-hole San Fernando and the clutches of Carmen Reja. This Dias was the macho eccentric whom Wallace had met two years previously, the boat-builder and trader famous for his harem of women making fancy hammocks. But Dias was a generous friend to both tall and lonely English naturalists. It took a week to ascend the Atabapo to Javita. Spruce was so weak that he had to be carried in a hammock across the portage road to Pimichin. He recovered his own boat at Tomo, and his Indians got him back to San Carlos de Río Negro at the end of August.

Spruce had commissioned one of Dias's famous hammocks, gaudily fringed with a galaxy of bird feathers. He took delivery of the gorgeous item and sent it to the Countess of Carlisle, as thanks for the assistance that she and the Earl had given him before he left for Brazil. He explained in a covering letter: 'I hope that [it will] be found worthy of a place in the Museum at Castle Howard, and... from the beauty of its fabric to occupy a place alongside the noblest works of art there.' His hope was realized. The hammock is still on view in a display case in the Long Gallery of Castle Howard, the Yorkshire mansion on whose estate Spruce was born and was to die. It is a magnificent object, about 1.8 metres (6 feet) long, with feathers intricately woven into floral patterns that look like satin embroidery – and the colours are still as bright as when they were on Amazonian birds [PL. 74]. In his letter Spruce told how it had been made by Indian girls working for Senhor Dias, that it took a girl over four months to make, and that the feathers were mostly tiny iridescents from hummingbirds, but also from toucans, parrots, 'and especially the beautiful (and now very rare) Cock of the Rock', all glued onto fibre of the prickly tucumã palm with sticky sap of the maçaranduba 'cow tree'.

Spruce spent two-and-a-half months at San Carlos, collecting, arranging his affairs and regaining some strength after the malaria. He left for good on 23 November 1854. He did so with few regrets, because the town's Indians were bullied by the few whites and mestizos (known ironically as *racionales*) and were therefore surly and rebellious – and hopelessly addicted to rum.

Spruce's ordeals were by no means over. His boat was manned by an indigenous Baré pilot, Pedro Deno, with his two sons and two other Indians. Their first night was at the pilot's little riverbank farm. The boat was tethered and they all clambered up to the hut, which had a cane-distilling shed alongside. Spruce bought and ate a caiman's forequarters, and retired to his hammock slung in the shed. He could hear the Indians getting noisily drunk on rum, but paid little attention because 'nothing is more tiresome than the conversation of these people when intoxicated'. But he had learned the Baré people's dialect of the Arawak language, so he noticed that they kept mentioning *heinali* – meaning 'the man' or him. 'I could not help listening attentively to what they said, and it was well I did so.' The pilot's disreputable son-in-law, Pedro Yurebe, had the idea of paying off his debts by getting Spruce to engage him as a boatman, paid in advance for the journey to Manaus; but after three or four days, when Spruce was asleep, they would all take the montaria, escape up a tributary and return to San Carlos – thus shirking the long and tedious voyage for which they had been paid. They all agreed to this plan. But the plot then thickened. Yurebe asked whether 'the man' had much merchandise with him. They exclaimed, in Arawak, 'He has plenty! He has everything!' – not realizing that his boxes were full of plants and drying paper. Yurebe said: 'Then we must not leave him without carrying off as much as we can of his goods, and for this purpose it will be necessary to kill him.' This was also approved and its consequences discussed at length. The drunken conversation continued, until Yurebe declared: 'Why should we not kill the man now?' Spruce was a sick man, with no family in San Carlos. So no one would be surprised to learn that he had died, and that they had buried him a day later as was the custom. They would leave some of his belongings so that these would appear undisturbed, and boldly return to San Carlos to report the Englishman's demise. Spruce of course 'listened to all this with breathless attention', but he doubted whether they would actually commit the murder, until he heard them lashing themselves into a fury against all white men in general and reciting the many injuries they had received – not, of course, from Spruce himself.

Spruce had diarrhoea, so had to leave his hammock two or three times. After midnight, he heard them agree that Yurebe was going to strangle him in his hammock as soon as he fell asleep again. The fires had gone out and the cabin was lit only by dim starlight. Spruce kept still, but with his feet on the ground so that he could spring up if attacked. He heard an Indian whisper that he was asleep, so that the time was ripe. 'I got up and walked leisurely towards the forest as if my necessities had called me thither again; but instead I... walked straight down to the [boat], unlocked the door of the cabin, which I entered.' He put a bundle of drying paper in the doorway, laid out his loaded double-barrelled shotgun and a machete and knife, 'and thus awaited the attack which I still expected would be made.' He heard the Indians angrily wondering why he did not return to his hammock. 'It may be imagined in what state of mind I passed the rest of the night, never allowing my eye and ear to relax their watchfulness for a moment.' Dawn finally broke, but the danger was not over. Pedro Yurebe duly asked to join the crew, but Spruce – with his gun beside him – refused.

Even though this troublemaker was left behind, 'I took care throughout the rest of the voyage that the Indians should never approach me unarmed, and I never spent a gloomier time.' After six days they reached the mouth of the Uaupés where Spruce met two of his old trader friends, and these lent him four men with whom to continue his journey downriver to Manaus. He promptly dismissed the Baré crew, who kept their pay and made their way home. Spruce had been saved thanks to his knowledge of Arawak and his lonely courage.

10

BATES ON THE AMAZON

In October 1852, after his four-month expedition up the Tapajós, Henry Walter Bates decided that he could no longer afford the luxury of chartering his own boat. He would henceforth remain in towns, work in their surrounding forests, and travel only occasionally as a passenger on trading boats. This was a sensible and pragmatic decision. The Amazon is the world's richest ecosystem and its insect life is so unbelievably prolific that Bates, who was above all an entomologist, would never fail to find thousands of new species within reach of his chosen bases. But it would give him little of the excitement, explorations, visits to indigenous peoples – and near-lethal malaria – of his more adventurous companions Wallace and Spruce. So Bates stayed in Santarém – and in the event did not leave for a further two-and-a-half years. Apart from two brief excursions, he was there throughout 1853 and 1854 and until June 1855.

Bates loved many aspects of Santarém. As we have seen, he was very pleased with his house and its view and verandah, and with his collector-helper José whose family looked after them both so well. There were none of the insect pests of other parts of Amazonia – no pium black-flies, no sand-flies, no mutuca horse-flies and no mosquitoes at night. The weather was glorious. The dry season from August to February had cloudless skies, but its dry heat was tempered by fresh breezes. Even the rainy season was tolerable, with sunshine between showers and occasional storms. The streets were always clean and dry, even at the height of the wet season. Also, unlike in every other town up the Amazon, food was plentiful – albeit expensive apart from excellent beef from the savannah pastures around the town. Fresh fish was landed each evening: there was always a rush to buy it from the fishermen. 'Very good bread was hawked round the town every morning, with milk, and a great variety of fruits and vegetables.' Another attraction was 'delicious bathing in the clear waters of the Tapajós' from the sandy beach below Bates's house.

The only drawback for Bates was a lack of social life. Garrulous old Captain Hislop, who had so entertained Wallace and Spruce in 1849, was fading after being knifed during a robbery attempt. Bates missed the lively and good-humoured atmosphere of towns like Cametá on the Tocantins, where plain-living mamelucos formed the bulk of the population and mixed amicably with both whites and Indians and where he made friends with the distinguished politician Ângelo Correira, or Manaus with the hospitable Henrique Antonij, or even remote Ega on the Solimões where a few elderly whites were so friendly. In Santarém, by contrast, 'I did not find... the pleasant, easy-going, and blunt-spoken country folk that are met with in other small towns of the interior.' The white Brazilians and Portuguese were pompous and pretentious. These were either merchants and shopkeepers, the owners of slave-worked plantations, or civil and military authorities. Their manners were stiff and formal. There was none of the hearty hospitality usual along Amazonian rivers. Instead there was 'much ceremony in the intercourse of the principal people with each other and with strangers.' These worthies called on one another at midday, and visitors were expected to wear black dress coats 'regardless of the furious heat which rages in the sandy streets of Santarém' at that hour. All gentlemen, and most ladies, wore gold watches and chains, and there was much snuff-taking from gold and silver snuff boxes. A reception room in each house had a quadrangle of cane-bottomed sofas and chairs, all lacquered and gilded. Compliments were passed and 'in taking leave, the host backs out his guests with repeated bows, finishing at the front door.' Bates of course had none of these clothes or accessories, so he seldom took part in this social formality. Parties were rare, because the men were fully occupied with their businesses and spent their leisure time in billiards and gambling rooms, leaving their wives and daughters shut up at home. The English visitor did get invited to one ball, but he was dismayed to find that the men sat on one side of the room with the ladies opposite them, and dancing partners were allocated by numbered cards distributed by a master of ceremonies. The penniless young foreign naturalist was not much of a catch, so did not draw the most attractive ladies. He deplored this 'old, bigoted, Portuguese system of treating women, which stifled social intercourse and wrought endless evils in the private life of Brazilians'.

Santarém had fewer religious ceremonies than other Amazonian towns, partly because its vicar lacked religious zeal. The people therefore had more lay amusements than 'the processions and mummeries of the saints' days.' Young people were very musical, with flutes, violins, guitars and small four-stringed violas. Bates enjoyed hearing these musicians playing French and Italian marches and dance music, and singing the latest songs, 'in the cool and brilliant moonlit evenings of the dry season'. But although Bates was in his twenties, he does not appear to have been included in these youthful entertainments. Nor did he take part in the big processions at Carnival, Easter, and the Eve of São João's Day. He was not invited into 'the more select affairs got up by the young whites, and coloured men associated with whites', when thirty or forty of them dressed in uniforms 'in very good taste... as cavaliers and dames', and visited the larger houses for some dancing, pale ale and sweetmeats. At Christmas black people mounted a grand semi-dramatic display in the streets. And once a year the Indians had their turn. About a hundred detribalized Mundurucu staged a torchlit parade, performing hunting and mythological dances, with the men in magnificent feather headdresses and tunics, the chief carrying a sceptre of orange, red and green parrot feathers, the women naked to the waist and the children fully naked but all decorated in red annatto (urucum) body paint. Men carried bows and arrows, women had their babies in baskets and children brought pet monkeys, coatis or tortoises. This ingenious procession was 'a great treat... got up spontaneously by the Indians... simply to amuse the people of the place.'

Bates did take an interest in children's education. Santarém had a state-funded primary school for boys and one for girls, then a secondary school with Latin and French classes. It was surprising 'how quickly and well the little lads, both coloured and white, learn reading, writing, and arithmetic... [with] a quickness of apprehension that would have gladdened the heart of a northern school-master.' Traders and planters wanted their sons to progress to the Lyceum or the Bishop's Seminary in Belém. Bates was honoured to be invited to be an examiner of this entrance test one year – the examination's managers were not aware that he had left school at thirteen and was largely self-taught.

Three developments changed the lives of the small population of settlers in Amazonia during the 1850s. One was the separation of the province of Amazonas from that of Pará – but this affected Manaus more than Santarém, which remained in Pará. A second was the advent of steam navigation. The industrialist Irineu de Sousa, Viscount of Mauá, in 1852 founded the Companhia de Navegação a Vapor do Amazonas ('Amazon Steam Navigation Company'). After a shaky start, the new service eventually meant that passengers and freight were no longer at the mercy of seasonal winds and currents or the lack of Indian paddlers. Steamers took a few days for journeys that had previously taken weeks or months, and both sexes no longer travelled with their hammocks slung in a tolda above a hold full of fish or forest produce. Bates noticed how Santarém's stuffy 'customs changed rapidly... after steamers began to run on the Amazons (in 1853), bringing a flood of new ideas and fashions into the country.' The service also revolutionized Bates's own communications. In March 1853 he wrote to Stevens that the Manaus steamer had already made two voyages – albeit heavily subsidized by the government since its freight income covered only a sixth of its monthly costs. He sent a stream of collections down on the steamer; and received letters, payments, magazines and reference books on it. He also planned to use the service to return up the Amazon-Solimões to Ega – since getting there in his own boat or as a passenger in someone else's was '*an utter impossibility*'.

The third change was the start of what was eventually to become the gigantic Amazon rubber boom. Wallace and Bates had noticed communities tapping rubber on the Tocantins in 1848; and Bates mentioned it again on the Tapajós in 1852. When Spruce started up the Rio Negro in 1851 he tried in vain to get its inhabitants to extract rubber 'but they shook their heads and said it would never answer'. Everything changed from inventions in faraway Scotland, England and the United States. By mixing sap from *Hevea* trees with naphtha and then heating it by 'vulcanization' the sticky latex became the hard, elastic and versatile product we know as rubber. Thus demand rose rapidly. When Spruce returned to Manaus in 1855, 'all the way down the Rio Negro the smoke was seen ascending from recently opened seringales [*seringal* was the Portuguese for a rubber stand].... The extraordinary price reached by rubber in Pará in 1853 at length woke up the people from their lethargy, and when once set in motion, so wide was the impulse extended that throughout the Amazon and its principal tributaries the mass of the population put itself in motion to search out and fabricate rubber. In the province of Pará alone... it was computed that 25,000 persons were employed in that branch of industry. Mechanics threw aside their tools, sugar-makers deserted their mills, and Indians their roças....' The governing class at first deplored this labour rush, because it meant that sugar, rum and manioc were in short supply; but official attitudes changed when export profits started gushing in.

The diligent botanist Spruce identified eight species of rubber tree on the Negro and lower Amazon. On the Uaupés he himself discovered a long-leaved species that he called *Siphonia lutea*. He was told about another species in forests beyond Manaus that was said to yield more milk: and it was indeed later found that the best rubber came from *Hevea brasiliensis* trees far to the southwest, on the Purus and Juruá tributaries of the upper Amazon. Spruce then noted that rubber latex was best extracted during the dry months, when the trees were not flowering or fruiting; and he described how rubber was dried by smoking it onto a paddle in thick acrid smoke from burning certain palm fruits. A paper that Spruce sent home was the earliest publication solely about the rubber trade: 'Notes on the India-rubber of the Amazon', in *Hooker's Journal of Botany* in 1855.

At Santarém, Bates resumed his daily routine of collecting and preparing his specimens. As the months went by, he tended to change his collecting location. About 8 kilometres (5 miles) west of the town towards the Tapajós was a little bay called Mapiri. This was reached by a delightful walk along a sandy beach (before the river rose to flood it, from April to July) and past a belt of woodland studded with occasional huts. Then came shallow lagoons covered in water lilies and surrounded by dense thickets. These pools were frequented by white egrets (*Casmerodius albus*), dusky-striped herons (*Butorides striatus*), and masses of smaller birds. Around these, rosy-breasted troupials (of the *Icterus* genus) – which reminded him of British starlings – fed on insect larvae in the moist soil. Bates's daily walk then took him along beaches overlooked by a beautiful grove of trees: in April these were covered in blossom and their trunks thickly clothed in *Epidendrum* orchids profusely blooming with large white flowers. This small place was home to four species of kingfisher, and several hummingbirds, 'the most conspicuous of which is a large swallow-tailed kind (Eupetomena macroura) with a brilliant livery of emerald green and steel blue.' Emerging from this grove was another long stretch of beach with high rocky ground and a broad belt of forest beyond. Finally, rounding a projecting bluff, came the bay of Mapiri, its shores wooded, and with a line of clay cliffs on the far side. It was a beautiful place, constantly fruitful for the collector. There were fine views: southwards to hills clothed with rolling forest, and west across the immense expanse of the Tapajós whose far shore was visible only as a thin grey line on the horizon.

During the rainy season of 1852–1853 butterflies, moths and other Hymenoptera were scarce, so Bates concentrated on small beetles – he was pleased that Stevens wrote recommending that he do just that. In March 1853 he told his agent that many of the beetle species he was sending were *new* – they did not appear in his two reference books, one by the French coleopterist Jean-Charles Chenu and the other the catalogue of the Natural History Museum. He commented that to gather these beetles 'they require *very close* searching for, most of them being very quiet, feeding on the under side of the leaves of spiny palms in swamps.' He listed new species and genera 'which, together with other things, I find at roots of trees on the borders of swamps. I have a unique specimen of a new species of Metopias, *Gory*, with spined thorax and elytra'.

By August 1853 Bates wrote that he had twenty-seven times made a 10-kilometre (6-mile) walk 'over scorching, sandy desert' to another locality. The reward was an 'astonishing number of Coleoptera there; they run small, but a great many Longicornes, Clerii, &c. unfortunately mostly unique specimens: of Ibidion alone I number eighteen species now, and thirty-seven species of Clerii.' (He kept one example of each species for his personal collection, so that 'unique specimens' meant that he had no duplicate to send to England.)

Resting in the shade during the midday heat, Bates used to amuse himself by watching a pale-green species of sand wasp (*Microbembex ciliata*) at work. Females dug little tunnels into a bank, throwing up small jets of sand. 'The little miners excavate with their fore feet...; they work with wonderful rapidity' digging out continuous streams of sand. After making a sloping gallery, 5 to 8 centimetres (2 or 3 inches) deep, 'the busy work-woman flies away; but returns... with a fly in her grasp, with which she re-enters her mine.' The fly had been numbed by her sting. She then lays an egg on the victim and carefully closes the entrance. The fly of course 'serves as food for the soft, footless grub soon to be hatched from the egg.' These sand wasps make a separate chamber for each egg. Higher up the Amazon, Bates observed a larger but related species, *Stictia signata*, which flew long distances to catch the blood-thirsty mutuca horse-flies as food for its egg – thus doing humans a great service. Bates once had this big wasp fly straight for his face, in order to seize a mutuca that was about to bite him.

Another favourite collecting place was an inlet called Mahicá, on the main Amazon about 5 kilometres (3 miles) east of Santarém. This was an area of level pasture ringed by a wall of tall forest. A few caboclos grazed oxen on the meadows. These people were 'wretchedly poor' and Bates was shocked that it did not occur to them to fence some areas for planting kitchen gardens – although he did admit that it was difficult to find fencing-wood that would not be destroyed by termites. Woods surrounding this pasture 'teemed with life; the number and variety of curious insects of all orders which occurred here was quite wonderful.' Towards the river, the trees grew on a bed of stiff white clay. Local people made every sort of coarse pottery and kitchen utensil from this, a form of tabatinga marl that occurs in pockets all over Amazonia. Bates was delighted by a symbiotic relationship between man and insect: scores of shallow pits left by clay extraction were very attractive to mason bees and wasps. He spent 'many an hour watching their proceedings'. The most conspicuous species was a large yellow and black wasp, *Sceliphron fistularium* [PL. 25]. This came to a clay pit with a loud hum, took two or three minutes to knead a small round pellet, and flew off with this in its mouth. It used these pellets to build a pouch-shaped nest, hanging from a branch or other projection. One wasp was so intent on building a nest on the handle of a chest in Bates's boat that it did not mind him watching it closely through his magnifying glass. 'Every fresh pellet was brought in with a triumphant song, which changed to a cheerful busy hum when it alighted and began to work.' It used its mouth and mandibles to smooth the clay around the rim of its little cell, and then patted the sides with its feet. 'The gay little builder' took a week to finish. Whenever Bates opened one of these wasp's nests, he found it stocked with small spiders of the genus *Gasteracantha*, 'in the usual half-dead state to which mother wasps reduce insects which are to serve as food for their progeny'. There were other mason wasps, such as a large black species, *Trypoxylon albitarse*, which 'makes a tremendous fuss whilst building its cell'. If two or three of them chose to do this in crevices in a dwelling 'their loud humming keeps the house in an uproar'. A smaller variety, *T. aurifons*, made nests shaped like tiny carafes, with round bodies and rimmed openings.

To Bates, the most numerous and interesting clay artificers were worker bees of the species *Melipona fasciculata*. These small stingless bees live in immense colonies. They collect pollen as do other bees, but great numbers of their workers also gather clay for their hives. 'The rapidity and precision of their movements whilst thus engaged are wonderful.' They first scrape the clay with their jaws, then use their front feet to form pellets held by their large foliated hind shanks, and fly off with as much as they can carry. Bates discovered how they used the clay: to wall-off crevices in trees or riverbanks in which they made their honey combs. Their liquid honey is aromatic, but they cannot be domesticated like other honey bees. Having no stings, when Indians raid their combs their only defence is for hundreds to smother the robbers – particularly in their hair. Bates commented that 'these tiny fellows are often very troublesome in the woods, on account of their familiarity; for they settle on one's face and hands and... get into the eyes and mouth or up the nostrils.' He did not appreciate that they wanted the salt in human saliva and tear ducts. This is why they are now known as 'sweat bees'. They are still common, and their attentions are a harmless minor irritant.

Bates's third collecting area near Santarém was due south, away from the large rivers and towards the low conical hill Irurá. He went to this region of inland streams flanked by rich woods once or twice a week throughout the three years he spent at Santarém. 'These forest brooks, with their clear cold waters brawling over their sandy or pebbly beds through wild tropical glens, always had a great charm for me.' Lofty forests beside the streams and on the flanks of low hills contained a great variety of trees, notably colossal

Brazil-nut (*Bertholletia excelsa*), piquiá (*Caryocar villosum*), with a fruit much appreciated locally but which Bates found as boring as raw potato, and immensely tall tonka bean trees (*Dipteryx odorata*), whose seeds were exported to Europe for scenting snuff. In the open spaces, the air was wonderfully transparent and there were sweeping views, to the white glistening shores of the Tapajós and inland to forested hills.

Excursions to Irurá 'had always a picnic character'. They left at dawn, with two boys, black and Indian, carrying the day's provisions – some beef or fried fish, farinha and bananas, with a cooking kettle and plates. The faithful José took two guns, ammunition and game-bags, while Bates himself had the insect-collecting apparatus of a net and 'a large leathern bag with compartments for corked boxes, phials, glass tubes, and so forth'. They would halt at a clearing reasonably free of ants and close to a stream, spend the morning hunting in different directions through the woods, take their well-earned meal on the ground, with banana leaves as a tablecloth, and rest for a couple of hours in the great heat of the afternoon. The boys would sleep, but Bates liked to lie and watch birds – flocks of glossy black anis (*Crotophaga ani*) calling to each other as they moved from tree to tree, or a toucan peeping into crevices of branches – or jacuaru lizards (*Tupinambis teguixin*) 'scampering with great clatter over the dead leaves' as they chased one another. By four o'clock the midday heat was mitigated and, after some more collecting, they walked home. Being so close to the equator the sun always set at six, but the little party was sometimes benighted while crossing the campo – not a problem on moonlit nights. 'After sunset the air becomes delightfully cool, and fragrant with the aroma of fruits and flowers. The nocturnal animals then come forth.'

At the end of August 1853 Bates got a ride in a trader's boat for over 20 kilometres (12 miles) up the Tapajós back to Alter do Chão. This was a 'most wretched starved, ruinous village', but it lay in a lovely natural location – a deep bay, a belt of snowy sandy beach, towering pyramidal hills, wooded ridges and, most importantly for the naturalist, surrounding virgin forests that stretched into the untrodden interior. Bates spent three months at Alter do Chão, and in December sent Stevens a great consignment. In his accompanying letter, he showed that he was a master of every facet of entomology. Of butterflies, 'you will find many of the handsomest of the new species of last year, and some very remarkable and new things, good series of specimens, indeed I consider the Lepidoptera to be very fine; I took altogether about thirty-five new species of Diurnes....' He had carefully examined 222 species he collected; 'dissected a great quantity of specimens; found some laws in the variation of their neuration, &c.'; and did his best to classify them in the Linnaean system. Years later, the distinguished entomologist William Distant wrote that his friend Bates was unique in being a master of all three aspects of that discipline: he was a superb collector and field naturalist, 'a philosophical observer or recorder of evolutionary facts and arguments', and was also devoted to systematic classification. It was almost unknown for any one entomologist to pursue all three branches of entomology, because the methods and motivations were so different, and because those who excelled at one often disparaged the others.

When Bates and his young companions walked to Irurá, they crossed a campo savannah 'disfigured in all directions by earthy mounds and conical hillocks, the work of many different species of [termites]. Some of these structures are five feet [1.5 metres] high, and formed of particles of earth worked into a material as hard as stone.' As a passionate entomologist, Bates made a pioneering study of these amazing creatures. He was one of the first to appreciate the essential role of termites in hastening the decomposition of dead and decaying wood in hot countries. They are the cleaners and recyclers of tropical rain forests. He examined their narrow tunnel galleries that criss-crossed the ground and snaked all over dead wood; and he peered into several hundred of their rock-hard colonies. Everywhere, he found 'a throng of eager, busy creatures' transporting dead wood to be rendered-down, or grains of earth for their constructions.

Among ants, by contrast, workers are by far the most numerous part of a community; these are undeveloped females, and the functions of different categories 'do not seem to be rigidly defined. The contrary happens in the Termites, and this perhaps shows that the organization of their communities has reached a higher stage, the division of labour being more complete. The neuters in these wonderful insects are always divided into two classes – fighters and workers; both are blind, and each keeps to its own task; the one to build, make covered roads, nurse the young brood from the egg upwards, take care of the king and queen...: the other to defend the community against all comers.' Ants progress from grub, to pupa in a membrane, and then to adults. Most termites, on the other hand, are neuters that emerge from the egg fully developed in shape. Bates was amazed by the 'principle of division of labour, the setting apart of classes of individuals for certain employments, [which] occurs only in human societies in an advanced state of civilisation.' Such differences by function are not found among higher animals. 'The wonderful part in the history of the Termites is, that not only is there a rigid division of labour, but nature has given to each class a structure of body adapting it to the kind of labour it has to perform.' The few males and females developed eyes and wings, so that they could fly to start new colonies. Neuter workers are smooth and rounded, with mouths adapted for working materials. Soldiers had large heads provided with special organs of offence or defence – horny pikes, tridents, or differently shaped jaws – which varied between species.

Whenever a termite mound or tunnel was disturbed, the first to be seen were workers, but these quickly vanished into their labyrinth of tunnels; and the soldiers appeared. Bates admired their suicidal bravery. When a hole was made, its edges 'bristled with their armed heads as the courageous warriors ranged themselves in a compact line.... They attacked fiercely any intruding object, and as fast as the front ranks were destroyed, others filled up their places. When the jaws closed in the flesh, they suffered themselves to be torn in pieces rather than loosen their hold.' Despite the bites, Bates dug into dozens of termite hills. He identified and drew soldiers from eight different species. He also reached large chambers at the heart of mounds, where the 'royal couple' of a very large king and the perpetually gravid queen were closely guarded by a detachment of workers. As fast as the queen laid eggs, these were carried off to minor cells dispersed throughout the colony. After months of daily study, Bates established that winged termites were males and females in equal numbers, that when they flew off to create a new colony some of them shed their wings and became its king and queen, that soldiers and workers were adults and not larvae – as supposed by other entomologists, and that the different job-functions did not develop by having separate food or conditions.

Bates also recorded the different architecture and building methods of various termite species. Each used distinct particles, its own method of compacting and cementation, and a unique shape of mound. But, as he wrote to Stevens, 'a large hillock is always an agglomeration of many very distinct species which build with very different materials.' Of tree-dwelling termites, 'some are wholly subterranean, and others live under the bark or in the interior of trees: it is these latter kinds which get into houses and destroy furniture, books, and clothing.' He found some nests without a king and queen, just workers bringing eggs from an overstocked termitarium; whereas in others the king and queen were only starting to reproduce or create a mound. He also watched the elaborate preparations when winged termites were about to fly to a new location: 'The workers are set in the greatest activity, as if they were aware that the very existence of their species depended on the successful emigration and marriages of their brothers and sisters.' The exodus took several days. Flying termites were 'much attracted by the lights in houses, and fly by myriads into chambers, filling the air with a loud rustling noise, and often falling in such numbers that they extinguish the lamps.' Innumerable enemies – bats, goatsucker birds, spiders, ants and reptiles – devoured most of the termites during a migration. 'The waste of life is astonishing.'

Bates's intensive study of termites was pure scientific research that earned him nothing. He wrote, modestly, that 'very little, up to that date, had been recorded of the constitution and economy of their communities' in South America. In April 1854 he sent six boxes of insects, including many termites, back to England. He had great difficulty preserving termites from mould during a drenching rainy season, so packed them in spirits. He hoped that the Natural History Museum would want to buy them. Bates asked Stevens to get his termite species professionally drawn and then submit his notes for publication in the *Transactions of the Entomological Society*, where 'they would form a splendid paper; in fact, I flatter myself, would attract much the attention of naturalists.' In 1854 and 1855 he also sent his observations of termites to Professor Westwood at Oxford, and they were published in German by Dr Hagen in the *Linnaea Entomologica* journal. Bates later summarised these findings in his book *The Naturalist on the River Amazons*; but they were omitted in the abridged second edition of 1864.

In August 1853 Bates was complaining to Stevens that it was too long since he had received any 'papers, magazines, entomological notes and books, which would be a great treat to me now.... I certainly could not exist in these deserts without such things.' But by the following January he cheered up on receipt of two parcels from Stevens that had been languishing in the customs house in Belém, and he had got five catalogues of the Natural History Museum. He wrote to Stevens: 'I cannot give you an idea of the pleasure it caused me to receive so many cheering, valuable, and useful books and letters.' In another missive he repeated that they had 'been to me a continued intellectual feast, and given me lots of occupation.' He was particularly thrilled to have seven volumes of *Suites à Buffon*, 'especially the Hymenoptera part, and henceforward you may depend upon it the bees, ants, &c. will feel the effect...'.* Bates said that Stevens was quite right to have sent two volumes of catalogues from the Jardin des Plantes in Paris. He also loved getting copies of the *Illustrated London News.* And he asked for any first-rate monographs, such as Lacordaire's 'Phytophages' (plant-eating beetles) and F. Smith on Cryptocerus (ants), and more catalogues from the Natural History Museum (including zoology this time) and other major museums.

In September 1854 Bates sailed 300 kilometres (185 miles) up the main Amazon to Vila Nova (Parintins) at the eastern end of Tupinambarana Island. He recalled having

* *Suites à Buffon* was a journal adding to the thirty-volume natural history by the great eighteenth-century naturalist Georges-Louis Leclerc, Comte de Buffon.

found an astonishing number of 'new things' when he had spent a few days there five years previously. But this time he was disappointed. Vila Nova was a strange place, with much of the land around the little town seasonally flooded. Surprisingly, he spent five months there even though he admitted to Stevens that 'upon the whole I consider it a poor locality'. Of course he sent a collection resulting from his labours, much of it beetles; but he wished that there had been more. Finally he admitted that 'my health got very low in Vila Nova from bad food'. So he returned to salubrious Santarém and in April 1855 wrote that 'I never felt better, more disposed to work, or happier since I have been in the country.' For two months he turned his attention to birds. He was surprised to see how many species there were – particularly of hummingbirds, of which he shot or observed ten different species. He worked hard at mounting birds, and wrote to Stevens that local people told him that 'I have put them up very nicely'.

Bates now decided to return to Ega (Tefé), far up the Amazon-Solimões, and from there collect all along the upper river. The steamer service went upriver from Manaus to Peru every three months, and Bates decided to catch the June 1855 sailing. He commissioned dozens of insect-mounting boxes and bought two new guns – he had long since abandoned his zarabatana blow-gun as too difficult for a non-Indian. Thus prepared, Bates left Santarém, spent a week in Manaus, and reached Ega on 19 June. He was very grateful to the Brazilian government for heavily subsidizing this loss-making shipping service. It took only eight days of actual sailing to cover the 1,450 kilometres (900 miles) to Ega; whereas in 1849–1850 he had struggled for ninety-seven days on the same journey.

Within a couple of months, Bates was writing joyously to Stevens that he had already collected 2,600 chosen insects, as well as land shells and other prizes. He was reminded how the insect fauna of the upper Amazon was wonderfully different to that around Santarém. 'The forests are composed of quite a different class of trees.... There is no place on the lower Amazons that can at all be compared with Ega, in its exuberant fertility, ...teeming waters, and towering forests.' He immediately resumed his quiet routine, on every day except Sunday: 'up early, a walk, bath [in the lake] and breakfast; then out with my boxes [near the beach], selecting the new species and stowing away duplicates, until 9.30 a.m., when the sun is out hot and it is time to be off to the forest....' He concentrated on his favourites – beetles and butterflies. He was now so expert that he was constantly discovering new species and noting minute differences from beetles observed elsewhere. He wrote to Stevens: 'I… never return without bringing a species new to me; general average, four or five a-day new. Yesterday I got two new Lamellicorns… and one extraordinary and large Necydalis, quite new. Today got another new Longicorn. A few days ago I got a most splendid new Prionidae (i.e. new to science), being a Sternacanthus, of which there was only one species known before…. Ega is wonderfully rich in Erotilidae: in the woods, after a shower, you meet with great fellows, some nearly an inch long, gay in vivid red, black and yellow colours.' Curiously, Bates rarely collected a second example of these beetles: so his private collection was rich in 'beautiful unique specimens of new species'. He soon had 500 species just of longicorn beetles. Lepidoptera were abundant and, of course, more saleable to British collectors. 'In this department Ega is one of the finest districts in the world.'

Bates loved rambling on a path along the beach on the banks of Lake Tefé. 'The land above is high and covered with forest, and the beach is a grove of Araçá [fragrant fruit] trees: the limpid waters of the broad lake break gently on the sands: the trees overhead are full of gaudy coloured birds; the horned screamer (the noisiest bird in the world) yells from the forest [this is a large black bird, *Anhima cornuta*]; lonely swallows… are flitting about; but all along the beach is a succession of the most beautiful butterflies in the world. This morning I saw more species than constitute the whole English fauna in this department.' Bates also gathered micro-Lepidoptera, little butterflies and moths that most collectors neglected because they were too tiny. He reckoned that there must be several thousand species just of the Tineidae (now several families of leaf- and fruit-mining moths) and Tortricidae. A consignment he sent in July 1856 contained 158 specimens of these small butterflies, 'collected without giving any close attention to them, and whilst hunting for birds with a double-barrel shotgun slung at my left shoulder.' A familiar frustration was with beautiful big morphos – the most sought-after by collectors. Bates recorded three species near Ega, each with differently coloured females. But the flashy blue morphos glided high in the forest, rarely descending below 6 metres (20 feet), so that Bates could not possibly catch them in his net. He did not neglect other forms of insect. He sent specimens of workers of a new form of ant, and 'I have the winged female of a different species; they are all extremely rare; the neuration [nerve system] of the female is unlike any Myrmicidae or Scoliidae I have seen. …I consider it a very remarkable discovery.' Writing about his work at Ega a few years later, Bates proudly noted that 'the name of my favourite village has become quite a household name amongst a numerous class of Naturalists, not only in England but abroad, in consequence of the very large number of new species (upwards of 3000) which they have had to describe with the locality "Ega" attached to them.'

It is worth recounting Bates's description of catching one rare butterfly, just to illustrate his daily work during all those years of collecting. He wrote to a friend, from Ega in May 1857, that he had finally caught a beautiful *Zeonia* (now classed as *Chorinea*) in forests that he had traversed repeatedly. After a month of unusually dry weather, 'in ascending a slope in the forest by a broad pathway mounting from a moist hollow, choked up with monstrous arums and other marsh plants, I was delighted to see another of what had always been so exceedingly rare a group of butterflies; it crossed the path in a series of rapid jerks, and settled on a leaf close before me.' He captured it and two other zeonias. But he found none elsewhere, 'for it very often happens that a species is confined to a few square yards of space in the vast forest, which to our perceptions offers no difference throughout its millions of acres to account for the preference. I entered the thicket... and found a small sunny opening, where many of the Zeonia were flitting about from one leaf to another, meeting one another, gambolling, and fighting; their blue transparent tinge, brilliant crimson patch, and long tails all very visible in the momentary intervals between the jerks in their flight.' In that one thicket of 17 to 25 square metres (20 to 30 square yards), 'as far as the eye could reach, the leaves were peopled with them.' Bates rapidly caught perhaps 150 of the rare insects – although the majority 'fell to pieces in the bottom of the net, so fragile is their texture.' He returned on the following days. 'On the second day, the butterflies were still coming out [of their pupae]; on the third, they were much fewer, and nearly all worn; and on the fourth day, I did not see a single perfect specimen, and not a dozen altogether.' One species was later named *Zeonia batesii* after him [PL. 23].

As so often, Bates was torn between the joy of his profession and the loneliness of his private life. He poured out his thoughts to his agent Stevens – who promptly sent it all for publication in the *Zoologist*. 'When you consider the great pleasure there is in this, and at the same time the liberty and independence of this kind of life – the tolerable good living (turtle, fresh fish, game, fowls, &c.) – the suavity of the climate, &c., you will readily understand why I am disinclined to return to the slavery of English mercantile life.' (Bates recalled his hard work as a hosiery apprentice and in a brewery.) 'One great privation, however, I suffer and feel acutely – the want of frequent receipt of letters, books, newspapers and magazines.' It was four months since he had received anything written in English. He even hankered for 'stirring news' of the Crimean War.

Bates was keen to collect further up the Solimões but, as usual, this was easier said than done: one steamer was too full to take him and his luggage; porterage along a forest trail was out of the question, even to the next village; and westerly winds were wrong for sailing. So it was not until 5 September 1856 that he finally embarked on 'our neat little

Upper Amazon steamer' *Tabatinga* on the first of his excursions west of Ega. He took all supplies for a three-month journey, apart from meat and fish: fifteen large packages of collecting equipment, hammocks, cooking pots, crockery and so forth. Although no one connected mosquitoes with malaria, every passenger on the upper Amazon slept in a mosquito tent because 'without it existence would be scarcely possible'. Mosquito nets were not yet invented – Bates's tent was made of coarse calico, with sleeves at each end for hammock cords. He found that he could read and write or swing in his hammock in it, pleased that he had cheated the thirsty swarms of bloodsuckers that filled the ship's cabin. The 170-ton-burthen iron steamer sped along day and night, with a brilliant pilot, and lookouts for floating logs or sandbanks. The other passengers were 'mostly thin, anxious, Yankee-looking' Peruvians who traded hundreds of valuable Panama hats down the Amazon and brought back crockery, glass and other heavy goods.

Bates disembarked at Tonantins, a village at the mouth of a river of that name, on the north bank of the Solimões roughly halfway between Ega and the Peruvian frontier at Tabatinga. He spent nineteen days at Tonantins, a muddy hamlet of twenty 'hovels', hedged by lofty sombre forest, surrounded by swamps, covered in veritable clouds of pium black-flies, and with air that was 'always close, warm and reeking'. But there was a continual hum and chirp of insects and birds. The weedy ground beside the village teemed with plovers, sandpipers, striped herons and scissor-tailed flycatchers.

Collecting – of monkeys, birds and insects – was good. Bates was delighted to catch a new species of butterfly, later named *Callicore excelsior* 'owing to its surpassing in size and beauty all the previously-known species of its singularly beautiful genus. The upper surface of its wings is of the richest blue... and on each side is a broad curved stripe of an orange colour.' He also made the same important observation that Wallace had noted elsewhere: that many species could not cross the broad river and had thus evolved, during the latest geological era, slightly differently from those on its far side. This modification was particularly true of insects; but it also occurred with the rare and shy uacari monkey (*Cacajao calvus*) found only here on the north bank of the Solimões between its Tonantins and Japurá tributaries. These small monkeys are bald, with bright scarlet faces and an endearing quizzical expression. Their almost-tailless bodies are covered in long straight shining whitish hair, like an angora goat's, and they live in small bands in the tops of tall trees. Most die soon after captivity, so that they are a great rarity in zoos.

At the end of November, Bates left Tonantins on a trading schooner returning to Ega. After four days they reached the southern Jutaí tributary and Bates had a couple of days there – while the ship's owner traded for salt-fish – before moving to a place called

Fonte Boa. Bates was stuck for seven weeks in this dilapidated village, which was famous for being the region's 'head-quarters of mosquitoes'. At night he was protected in his calico hammock-tent; but the bloodsuckers were also active during the day in dark huts – swarms of them 'settled by half-dozens together on the legs'. These tormentors' insatiable thirst for blood and their itchy stings spoiled all comfort. Sadly for the collector, the plague was even worse in the forest, where a different, larger species of mosquito followed Bates in a little cloud at every step of his woodland ramble. Maddeningly, 'their hum is so loud that it prevents one hearing well the notes of birds'.

However, despite the damp, mud and insects, Bates was in good health and greatly enjoyed himself at Fonte Boa. Broad paths through rich forest led to delightful clearings. 'In every hollow flowed a sparkling brook, with perennial and crystal waters. The margins of these streams were paradises of leafiness and verdure; the most striking feature being the variety of ferns, with immense leaves, some terrestrial, others climbing over trees, and two at least arborescent.' Forest trees were gigantic. 'Birds and monkeys in this glorious forest were very abundant; the bear-like [parauacú] *Pithecia hirsuta* being the most remarkable of the monkeys, and the Umbrella chatterer and Curl-crested toucans amongst the most beautiful of the birds.'

Indians and caboclos from Fonte Boa farmed little plantations beside these streams. They greeted Bates cheerfully and were pleased when he offered to join them for meals, adding the contents of his provision-bag to their dinner and squatting among them on a mat.

On 25 January 1857 the steamer returned downriver, and whisked Bates back to Ega in a mere sixteen hours. Seven months later, he boarded the *Tabatinga* again, this time to sail 650 kilometres (400 miles) upriver to São Paulo de Olivença. They arrived after only five days, on 10 September. The town was built on a high bluff of tabatinga clay, but was surprisingly damp – Bates had difficulty protecting his collections from mould. It had once been a thriving mission, but was now a run-down village. The 500 inhabitants were mostly mamelucos or Ticuna and Kulina indigenous peoples. Bates was shocked by the morals of the few whites at São Paulo de Olivença. He wrote to Stevens that life there 'is an almost perpetual orgy. I never saw anything so disgusting in the course of my travels.' He partly blamed the priest, who 'spent his days and most of his nights in gambling and rum-drinking, corrupting the young fellows and setting the vilest example to the Indians.' Two

of the town's officials, the magistrate Geraldo and the Director of Indians Antonio Ribeiro, took against the straight-laced visitor because he would not join in their drinking bouts. These took place every third day. They started early in the morning with an explosive mixture of cachaça and ginger. This made Bates's neighbour Geraldo stand outside his house raving about foreigners and gesticulating against him in particular. (In the evening, when sober, Geraldo usually came to offer his humblest apologies.) The wives of São Paulo's worthies were also 'hard drinkers and corrupt to the last degree'. Wife-beating was rampant. At night Bates found it best to lock himself indoors 'and take no notice of the thumps and screams which used to rouse the village... throughout the night'. Bates's only companions proved to be two free black men: the deputy sheriff José Patricio, 'an upright, open-hearted and loyal' man, and the tailor, 'a tall, thin, grave young man' known as Master Chico (short for Francisco). By coincidence, Bates had met the young tailor Chico in Belém, where he used to live with a widow known as Tia ('Aunt') Rufina. This admirable lady was born a slave, but was allowed to trade in the market. So she earned enough to purchase first her own and then Chico's freedom, and prospered so much that she bought a large house. She used to care for Bates's belongings when he was away on a trip, and was totally trustworthy, gentle, cheerful and financially generous to the Englishman. Chico neither drank, smoked nor gambled, and was as shocked as Bates 'by the depravity of all classes in this wretched little settlement'. He would give a special knock on Bates's shutters, to show that it was not another obstreperous drunk. The two young men then spent long evenings working and conversing: Chico was invariably courteous and well worth listening to. Of all the racial groups in the Brazilian Amazon, Bates admired and liked free black people the most.

Bates spent five months at São Paulo de Olivença – but he wrote that five years would have been insufficient to do justice to its treasures of zoology and botany. Although he had by now spent a decade in Brazil, he was as excited as a newcomer by its beautiful forests. One of his favourite walks was through the forest to 'a cool sandy glen, with a brook of icy-cold water flowing at the bottom. At mid-day the vertical sun penetrates into the gloomy depths of this romantic spot, lighting up the leafy banks of the rivulet... where numbers of scarlet, green, and black tanagers and brightly-coloured butterflies sport about in the stray beams.' One bird was 'the most remarkable songster by far of the Amazonian forests'. This was the musician wren, *Cyphorhinus arada,* whose lovely music sounded to Bates like a flageolet flute and to others a violin – this was the songbird that so enchanted Spruce when he was near the lower Amazon. This songster's music started with a few tender notes, but then stopped abruptly or made 'a clicking noise like a hand-organ out of wind'. São Paulo's glorious forest was full of sparkling brooks and bubbling springs; Bates could not find a local helper, but on his solitary walks he often bathed for an hour at a time in the bracing waters of a swift-flowing stream – 'hours which remain amongst my most pleasant memories'.

In all these years of walking in tropical forests, it never occurred to Bates that he was in any danger. As with his two friends – and indeed Humboldt, Waterton, Schomburgk and others before them – he stressed the beauty, healthiness and relative safety of the forest. Notions of 'green hell', with man-eating piranha, jaguars, venomous snakes, aggressive peccary, huge black caiman, and Indians firing poisoned arrows, came later in the overheated imaginations of fake 'explorers'. Bates often saw snakes, but he knew that most were harmless. Soon after arriving in Belém he became entangled in the folds of a wonderfully slender 1.8-metre-long (6-foot) dryophis (called locally a cobra-cipó or liana snake; probably *Oxybelis fulgidus*); and near the Tapajós he chased a huge anaconda as it glided through the forest. The three most poisonous Amazonian snakes are vipers (particularly surucucú (*Lachesis muta*) and jararaca (*Bothrops jararaca*)), rattlesnakes (*Crotalus durissus*), and corals (*Micrurus corallinus*). The vipers and rattlesnakes are big snakes with long fangs, camouflaged to blend into a carpet of dead leaves (there is now – but not in Bates's day – a serum to counteract their lethal venom). By contrast, corals are brightly coloured in black, red and yellow bands like old school ties (and there is no antidote to their intense poison). However, corals rarely strike and their fangs are short. Bates trod on a jararaca viper near Belém, but it was killed by an Indian boy with a machete before it could strike him. On an excursion from Ega he encountered another jararaca, with its 'hideous, flat triangular head' reared back ready to strike; but another Indian shattered this one with a blast of shot – to Bates's dismay as he wanted the specimen. Some local people were convinced that these vipers jumped up in unprovoked attacks. But Bates refuted this: 'I met, in the course of my daily rambles in the woods, many Jararacas, and once or twice very narrowly escaped treading on them, but never saw them attempt to spring.'

Because he could not find an Indian helper, which he admitted was essential for locating and catching birds, Bates concentrated on butterflies. He immediately saw that most of the conspicuous diurnal Lepidoptera were quite different from any he had seen elsewhere.* He sent Stevens an eight-page report on his months at São Paulo de Olivença, which the *Zoologist* published as a separate paper; and most of this consisted of very detailed descriptions of butterflies. At that time of no field or coloured photography, he had to describe verbally all the colours and wing patterns. Bates recorded that he

* 'Diurnal Lepidoptera' means daytime-flying insects with four wings covered in fine scales: in other words, butterflies.

recovered 5,000 specimens of insects from that locality, 'amongst which there were 686 species new to me of all orders; 79 being new species of Diurnal Lepidoptera.'

São Paulo de Olivença was the eastern limit of the Ticuna indigenous people. The Ticuna are an independent cohesive nation, with a complicated language and some customs unrelated to those of any other Indians. In colonial times they had migrated from remote lakes and rivers to occupy hundreds of kilometres of the main Amazon. They were filling a vacuum left by the departure of the sophisticated Tupi-speaking Omagua, who were extinguished by disease, slavery and migration upriver into Spanish Peru. The Ticuna are today one of the largest unified indigenous peoples of Amazonia, with tens of thousands occupying a swathe of forests and rivers in Brazil and Peru.

Bates of course knew nothing of the history of the Ticuna when he stumbled on one of their malocas in the forest. When this solitary white man first appeared among them they scampered off to hide in the trees; but on subsequent visits they became friendlier and he found them harmless and good-natured. Their maloca was a large oblong, but with rafters, palm-thatching and tree piers so uncoordinated that each part seemed to have been done by different builders. There were the usual clusters of hammocks and cooking fires for each family, and at one end a raised mezzanine platform built of split palm stems. The Ticuna excelled in pottery, notably huge broad-mouthed jars decorated with a lattice of different coloured dyes, to hold *tucupi* sauce (pepper and manioc) or manioc beer, as well as skilfully woven baskets and bags knitted from bromeliad fibres. Both sexes tattooed their faces with a scroll or rows of short lines on each cheek. In their huts they wore no clothing, just bracelets, anklets and garters of tapir hide.

The Ticuna loved frequent festivals. Some of their drinking bouts were to placate the mischievous jurupari ('devil'), or to celebrate weddings, fruit festivals, children's puberty or (according to the rather puritanical Bates) 'a holiday got up simply out of a love of dissipation'. At puberty, girls were secluded for a month on the maloca's platform, with minimal diet. They then had some head hair ceremonially plucked by older women.* At festivals, chiefs and elders wore spectacular headdresses made by fixing toucan breast-feathers to a web of bromeliad twine, with macaw tail-feathers standing up dramatically behind. But the Ticunas' most striking adornments are masked dresses. These are full-length shifts made of the thick, whitish inner bark of some trees of the fig family – a substance whose fibres are so regularly interlaced that it looks like artificial cloth. The tubular gowns cover the head, with eye holes, ears on either side made of bark-cloth on

* In the twentieth century, a Colombian travel company called Green Hell Tours had to be stopped from bringing tourists to gawp at the ritual of 'plucking the virgins' hair'.

round reeds, and human or animal features. Grotesque faces are either painted onto the bark-cloth, in vivid yellow, black and red dyes, or they are solid masks of balsa wood. They depict monkeys, fish, birds or other animals, and 'the biggest and ugliest mask represents the Jurupari'. Bates drew such a scene of wonderfully exuberant masked celebrants, and it appeared in his book [PL. 53]. He himself is at one side, with sideburns and a checked shirt, being served manioc brew by a slim naked maiden. Bates was less enraptured by this ceremony than Wallace had been with that of the Tukano on the Uaupés. He wrote that the Ticuna 'go through their monotonous see-saw and stamping dances accompanied by singing and drumming, and keep up the sport often for three or four days and nights in succession, drinking enormous quantities of caysúma [manioc beer], smoking tobacco, and snuffing paricá powder.'

Henry Bates had seen villages of relatively unspoiled Indians only twice – these Ticuna, and the Munduruсu on the Cupari tributary of the Tapajós – and both were small groups far from the heartland of their peoples. But he gave much thought to Indians, even though he was in no sense an anthropologist and was baffled by the indigenous psyche. He was well aware of the differences between indigenous peoples, but he judged them by the work-ethic and values of Victorian England. At the base were nomadic peoples like the riverine 'debased' Mura, whom he regarded as indolent, dirty, untrustworthy and with shabby hovels on sandbanks.

Also in his lowest stratum were the Kaixana (or Kayushana), a forest people inland from Tonantins. On a collecting walk when he was there in the previous September, Bates had stumbled across a secluded conical hut with a very low doorway in the gloom of the forest. This was the home of a family of timid Kaixana. They were smeared with mud and black genipapo dye, wore only a form of bark-cloth apron, and the man's 'savage aspect was heightened by his hair hanging over his forehead to his eyes.' They had no villages, just scattered small huts, they hunted only with blow-guns, and even their festivals were little more than some drinking. The men played pan-pipes of lengths of arrow-grass, with which they wiled away the hours, 'lolling in ragged bast hammocks slung in their dark, smoky huts.' The Arawak-speaking Kaixana were harmless, peaceable people, living in forests and lakes between the lower Japurá and Içá rivers. To Bates 'their social condition is of a low type' little better than that of forest animals; and they are sadly now extinct.

Somewhat higher in Bates's estimation were the Marawá, another Arawak-speaking people near the Jutaí (a southern tributary of the Solimões upstream of Ega). Bates was paddled for a day in a canoe to reach their cluster of huts. His party was received in a frank, smiling manner. One young Indian, 'a fine strapping fellow nearly six feet [1.8 metres]

44

45

44 A rectangular maloca of the Uaupés river peoples. Wallace and Spruce were both thrilled by these great communal dwellings, whose triangular facades were decorated with geometric paintings.

45 A circular maloca of the Yekuana Maquiritare (Mayongong) of the upper Orinoco. Spruce was impressed by the efficiency of these people when he visited them in 1853–54.

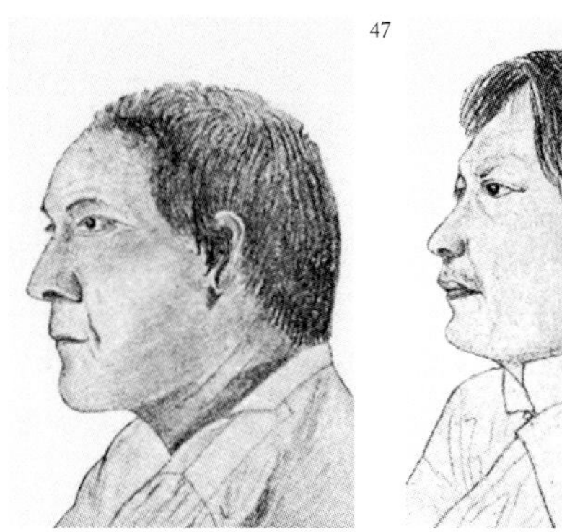

Spruce's drawings were possibly the first portraits ever made of individual Brazilian Indians.

46 Chief Callistro of the Tariana at Iauaretê on the Uaupés, 1852.

47 Chief Ramon Tussari of the Yekuana Maquiritare of the Cunucunuma river, 1854.

48 Spruce's drawing of two wretched Maku (Hupdu) girls, aged nine and sixteen, seized from their nomadic forest people and held in Marabitanas fort before being sent to servitude in Manaus.

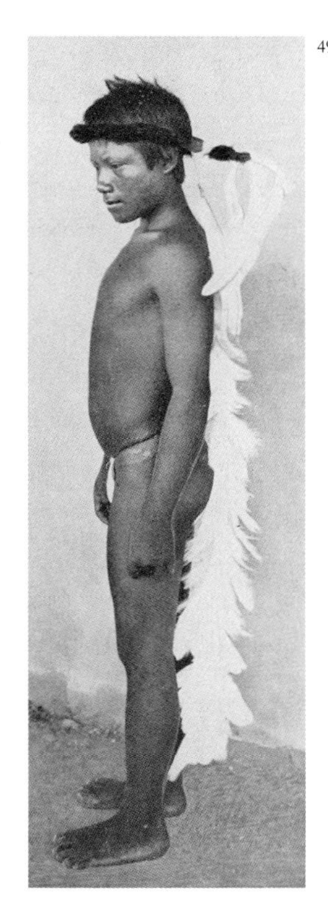

49 Desana men, of the Uaupés river, had long hair hanging down their backs, tied with monkey hair and decorated with a cascade of egret or heron feathers.

50 A Tukanoan man in full ceremonial regalia, with a coronet of toucan and eagle feathers, a prized quartz cylinder around his neck, a long apron and garters with rattles below his knees.

51 Cigars of wild tobacco were too hot, so Tukanoan peoples made elegant holders for them. Wallace drew this artifact, and Spruce sent one to the museum at Kew Gardens.

52

53

52 Boys of a Uaupés people perform a dance, each holding a maraca gourd rattle and with a band of nut rattles around his ankles.

53 Bates gatecrashed a Ticuna wedding, near São Paulo de Olivença on the Solimões. Some elders wear fantastic inner-bark masks and gowns. A girl serves Bates manioc beer, but the drawing makes him appear more portly than he was from his meagre diet.

54

55

57

Manioc was the staple food of indigenous peoples throughout Amazonia.

54 Lethal cyanide had to be leached from manioc before it could be eaten. This was done in a basketry tube called tipiti, which squeezed out the poison as it contracted. Here a Maquiritare man weaves a long tipiti.

55 Before leaching, manioc roots were grated and soaked into pulp. Quartz-studded graters were prized trade goods. This woman is fixing quartz teeth into a grater, while feeding her baby.

56 Manioc is roasted into beiju pancakes or farinha granules in a broad pan, itself another valuable trading artifact.

57 The piaçaba palm grows abundantly in the upper Rio Negro. Its fibres were packed into conical bales and formed one of the region's main exports. They were used to make rope and broom bristles.

58 Only indigenous men had the skill to shoot swimming fish, using arrows with barbed detachable heads. Travellers were fed by their Indians' fishing and hunting.

59

60

59 A camp of cinchona-bark collectors, on the slopes of the Ecuadorian Andes. They wastefully felled the trees before removing the bark for its quinine. These casqueros helped Spruce in his quest for cinchona.

60 Spruce's team would have looked like this group of horsemen, on high páramo near Riobamba and below snow-clad Mount Chimborazo.

61

61 Spruce's hundreds of pressed plants are highly valued by the herbarium of the Royal Botanic Gardens, Kew. This specimen is the powerful hallucinogenic used by shamans of different peoples, known as caapi, ayahuasca or yagé. Spruce identified and studied this potent plant and gave it the botanical name *Banisteria* (now *Banisteriopsis*) *caapi*.

62

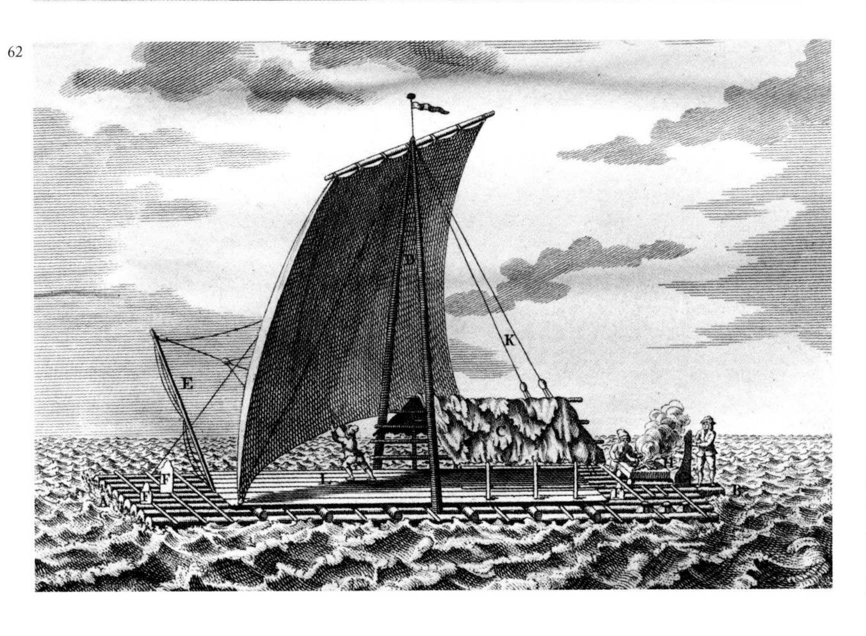

62 A balsawood raft used in Ecuador. Spruce chartered such a craft to bring his precious cinchona plants and seedlings down the Ventanas (now Babahojo) river to the port Guayaquil in 1860.

63

64

65

66

63 Wallace in 1869, aged 46.

64 'Bates of the Amazon' as he was affectionately known, when he was running the Royal Geographical Society.

65 Spruce in his seventies, when he was crippled but still working on his botanical collections.

66 Spruce lived in a small house, now called Spruce Cottage, in Coneysthorpe on the Castle Howard estate in Yorkshire.

high, with a large aquiline nose' showed Bates various interesting things: how to look fierce with plugs in his lip and ears, a butterfly chrysalis (something that the lepidopterist knew well), and the preparation of mildly hallucinogenic ipadu (coca) powder. These Marawá are also now extinct.

Near the top of Bates's pecking order were the Tupi-speaking Kocama, 'a shrewd, hard-working people' who lived on the upper Solimões river in Brazil and Amazonas in Peru. These were the only tribe who willingly volunteered to work as boatmen. Kocama had manned the boat in which Bates reached Ega in 1850, and 'it would be difficult to find a better-behaved set of men in a voyage than these poor Indians'. They worked tirelessly paddling and winching the boat upriver, always in good humour. Unusually for Indians, the Kocama had 'provident habits': each wanted to own a locked wooden trunk, in which he stored his 'clothing, hatchets, knives, harpoon heads, needles and thread, and so forth.'

Other peoples in this part of Brazil whom Bates admired were the Arawak-speaking Pasé and the Yuri subgroup of Ticuna. Both were 'peaceable, gentle and industrious people, devoted to agriculture and fishing, and have always been friendly to the whites.' The Pasé were 'a slenderly-built and superior race of Indians'. When he was first at Ega, Bates was paddled for about 30 kilometres (20 miles) up forested canals to visit the house of a Pasé chief. He was impressed by the efficiency and cleanliness of the farm and establishment, the warmth of Indian hospitality and the dignity of his aged host and his wife. But all these peoples were dying out, from influenza and consumption (tuberculosis) – pulmonary diseases against which they had no inherited immunity. Far from showing no emotion, the Pasé deeply mourned their dead relatives and lamented the extinction of their people. Their demise as a coherent tribe was hastened by intermarriage into settler society. As Bates noted, they were regarded as the most advanced of all Indian nations, and their 'industrious habits, fidelity, mildness of disposition, ...docility and, it may be added, their personal beauty, especially of the children and women' made them sought-after wives, concubines or household servants. In the twenty-first century, these three 'advanced' peoples – Kocama, Pasé and Yuri – are, indeed, virtually extinct.

Bates's judgement of these and other detribalized indigenous peoples was harsh. He felt that their goodness arose more from 'the absence of active bad qualities, than in the possession of good ones'. He generalized that all Indians were phlegmatic and apathetic, lacking imagination or intellectual curiosity, and never 'stirred by the emotions: love, pity, admiration, fear, wonder, joy and enthusiasm.... This made Amazonian Indians very uninteresting companions anywhere.' Bates tried in vain to discuss the geography of the

Amazon river, the sun and stars, and a supreme God with his indigenous companions. He concluded that their dullness was the result of living in a wilderness, concerned only with daily subsistence, and lacking a written language or a leisured class who could hand down acquired knowledge to the next generation. This observation was unfair, since Bates never ceased to admire Indians' intense knowledge of their environment and ability to locate birds or move through forests – even if they conducted no scientific research as he understood it. He also marvelled at the skill of indigenous boatmen, despite never having to navigate fearful rapids as Wallace and Spruce had done. In fairness to Bates, the disciplines of anthropology (the study of man) and ethnography (human societies) scarcely existed: J. C. Prichard's seminal *Natural History of Man* had been published only in 1843. Also, Bates's generalizations were based on fleeting visits and contacts, generally with Indians near the settler frontier rather than in uncontaminated villages where their society could function in all its glory.

In January 1858 Bates finally succumbed to malaria, as Wallace had done on the Uaupés in 1851 and Spruce on the Orinoco three years later. Bates fell seriously ill, with 'ague' or intermittent attacks that left him 'with shattered health and dampened enthusiasm'. Bates guessed that his illness was the culmination of several years of gradually deteriorating health. 'I had exposed myself too much in the sun, working to the utmost of my strength six days a week, and had suffered much, besides, from bad and insufficient food.' He sensibly took quinine, of which he had been carrying a small phial for seven years without having had to use it. But Bates had curious theories about medicine. He worried that if he 'indulged' his feeling of lassitude – a symptom of every form of malaria – it would result in 'incurable disorders of the liver and spleen'. So he made a great effort to rouse himself. 'Every morning I shouldered my gun or insect-net, and went my usual walk in the forest. The fit of shivering very often seized me before I got home, and I then used to stand still and brave it out.' The friendly captain of the steamer *Tabatinga* 'was shocked to see me so much shattered', so he persuaded Bates to return with him from São Paulo de Olivença to Ega. Bates did this, in February 1858. He had hoped to collect 'yet unseen treasures' far up the Amazon and deep into Peru, but had to abandon this plan. Although the malarial spasms ceased, his general health remained too weak. Despite this, he spent a further year in Ega, still collecting diligently.

During that final year in Ega the thirty-three-year-old naturalist purchased two Indian children. He bought them from a trader on the Japurá, the river that enters the

Solimões opposite Ega and is called Caquetá upriver in Colombia. There was clearly an illegal traffic in indigenous children on the Japurá, just as there was on tributaries of the upper Rio Negro as witnessed by Wallace and Spruce.* Bates's boy was about twelve, dark-skinned and speaking a language that no Indians in Ega could understand: he might therefore have been a Tukano-speaking Kueretú, or an Arawak-speaking Yumana. After some months of sulky silence, the boy learned to speak Portuguese. Bates called his 'little savage' Sebastião, and found him invaluable in the forest, finding birds that fell into thick undergrowth, or shinning up trees to gather fruit. In a matter-of-fact manner, Bates wrote that 'the use of these Indian children is to fill water-jars from the river, gather fire-wood in the forest, cook, assist in paddling the montaria in excursions, and so forth.' He treated his little captive well, eventually taking him down to Belém where he became an apprentice to José (who had been Bates's faithful assistant in Santarém in 1853–1855, and was now a goldsmith in the provincial capital). Bates's other purchase was delivered to his door by an old woman one rainy night. It was a little girl, 'thin and haggard, drenched with wet and shivering with ague [malaria].' She was a Miranha and called Oria by her people. Bates and José 'took the greatest care of our little patient', giving her quinine and nourishing food, and engaging a Miranha woman to nurse her. Bates was paternally smitten by the child, who was always smiling and full of talk, loved being carried to the river to bathe, asked for fruit, or taking objects in the room for playthings. But their ministrations were in vain and she sank rapidly. During her final weeks, 'it was inexpressibly touching to hear her' repeating verses she had been taught in her village. Bates had Oria christened before she died, and insisted that she be buried with the Catholic church's 'little angel' ministrations, as accorded to non-indigenous children.

Bates finally left Ega on 3 February 1859, en route for England, and reached Belém by steamer six weeks later. He found Belém much changed in the eight years since he had last been there. He was amazed and touched by the warm welcome he received from old friends, both Brazilian and foreign. The doyen of foreign merchants, George Brocklehurst of R. Singlehurst & Co. – who was Bates's shipping agent – had him to stay and was most hospitable. Bates was also surprised to find himself something of a celebrity because he had devoted so many years to scientific study of the fearful jungle wilderness. The town was greatly improved from the 'weedy, ruinous, village-looking place' he and Wallace had first seen in 1848. Its population had grown to 20,000, largely from European immigration. Its main streets were paved and lined with trees, dilapidated houses had

* Almost forty years previously, the German botanist Martius had been shocked to see such covert slavery rampant on the Japurá – although, like Bates, he also acquired young Indian captives, to save ten of them because they were stricken with malaria.

been renovated and boasted elegant balconies, swamps were drained, and magnificent new rides were created. There was now public transport of sixty horse-drawn cabriolets, a public library, printing presses and four daily newspapers. People were as pleasure-loving as ever, but their parties were more European-inspired and less religious. The main downside was a huge increase in the cost of living, and a shortage of labour. But to the naturalist Henry Bates, the worst change was the removal of the 'mantle of shrubs, bushes, and creeping plants' that used to cover open spaces, and deforestation to make way for ugly roads through 'once clean and lonely woods'. Some of his beautiful collecting grounds had been obliterated.

On 2 June 1859 Bates left Brazil, for good. He sailed on an American trading ship, via the United States. He prudently divided his collections into three shipments on different vessels – to avoid a disaster such as had befallen his friend Wallace. As he sailed down the Pará river he 'took a last view of the glorious forest for which I had so much love, and to explore which I had devoted so many years.... The saddest hours' were when they entered the Atlantic Ocean and he severed 'the last link... with the land of so many pleasing recollections... a region which may be fittingly called a Naturalist's Paradise.'

He was consoled, of course, by the thought of returning to his beloved England. Its defects – grey twilights, murky atmosphere, 'factory chimneys and crowds of grimy operatives, ...union workhouses, confined rooms, artificial cares, and slavish conventionalities' – were more than compensated by 'civilised life, where feelings, tastes, and intellect find abundant nourishment'.

11

SPRUCE CLIMBS TO THE ANDES

In a world devoid of means of communication apart from haphazard word of mouth, the three English naturalists often narrowly missed seeing one another. As we have seen, in October 1849 Bates stopped at Santarém a few days after Wallace had left for Monte Alegre; later that year Bates, Wallace and Spruce were each at Óbidos a few days apart; and in June 1852 Wallace on his return journey reached Santarém a mere week after Bates had sailed up the Tapajós. Now in 1855 Bates and Spruce just missed a meeting in Manaus. Spruce spent the first three months of the year in that town, organizing the despatch of his magnificent collections from three years up the Rio Negro. He then took the March sailing of the quarterly steamer upriver to his next destination, Peru. By bad luck Bates chose the next sailing from Manaus, in June 1855, to return to Ega. So the two lonely scientists missed the chance of intellectual conversation, traveller's tales and young men's chatter *in English* – which both of them craved desperately.

Like Bates, Spruce loved the voyage up the Solimões, in the luxury of a steamboat. He started by reaching Manaquiri in a mere ten hours – a journey that had taken seven arduous days four years earlier. It then took less than a fortnight to the Peruvian border at Tabatinga. He told his friend John Teasdale about the iron steamer *Monarca*. Its low-pressure engine delivered only 35 horsepower and was so large that it left very little space for cargo. Piles of firewood for this engine also took up much of the deck. The most common fuel was a handsome tree called pau mulato, so-named because it regularly shed its bark to expose new bark that turned from green to bronze to mulatto brown. This tree was abundant in the seasonally flooded igapó forest, and piles of it were gathered at regular intervals along the river. It burned well even when green, and it was easily accessible on both black- and white-water rivers. But a full load of this timber lasted the

Monarca for at most thirty-six hours. So half the time of the voyage was spent loading this firewood – as well as 'a cargo of mosquitoes always coming on board, uninvited'. The pau mulato tree meant much to Spruce, for two reasons. One was that he was travelling in a ship propelled by a wood named after himself: 'It was unknown to botanists until I sent specimens from Santarém, and Mr. Bentham has called it *Enkylista spruceana*' (the tree is still named after our hero, but its botanical name is now *Calycophyllum spruceanum*). The other reason was that the tree was a species related to *Cinchona*, a genus that includes 'fever-bark' trees yielding the malaria palliative quinine. Spruce did not know then, but cinchona trees were later to play a major role in his life and enduring legacy.

On the voyage, Spruce 'never tired of admiring the ever-varying forest panorama – the broad beaches densely clad with Arrow-reeds growing 20 or 30 feet [6 to 9 metres] high, behind which extended beds of slender and graceful willows (*Salix humboldtiana*), their yellow-green foliage relieved by the occasional admixture of the broad green-and-white leaves of *Cecropia peltata* (a tree of the nettle family), while beyond rose abruptly the lofty virgin forest' with its prodigious mixture of species. Then there was the river itself: noble, vastly broad, tawny brown, studded with innumerable islands, plenty of cranes and herons, never-failing caiman, and freshwater dolphins making playful somersaults.

One stop was at São Paulo de Olivença: considered by Spruce a miserable place, but popular with residents because its priest was a gambler who also indulged in every other vice of the Amazonian interior. This was the same vicar who would shock Bates when he was there two years later. The border fort at Tabatinga was equally decrepit. Spruce bought a souvenir from the local Ticuna indigenous people: a bark-cloth gown and grotesque mask, which is still in the Economic Botany collection of the Royal Botanic Gardens, Kew. The steamer voyage ended at Nauta, upriver of Iquitos in Peru.

It took Spruce two weeks to hire a couple of boats and Indian paddlers for the next stage of his journey, up the Huallaga – a major southern tributary of the Amazon. Dependent on sails and paddles again, Spruce took over two months on this voyage. He paused at two former mission villages, Lagunas and Yurimaguas, and had agreeable visits with their priests. The main event of this journey was to catch a herd of peccary swimming across the river. Spruce and his crew managed 'with considerable difficulty' to kill nine of these, with a gun and machetes. Peccary fight back bravely when menaced, so a wounded boar fiercely tried to climb into the boat to attack his enemies. He was killed just in time. The meat of these animals fed Spruce's men and local villagers for days.

Spruce reached his destination, Tarapoto, at the end of June 1855. Like Bates, he felt that he might do his best collecting by making occasional excursions from one base. He eventually spent almost two very happy years at Tarapoto. He had for some time been corresponding with the little town's 'principal inhabitant', a Majorcan Spaniard called Don Ignacio Morey, and it was an invitation from this gentleman that took Spruce to this remote part of Peru. Don Ignacio sent mules to bring the Englishman up from the Huallaga, then placed his well-furnished table at Spruce's disposal and secured an unoccupied house for him. The latter was just what he wanted: outside the small town, not on a street, and sufficiently isolated to be perfectly quiet. It was surrounded by a garden yielding sugar cane, sweet manioc, cotton, yams, broad beans and calabash gourds, with clumps of herbs such as capsicum, and some young yagua royal palms [PL. 35]. The house was on a bluff at the edge of a turbulent stream, of sparkling waters splashing over rocks, beyond which lay a village of two indigenous peoples. The sound of this stream reached Spruce 'in a soft murmur, often mingled with the less musical sounds of a cane-mill on the opposite margin; the squeaking of the cane-crushers; the shouts of the men who goad along the poor oxen or mules in their painful round; the grunting of pigs which chew the crushed canes as they are thrown out; and very often the laughter and playful screams of boys and girls bathing in the stream.' In the opposite direction Spruce looked over Tarapoto, a town founded seventy years earlier by Peruvian settlers, rapidly expanding to some 12,000 inhabitants. It included Quechua-speaking Lama people, also known as Motilones ('Shaven heads') because they shaved their crowns leaving only a fringe on the forehead.

Spruce was now at an altitude of some 450 metres (1,500 feet) so he was above the lowland rain forests and approaching more open Andean scenery. Most of the woods here were low, with patches of pasture for European domestic animals. Beside these fields 'or from amid the forest peep out the straw roofs of cottages, often accompanied by plantain gardens and by orange and other fruit trees.' The town was regularly built, with single-storey houses of thick adobe walls and palm-thatch roofs. It lay in a broad pampa, a wooded plain and rolling hills approached by a wide red-earth road, and in other directions more abrupt shale ridges and low mountains.

Spruce generally took a dim view of mestizos, known throughout the Peruvian Andes as *cholos*. He found them 'a very degenerate race, who have nothing about them of the European but a whitish skin; their ideas, modes of life, and language being still entirely Indian.' But he wrote to George Bentham that there were a few pleasant people with whom to converse. And his health – which had never been robust – was better than at any time in his six years of South American wanderings.

A typical excursion was northwestwards to a hill called Campana near Moyobamba. To Spruce's surprise there were almost no mule trails in this part of Peru: everything had to be carried on the backs of Indians, who were dwindling in numbers. For this month-long collecting excursion, he took six porters to carry his provisions (salt beef and fish) and pressing paper, and on the return his herbarium plants. Where there were no tracks, the party had to wade up icy streams. After a stay at the old colonial village of Lamas, they clambered over 'savage, rocky and wooded gorges, with rugged ascents and descents; and the torrents that traversed them must be crossed and recrossed' frequently because of cliffs on either bank. The reward for Spruce was good botanizing on the summits of unclimbed hills, covered in dense Amazonian forests with occasional natural grassy clearings.

As a diligent and dedicated botanist, Spruce collected in every part of the broad pampa around Tarapoto. Although most of this was cultivated, there were patches of virgin and secondary forest and – to his trained eye – 'much difference in the vegetation' in different parts of the plain. He then plunged further afield, in a series of often punishing expeditions. Three were to Mount Guayrapurina ('Where the Wind Blows') 20 kilometres (12 miles) east of the town. The mountain's name was apt: the wind was so strong that it almost blew him off precipitous trails, and it howled all night when he slept near the summit.

On one steep flank of this mountain Spruce was delighted to find his favourite plants. The rocks were 'densely clad with Hepaticae (especially Mastigobryum, Lepidozia, and Plagiochila)' as well as mosses, and at one clear stream 'the richest bit of fern ground I had seen in the world'. In an area with a diameter of less than 80 kilometres (50 miles), Spruce 'found 250 species of ferns and their allies, of which many were new, especially among the tree-ferns'. He found several hundred species of mosses and liverworts, which had been rare in the lowland tropical rain forests. And, as his editor Wallace reminded the reader, Spruce limited his collecting to species that he himself had never seen or that 'he believed to be unknown to European botanists'. Among the wealth of tree species, 'he could not possibly collect all he saw', so often had to omit 'many new species in favour of others which he believed to constitute new genera'. Spruce himself likened this to the thrill of exploration. It was 'the greatest of all pleasures to the naturalist… that of discovering new species, of dotting in (as it were) new islands on the map of nature and, in some cases of even peopling continents that appear to be deserts.'

Spruce had loved liverworts ever since he was a boy in Yorkshire. He was enthralled by the thousands of species of them in South America. 'They are so abundant and beautiful… that I think no botanist could resist the temptation to gather them. In equatorial plains, one set creeps over the living leaves of bushes and ferns, and clothes them with a delicate tracery of silvery-green, golden or red-brown; and another set, along with mosses, invests the fallen trunks of old trees in masses that you could not embrace with your arms. I have some species with a stem half a yard [45 centimetres] long, and others so minute that six of them grow and fruit on a single leaflet of an Acrostichum.'

He almost rejoiced that these tiny liverworts had no commercial value. 'It is true that the Hepaticae have hardly as yet yielded any substance to man capable of stupefying him or of forcing his stomach to empty its contents, nor are they good for food; but if man cannot torture them to his uses or abuses, they are infinitely useful where God has placed them...; and they are, at the least, useful to, and beautiful in themselves – surely the primary motive for every individual existence.' Spruce's love of plants was pure and passionate. He confessed that 'I like to look on plants as sentient beings, which live and enjoy their lives – which beautify the earth during life, and after death may adorn my herbarium. When they are beaten to a pulp or powder in the apothecary's mortar, they lose most of their interest for me.'

Other excursions were to gorges at the heads of surrounding rivers; down to tributaries of the Huallaga; and north to a range of mountains dominated by Cerro Pelado ('Bald Mountain') whose sharp sandstone ridges he climbed. For a month he based himself beyond Lamas to explore the slopes of the Campana ('Bell') mountains. Spruce wrote an anguished account of the difficulty of much of this travel, on rudimentary trails, through often dense vegetation, up steep hills or along sheer drops or slippery rocks. Local authorities employed no one to look after these backwoods paths.

For a while on his first Campana expedition, Spruce stayed in the isolated hut of an Indian called Chumbi. One day, tired of his diet of jerked beef and salt-fish, he sent Chumbi out to try to shoot a tasty curassow turkey. Creeping towards a bird, Chumbi was bitten by a deadly poisonous pit viper (*Bothrops bilineata*): he staggered back to the hut, but for three days was at death's door, in extreme pain, swollen and desperately weak. Spruce applied a series of exotic cures and, miraculously, the Indian survived. The botanist was well aware that, had Chumbi died, his family would have taken revenge on the foreigner who had sent him into the forest – a local priest confirmed that they would probably have poisoned the malign stranger. A few months earlier, Spruce had himself escaped being bitten. Ascending the Huallaga, he was collecting bark for the campfire when he heard rattling and 'saw that I had uncovered a large rattlesnake, who was raising himself up and poising his head for a spring at my leg, which was no more than two feet [60 centimetres] off'. Spruce ran off at speed – a strike by this *Crotalus durissus* could have been fatal. But, like Wallace and Bates, Spruce almost never mentioned danger from snakes during all his years in the Amazon forests.

Back in Tarapoto, Spruce was gathering specimen parts of an inga tree, when he was badly stung on his wrist by a hairy caterpillar. He had previously 'made light' of the stinging of large tropical caterpillars that abound in leguminous trees. Their venom was in long fascicled hairs, but the pain of the sting was usually no worse than that of a nettle. This one was different. The pain and irritation increased, so Spruce 'applied solution of ammonia pretty freely, and it proved so strong as to produce excoriation [peeling skin]'. Next morning, although the wound was inflamed and very painful, Spruce tied a rag over it and went into the forest as usual. It was a twelve-hour excursion, in cold rain. That evening, the hand and arm were badly swollen. 'That was the beginning of a time of the most intense suffering I ever endured. After three days of fever and sleepless nights, ulcers broke out all over the back of the hand and the wrist – there were thirty-five in all, and I shall carry the scars to my grave. For five weeks I was condemned to lie most of the time on a long settle, with my arm (in a sling) resting on the back, that being the easiest position I could find.' Poultices of rice and linseed did no good and 'the ulceration ran its course. At one time the case looked so bad that mortification seemed imminent', so Spruce contemplated getting his rustic neighbours 'to cut off my hand, as the only means of saving my life.' He blamed himself for impatiently applying too much ammonia, coupled with a 'chill from exposure to wet'. But he eventually recovered.

Spruce wrote to his mentor and agent Bentham about an unforeseen, but serious, problem in packing his specimens: there were no boards in Tarapoto. People used split canes (caña brava, *Gynerium saccharoides*) for everything from shelves to bedsteads. Their only need for a board was to make a door, and for this they would demolish an old boat. This was what Spruce had to do, in order to make crates for his plant specimens. He bought two old boats in ports down on the Huallaga river, then had to go again with a carpenter to cut these into pieces of the right size, 'which had to be conveyed to Tarapoto on Indians' backs, and afterwards laboriously adzed down into something like boards.' All this, with the time spent in finding Indian porters, then making boxes, 'left me little leisure for anything else for the space of near a month.' Even once the boxes were made, the task of transporting them from Tarapoto to London was daunting, not least from the danger of shipwreck in rapids on the Huallaga. Spruce wrote despairingly to Bentham that 'the difficulty, risk, and expense of getting plants from here all the way down to the mouth of the Amazon are so great, that I see my Tarapoto collections are not likely to repay more than the expense of collecting.'

Don Ignacio Morey and the other prominent Spanish resident of Tarapoto, Don Victoriano Marrieta, decided to travel to Ecuador, partly to trade in Panama hats made in Guayaquil. Spruce decided to go with them, because he had heard (correctly) that the forested eastern slopes of the Andes in Ecuador enjoyed immensely rich biodiversity. Travel overland to Ecuador on non-existent trails through scarcely explored mountains was out of the question, and crossing the Andes and descending to the Pacific coast of Peru, followed by a steamer voyage to Guayaquil, was considered unwise because of an epidemic of yellow fever. So the travellers decided to go by Amazonian rivers. There were months of delay, caused by a variety of problems. One was a revolution in Peru, between followers of the president, General Vivanco, and a rival called General Castilla. In remote Tarapoto, this involved bands of mutinous and pillaging soldiers. On one occasion Spruce hung a Union Jack to protect his house, at another he waited with his gun to fend off looters – as he had done in San Carlos de Río Negro three years earlier.

Finally, on 23 March 1857 Don Ignacio, Don Victoriano and Don Ricardo (Spruce) set off down the Huallaga, each in his own boat paddled by seven Indians. The canoes were open, because this was safer when shooting rapids. The first ordeal was a fearful *pongo* or narrows (now called Aguirre, after the crazed sixteenth-century traitor) in which Spruce's boat was almost swallowed up in a whirlpool. He had a large, handsome guard-dog called Sultan, that he had reared from a puppy and adored – 'there was not another such dog in all Maynas [province]'. But 'in the whirlpool, the horrid roar of the waters, which drowned our voices, and the waves, which splashed over us, so frightened the dog the he went mad!' For six days Spruce tried to save Sultan, but the dog refused food and water, and attacked terrified dogs, pigs, cows and people in riverbank villages. 'At length he began to snap at people in the canoe, and being worn almost to a skeleton, I... was obliged to shoot him.'

Every day of the long journey was 'impeded by swollen rivers and steeping rains'. The two weeks down the Huallaga were 'heart-sickening' and monotonous, with low forested banks almost totally uninhabited, and a plague of black-flies by day and mosquitoes by night. (This author descended the Huallaga a century after Spruce, and nothing had changed. But later in the twentieth century it became dangerous because of fighting between government forces and Sendero Luminoso ('Shining Path') terrorists allied to cocaine *narco-traficantes*.)

The travellers stopped at Lagunas near the mouth of the Huallaga (now a tidy Jesuit mission village) to fix tolda awnings to the boats, hire fresh crews of Indians and replenish provisions. They then entered the main Marañón/Amazon and sailed westwards up it for five days, to turn up its northern Pastaza tributary. Everyone here was terrified of the very warlike Huambiza indigenous people – one of the four divisions of the dreaded Auca nation (then known as Jívaro). There had been several short-lived and unsuccessful attempts to missionize the Jívaro. They loved fighting, particularly in inter-tribal vendettas but also against other intruders. They shrank their enemies' heads into infamous *tsantsa* trophies, by removing the skull, sewing up the lips and boiling the skin and hair in a shrinking herb. Spruce and his friends passed or entered ghost villages. These had been abandoned in such a hurry that food was still on the tables, being rapidly consumed by termites. Their boatmen were so scared that they tried never to land, and insisted on sleeping with the boats tethered to small islands in mid-river and with relays of watchmen throughout the night.

The ascent of the Pastaza took two 'wearisome and monotonous [weeks] with almost continuous rains, rarely any dry land to sleep on, and not a single village or settlement of any kind.' They had to change their crews of paddlers at the miserable village of Andoas, home to about sixty couples of peaceful Zaparoa-speaking Andoa people and a lonely and unhappy Peruvian governor. On 5 May they set off, their boats crammed with the flesh from two tapirs (as meaty as cows), armadillos, eighteen bunches of plantains, manioc, sweet potatoes, pineapples and pots of fermented yucca brew to keep the Andoa Indian crews happy. They were now technically in Ecuador, moving northwestwards up the meandering and empty Bobonaza river, a headwater of the Pastaza.

It took sixteen more days to reach a village called Pucayacu. After meeting the governor of this tiny settlement, they clambered down to sleep in their boats or on a beach. Spruce now endured what, apart from rapids, is the greatest hazard of Amazonian river travel: a flash flood. They had foolishly ignored distant thunder – the only warning that a surge of water might be rushing towards them. 'Then the storm burst over us, and the river almost simultaneously began to rise; speedily the beach was overflowed, the Indians leaped into the canoes; the waters continued to rise with great rapidity, coming in on us every few minutes in a roaring surge which broke under the canoes in whirlpools, and dashed them against each other.' The bignonia lianas holding the canoes finally broke. But the Indians in Spruce's boat bravely clung onto branches through the stormy night, 'using all their efforts to prevent the canoes from being smashed by blows from each other or from the floating trees which now began to career past us like mad bulls. So dense was the gloom that we could see nothing, while we were deafened by the pelting rain, the roaring flood, and the crashing of the branches of the floating trees as they rolled over or dashed against each other; but each lightning-flash revealed to us all the horrors of our position.' At one moment the rising water pushed Spruce's boat into overhanging prickly bamboos and threatened to swamp it until they managed to cut away the matted foliage. 'Every hour thus passed seemed an age, and the coming of day scarcely ameliorated our position.... Assuredly I had slight hopes of living to see the day, and I shall for ever feel grateful to those Indians who... stood through all the rain and storm of that fearful night, relaxing not a moment in their efforts to save our canoes.' The raging river did not start to subside until the middle of the following day. By then, 'we were wearied to death, and myself in a high fever'. The two Spaniards and their Indians had been equally surprised by the flash flood. They spent the pitch-black night lost, trying to grope away from the raging waters, 'and when day broke it found them half dead with cold, and their clothes and bodies torn and wounded by prickly bamboos and palms.' They then unloaded the boats. But while they were doing this, the bank to which they had originally been moored crashed into the river with several large trees – another close shave for the travellers.

The two gentleman-traders from Tarapoto continued overland up into the Andes, to seek their return cargo of Panama hats and other goods from Guayaquil. But Spruce remained for three weeks at Pucayacu. By 10 June he had lightened his mass of luggage and set off with seven reluctant (but well paid) Indian porters, each of whom had a boy or young woman to carry his food. It took three weeks to climb past the village of Canelos to the town of Baños on the Andes altiplano. This was a terrible trek, with almost continuous rain and cold, nights huddled under dripping palm roofs, muddy trails, steep climbs to mountain passes, and repeated crossings of rivers and streams. Spruce's only consolation for the delays and discomforts was 'finding myself in the most mossy place I had yet seen anywhere. Even the topmost twigs and the very leaves were shaggy with mosses, and from the branches overhanging the river depended festoons of several feet in length, composed chiefly of Bryopterides and *Phyllogium fulgens*, in beautiful fruit.' Whenever the hardships seemed overwhelming, Spruce thanked heaven for being able to contemplate a simple moss.

Years later, his friend Alfred Wallace used Spruce's journals, his article in the *Revue bryologique*, and his later book on liverworts, *Hepaticae Amazonicae et Andinae*, to describe the cloud forest of Canelos through which he was passing. He was sure that it had 'the honour of being the richest cryptogamic [of non-seed-bearing plants] locality on the surface of the globe. The trees even... seem to serve no other purpose than to support ferns, mosses, and lichens. The epiphytic ferns, which are the most abundant, are principally Hymenophylleae and Polypodium.... Among the ferns growing upon the ground there are some that attain a height which is almost gigantic: they belong to the genera

Marattia, Hypolepis, Litobrochia, etc.' After describing many more ferns, Spruce had no time to search slowly and scrupulously for the tiny liverworts that he loved. Despite this, he found some novelties, including an unpublished genus *Myriocolea irrorata*, which Wallace regarded as the most interesting that he had ever found.*

When Spruce and his porters reached the raging Topo river (a tributary of the upper Pastaza) it was swollen by pelting rain. They laboured for four days, trying to rig bridges from rock to rock across the torrent. They were drenched, and frozen at night; their food was totally exhausted; and the demoralized porters seemed about to abandon Spruce. In the end, the naturalist had to leave all his baggage behind, in leather chests under a cover of leaves. The men crept one-by-one across spindly wet bamboos that bent perilously into the boiling waters, knowing that one slip might be fatal. 'One of the middle bridges was so long (at least 40 feet [12 metres]), that [its] three slender bamboos almost broke under the weight of a man, even unloaded, and it was... impossible to get my boxes across.' Just after the last man got over the Topo its waters rose higher and swept away their makeshift bridge. 'Those who have escaped from death by hunger or drowning may understand what a load was taken off my heart when we had all got safely across the Topo, although I had been obliged to abandon so many things which to me were more valuable than money.'

There were more travails, such as wading for a mile 'through fetid mud in which grew beds of gigantic horse-tails 18 feet [5.5 metres] high'. Then there was a giddy descent to another headwater of the Pastaza, down a 'ladder' that was merely a notched pole from a rock overhanging the river by 150 feet (45 metres) to which 'each as he descended... clung like grim death'. There were eight ascents and descents of mountain ridges, then 'a long puddle-hole' where a path of slender poles was covered in water, so that Spruce slipped off into waist-high floods. They finally reached the relative comfort of Baños on 1 July 1857 – 103 gruelling days after leaving Tarapoto. Spruce was 'much fallen in flesh, and my thin face nearly hidden by a beard of three months' growth.' The rains continued unabated and it was cold, so Spruce was wracked with catarrh, and coughed blood.

His first concern was to try to recover his baggage from the far bank of the Topo. He engaged porters, but they had to wait for eleven days for the waters to subside, during which Spruce sent them food. When they did finally get to his trunks their leather was rotten and full of maggots; but fortunately much of the contents, including his precious

* Spruce illustrated this new genus with a plate in his later monumental book, *Hepaticae Amazonicae et Andinae* (1885). This species/genus was not rediscovered until 2002, and it is known from only two localities – the Topo river (where Spruce found it) and another tributary of the Pastaza. It is thus considered critically endangered.

books, were intact. And a few weeks of feasting on beef, mutton, bread, beans and fruit restored him, in time for his fortieth birthday on 10 September 1857.

During Spruce's three weeks on the Pastaza he observed the Andoa indigenous people's uses of hallucinogenic plants. This magical property fascinated him. He had already seen drugs taken by Baré Indians at São Gabriel da Cachoeira in 1851, Tukano-speaking peoples of the Uaupés in 1852, Baniwa on the upper Rio Negro in 1853, Maquiritare on the Cunucunuma of the upper Orinoco in 1853–1854, and Guahibo at Maipurés on the middle Orinoco in 1854. Based on these experiences, Richard Spruce was the first serious botanist to publish a detailed description of narcotics and stimulants. He knew that from the sixteenth century onwards, travellers and chroniclers had mentioned powerful drugs used by shamans in their incantations. These were inhaled as smoke or snuff or taken in drinks, and they produced intoxication or even temporary delirium. But no one had properly analyzed the plants themselves. So Spruce, 'having had the good fortune to see the two most famous narcotics in use, and to obtain specimens of the plants that afford them' their powers, decided to be the first to 'record his observations on them, made on the spot'.

The most famous and potent of Amazonian hallucinogenic plants is the liana known as *caapi* to peoples of the upper Rio Negro and Orinoco, *ayahuasca* on the eastern foothills of the Andes in Ecuador and Peru, and *cadana* in the Tukanoan language of the Uaupés. Spruce brilliantly established that these were all the same plant (he did not know that it is also called *yagé*, in the Colombian Amazon). Belying its mind-blowing power, caapi is an inconspicuous-looking vine, with a stem the thickness of a thumb, oval leaves and delicate little yellow blossoms. The Indians grew it in their roça cultivated clearings near rivers. Spruce recalled seeing 'about a dozen well growing plants… twining up to the tree tops' at the edge of a clearing on the Rio Negro. 'It was fortunately in flower and young fruit; and I saw, not without surprise, that it belonged to the order Malpighiaceae', a group of many species of handsome little trees found throughout Amazonia. He correctly realized that this was the only malpighiad that was a narcotic, and that it was undescribed. So he gave it the botanical name *Banisteria caapi* (this has recently been slightly changed, to *Banisteriopsis caapi*, because it was found to belong to a closely allied genus of this family). When used by shamans (called 'medicine men' in nineteenth-century England, and pajé in Tupi) the lower part of its stem was beaten in a mortar, to which a variety of other plants and roots were added to enhance the potency, then sieved to separate woody fibre, and mixed in water. 'Thus prepared, its colour is brownish-green, and its taste bitter and disagreeable.' In his field notebook, now in Kew's archive, he surmises: 'Query? May not the peculiar effects of the Caapi be owing rather to the roots of the Haemadictyon (though in small quantities)....' This was a typically brilliant guess, both because each indigenous group *does* add other narcotic plants to caapi brews, and because *Psychotria viridis* root is one of the additives.

When on the Uaupés in November 1852, Spruce was invited to a Dabucuri festival – at which gifts were ceremonially exchanged. He had watched young men of this Tukano tribe running through the great maloca handing calabashes of caapi drink to others between their dancing. In a few minutes the hallucinogen took effect. 'The Indian turns deadly pale, trembles in every limb, and horror is in his aspect. Suddenly contrary symptoms succeed: he bursts into a perspiration, and seems possessed with reckless fury, seizes whatever arms are at hand... and rushes to the doorway, where he inflicts violent blows on the ground or the doorposts, calling out all the while, "Thus would I do to mine enemy (naming him by his name) were this he!" In about ten minutes the excitement has passed off, and the Indian grows calm, but appears exhausted.' Spruce fully intended to experiment on himself with the drug. He took one cup of the 'nauseous beverage'. But his hosts were too solicitous: a woman immediately appeared with a large calabash of caxiri manioc beer, 'of which I must needs take a copious draught... which [having seen it prepared with saliva to hasten fermentation] I gulped down with secret loathing.' No sooner had he done this than a huge lighted cigar, as thick and long as his forearm, was put into his hand. 'Etiquette demanded that I should take a few whiffs of it – *I* who had never in my life smoked a cigar or a pipe of tobacco.' Next came a large cup of pupunha palm wine. Unsurprisingly, Spruce wanted to vomit; but he just avoided this by lying in a hammock and drinking coffee. Brazilian traders told the Englishman that they had tried caapi. 'It made them feel alternations of cold and heat, fear and boldness. The sight is disturbed and visions pass rapidly before the eyes, wherein everything gorgeous and magnificent they have heard or read of seems combined; and presently the scene changes to things uncouth and horrible.' One friend told Spruce that after a full dose of caapi he saw a rapid panorama of all the marvels he had read in the *Arabian Nights*, 'but the final sensations and sights were horrible, as they always are.'

Some 1,000 kilometres (600 miles) southwest of these Tukano, the Andoa people of the Pastaza called the drug ayahuasca – a Quechua (Inca) word meaning 'dead man's vine', because their shamans used its trances to converse with the dead. Spruce learned that shamans drank it when they had to adjudicate in a dispute or answer an embassy, or to discover an enemy's plans, tell whether wives were faithful, or learn who had bewitched a sick man. All who took it 'at first feel vertigo; then as if they rose up into the air and were floating about. The Indians say they see beautiful lakes, woods laden with fruit, birds of brilliant plumage, etc. Soon the scene changes; they see savage beasts preparing to seize them, they can no longer hold themselves up, but fall to the ground.' When a participant awoke from his trance, he had to be held by force in his hammock or he would jump up, grab his weapons and attack the first person he encountered. He then became drowsy and fell asleep. 'Boys are not allowed to taste ayahuasca before they reach puberty, nor women at any age: precisely as on the Uaupés.'

Spruce made three important observations about ayahuasca/caapi. One was that some tribes used it extensively whereas others did not. For example, for all peoples of the Uaupés it was essential in their rituals, whereas nearby Baré and Baniwa of the upper Rio Negro did not touch it; but north of them it was used by some peoples of the Orinoco and Colombian llanos. Another discovery was that shamans in different tribes used the hallucinogen for distinct purposes. A third was that indigenous peoples did *not* use plants as curative medicines in the way that Europeans and Asians did. 'Although the medicine-man doses himself with powerful narcotics [to prepare for and perform his rituals], no drug whatever is administered to the patient.' Spruce saw people rubbing plant juices on swellings or using plants as plasters on wounds, but not swallowing them. An exception was powder from some tree barks that was taken as a febrifuge to reduce malarial fevers – which was to play a pivotal role in Spruce's later career.

After Spruce's time countless outsiders, from scientists to hippies, experimented with plant narcotics – and taking these is still a fringe type of Amazon tourism. The outstanding twentieth-century botanist to study these fascinating plants was the Harvard professor Richard Evans Schultes, as well as his brilliant pupils Timothy Plowman and Wade Davis. All recorded their extreme reactions and visions from the drugs. Schultes confirmed that shamans used ayahuasca/caapi 'for prophetic and divinatory purposes, a narcosis characterised, among other strange effects, by frighteningly realistic coloured visual hallucinations and a feeling of extreme and reckless bravery.' He gave full credit to Richard Spruce for the first 'precise determination of the botanical source of the drug..., a description of the species from living specimens... recorded with his wonted preciseness in 1852.' Spruce was also the first to understand its genus and name the species. Schultes declared, in the late twentieth century, that 'in modern phytochemistry it is undoubtedly Spruce's pioneer work on the malpighiaceous narcotics of South America that commands prime attention.... We find it necessary... for proper chemical studies... to study and re-evaluate Spruce's meticulous field observations.'

The other hallucinogen studied by Spruce was the snuff called paricá in Brazil and niopo (or yopo) in Venezuela. Spruce first saw it being used by an old Guahibo man at Maipurés on the middle Orinoco in 1854. The Guahibo roasted niopo seeds, then ground them into powder with a small pestle of bow-wood on an oval wooden platter with a handle – roughly the shape and size of a hand mirror. (Spruce purchased this apparatus from the Indian, and sent it to Kew Gardens where it is now in the Economic Botany collection.) The snuff was inhaled through Y-shaped tubes, made from leg bones of herons or other wading birds. The stem of this Y was in the snuff-box and the two arms inserted into both nostrils. The snuff was forcibly inhaled, which 'thoroughly narcoticised' the user if taken in sufficient quantity; 'but this endures only a few minutes, and is followed by a soothing influence which is more lasting.' The old man also had a piece of caapi narcotic, which he chewed while inhaling the niopo. He happily said to the Englishman, in broken Spanish: 'With a chew of caapi and a pinch of niopo, one feels so good! No hunger – no thirst – no tired!'

Half a century earlier, Humboldt had mentioned Guahibo and related Otomac of the Venezuelan/Colombian llanos taking niopo snuff, and there have been plenty of observations of other Indians such as the Mundurucu inhaling paricá (its Brazilian name). From Humboldt, Linnaeus and other sources, Spruce decided that the seeds came from an acacia-type tree of dry forests, which he called *Piptadenia peregrina* (now *Anadenanthera peregrina*). He had seen and collected this tree quite often on the lower Amazon and then on the Januária tributary of the Negro near Manaus. Spruce recognized a problem about this snuff. It seemed uncannily similar to powders inhaled by Yanomami and other peoples, but their snuff came not from *seeds* but from the *bark* of virola, a tree of the nutmeg family.*

A century after Spruce, his great admirer Richard Schultes solved this mystery. He had seen Indians of southern Colombia – including peoples of the upper Uaupés/Vaupés – using hallucinogenic snuff that they called *ya-kee*, derived from various species of *Virola* bark. Then the Yanomami were found to use this same snuff, which they called *epena*. Wade Davis explained that when Schultes had these hallucinogenic snuffs analyzed chemically he found that 'an astonishing 11 percent of it... was composed of a series of

* The Yanomami are the largest surviving indigenous nation living traditionally in rain forests. Some 35,000 of them still live in great *yano* huts, scattered throughout forested hills between Brazil and Venezuela. Spruce met one captive Yanomami (whom he called Guaharibo) on the Casiquiare, and he approached their territory when he explored the Pacimoni river; other travellers such as Schomburgk had fleeting glimpses of them, but the Yanomami were not extensively contacted until the mid-twentieth century.

potent tryptamines, including... arguably the most powerful hallucinogen known from nature. More curious than the strength of the snuff was the fact that the chemical constituents were both in kind and concentration almost identical to what had been found in *yopo* [niopo or paricá]. The source trees [*Virola* and *Anadenanthera*] were botanically unrelated. One drug derived from seeds, the other from inner bark. Yet somehow the Indians had recognized both plants and discovered ways to exploit their remarkable chemical properties.' The Kew botanist William Milliken and the French anthropologist Bruce Albert made an exhaustive study of Yanomami plant use. They found that these people knew all about the similarity observed by Spruce and explained by Schultes. The Yanomami went to great lengths to trade for seeds of *Anadenanthera peregrina* because they reckoned that its snuff was more powerful than theirs derived from *Virola* bark.

Hallucinogens divide very broadly into two types: those that induce fantastic baroque visions, and those that make the user feel that he is thinking with great clarity. The snuffs definitely have the second effect. Spruce reported that Indians inhaled niopo or paricá snuff 'to clear their vision and render them more alert' before embarking on a hunt. This author can confirm this. A Yanomami shaman blew a tube-full of epena up my nostrils and the effect was immediate: the wooden piers and the Indians themselves emerged from the gloom of the yano hut in sharp relief, and I was convinced that I could solve any problem.

Richard Spruce spent six months at and around Baños, 'a poor little place of about a thousand souls', at an altitude of some 1,800 metres (6,000 feet) and in agreeable Andean countryside. This was followed by two years of continuous explorations, during which his base was usually Ambato, a nearby larger town. When Wallace published Spruce's papers, he listed no less than *sixty* botanical excursions during those years. Some expeditions were from the town of Riobamba, others around Quito or near Ambato, as well as returns to Baños. All were on horseback, except when Spruce was climbing a volcanic mountain or returning to the cloud forest. Wallace wrote: 'His explorations covered a large extent of the surrounding mountains and forests, and he was often away from Ambato for weeks or months at a time.'

These years of glorious collecting were by a man who was now one of the world's finest and most experienced botanists. In addition to his encyclopaedic knowledge and indefatigable collecting, Spruce was famous for his orderliness and thoroughness. Even in the most primitive surroundings he kept his collecting materials, books, microscope, dried plants and stores of food and clothing as tidy as possible. Sir Joseph Hooker, who succeeded his father Sir William as Director of the Royal Botanic Gardens, Kew, recalled the excitement of the arrival of a consignment from Spruce, 'marvelling at the extraordinary fine condition of the specimens, their completeness for description, and the great fullness and value of the information regarding them inscribed on the tickets.' This was echoed by another great contemporary botanist, Daniel Oliver, who wrote that 'Mr. Spruce's specimens were most carefully collected, dried, and packed, extraordinarily so considering the difficulties of all kinds with which he had to contend.' Beautifully legible labels give precise locality, habitat and other pertinent information. Spruce's mentor and agent, George Bentham, praised not only the number of specimens – over 7,000 – that he collected, but also 'the number of new generic forms with which he has enriched science; for his investigation into the economic uses of plants...; for several doubtful questions of origin... of genera and species which his discoveries have cleared up; and for the number and scientific value of his observations made on the spot, attached to the specimens preserved.'

In mid-1859 there was a momentous change in Spruce's life. He wrote: 'I was entrusted by Her Majesty's Secretary of State for India with a commission to procure seeds and plants of the Red Bark tree, and I proceeded to take the necessary steps for entering on its performance.'

The 'red bark tree' that Spruce was asked to collect referred to a variety of *Cinchona* tree that contained the highest content of the malaria palliative quinine. Indigenous peoples of the region where Spruce was now active – southeastern Ecuador – had always known that bark from this tree was a miraculous febrifuge. When ground to powder and drunk in water, it did not totally cure malaria but it very greatly reduced its fever. In the early seventeenth century these Indians treated their malaria-stricken Spanish *corregidor* (magistrate) with this bark, which they called *quinquina*. A few years later in 1638 the wife of the Viceroy of Peru, the Condesa de Chinchón, caught malaria, and Jesuits rushed some of the bark to Lima to relieve her fever. There is some doubt about this romantic story, but it later inspired the father of taxonomy Linnaeus to name the tree *Cinchona* (he carelessly omitted the first 'h' of Chinchón). The Jesuits were smart businessmen, and they developed a lucrative trade in what became known as 'Countess's powder' or 'Jesuits' bark'.

Later in that century there was much malaria in Europe. It was particularly bad in Italy's Pontine Marshes and in swampy parts of Essex in England. No one made the connection to the mosquitoes that bred in such marshy areas, but it *was* known that quinine reduced malarial fever. Oliver Cromwell caught malaria in 1658, but he absolutely refused to touch 'the powder of the devil' because it had come from Jesuits; so the Lord Protector may have died of the disease. Ten years on, King Charles II also caught malaria; but he insisted that he be given the bark to cure his fever – despite the remedy's papist name. A few years later Sir John Evelyn wrote in his diary that he had visited the world's first heated greenhouse, in London's Chelsea Physic Garden, and saw among many rare annuals 'the Tree bearing the Jesuits bark' which had effected such remarkable cures of tertian fevers ('tertian' meant those that recurred every three days). How this tree came to be propagated in London is a mystery, because it was very difficult to move living plants or seeds around the world in sailing ships, and Jesuit missionaries kept a tight grip on their quinine monopoly. Also, Spanish authorities rigorously excluded all foreigners from their South American colonies.

The monopoly and export embargo were still in force two centuries later. So, although quinine was taken to allay malarial fevers throughout the eighteenth century, it was very expensive. Malaria, or 'swamp fever', struck George Washington's army in the War of Independence, and was to devastate the Union Army in the American Civil War. By the mid-nineteenth century malaria was a scourge in colonies of equatorial Africa and Asia. Quinine from cinchona bark was the only plant pharmaceutical from South America that was genuinely effective – others such as sarsaparilla or balsam (the combustible oil that set fire to Wallace's ship in 1852) were medicinally useless. So there were various attempts to remove cinchona trees and grow them in plantations outside their native Andes. For a variety of reasons, all these efforts failed.

The man who instigated the letter to Richard Spruce was another Yorkshireman, Clements Markham. Twelve years younger than Spruce, Markham was rather grander – his grandfather was Archbishop of York and his father Canon of St. George's, Windsor. Markham first visited Peru as a naval midshipman and was thrilled by the country and its Inca heritage. When his naval ship was trapped in the ice on a later Arctic voyage, he read William Prescott's recently published *History of the Conquest of Peru* and this fuelled his lifelong fascination with that country. So Markham left the Royal Navy, went to Boston to meet Prescott, and then spent ten months of 1853 travelling deep into Peru. This took him down into the Peruvian Amazon rain forests, where a priest at a place called Calisaya showed him another species of cinchona that also cured malarial fever. Back in England,

Markham wrote two books: one about his travels and the other a history of Peru and the Incas. He got a job in the civil service and soon transferred to the department handling dispatches from India and other British colonies. These reports impressed on Markham the terrible suffering from malaria in both India and tropical Africa, and the huge cost and difficulty of buying quinine derived from cinchona bark.

Various authorities, including Sir William Hooker of Kew Gardens, appreciated the urgency of getting cinchona trees out of their Andean habitat and trying to grow them in plantations in India. An official in the India Office suggested to Markham, still in his twenties, that he might lead an expedition to gather live specimens of trees from Peru. He took up the cause with his habitual energy and persuaded the authorities that his familiarity with Peru and its Spanish and Quechua languages outweighed his total lack of botanical skill. He gave himself a crash course in cinchona trees. He searched historical records of scientific travellers who had written about the subject – Charles-Marie de La Condamine from France, Hipólito Ruiz and José Pavón from Spain, and Justus Hasskarl from Holland. The Royal Geographical Society still has his personal notebooks from this time, in which he noted each type of this tree genus, with sketches of its identifying traits, details of preferred climate and soils, fever-reduction qualities, and botanical and vernacular names. He saw his mission as making quinine cheaply available not only to soldiers and explorers but also in every bazaar in India, so that 'countless multitudes will be saved from death or grievous suffering'. In April 1859 Markham's efforts succeeded in his being commissioned by the Secretary of State for India to organize expeditions to obtain the precious plants.

Markham suggested that there should be three ventures: Markham himself was to get yellow-bark *Cinchona calisaya* trees from southern Peru and Bolivia; a distinguished botanist, G. J. Pritchett, was to seek a 'grey-bark' variety (*C. nitida, C. micrantha* and *C. peruviana*) in north-central Peru; and the third mission was to Ecuador for 'red-bark' *C. succirubra* (now *C. pubescens*) as well as *C. condaminea* (now called *C. officinalis*). Hooker and Bentham enthusiastically named Richard Spruce as the finest British botanist active in South America and noted that, providentially, he was already in the Ecuadorian Andes. So the Secretary of State for India sent Spruce this third assignment.

Markham almost succeeded in his venture to Peru in 1859–1860. In a remarkably tough expedition he hacked down to the Tambopata river in the rain forests of south-eastern Peru, collected several hundred rare *Cinchona calisaya* seedlings, got them to the coast despite determined efforts of local authorities to thwart his robbery of their national treasure, and then circumvented attempts by customs officials to block the export of this special cargo. He loaded over 450 of them onto a steamer, crossed the Isthmus of Panama and reached England with his seedlings growing in glass-sided Wardian cases. Some 270 were still alive when his ship reached Southampton. But, instead of nurturing these at Kew, Markham immediately set off for India with them. A series of blunders and bad luck on the journey killed many saplings and, although a few were planted in a government plantation in the Nilgiri hills in southern India, none survived. Meanwhile, the Peruvians saw what Pritchett was up to in central Peru, and expelled him. Thus, the cinchona mission now depended entirely on Richard Spruce.

Markham had never met Spruce, who had by now been in South America for a decade. He left the letter of invitation, on India Office paper, with the British consul at Guayaquil when his ship stopped at that port. The consul conveyed it to the botanist in the sierra, and Spruce accepted. A later letter from the India Office, of November 1859, spelled out the contract: a salary of £30 a month, plus expenses of up to £500 to be drawn from the British consul. But he was to be frugal: 'You will endeavour to perform the services with due attention to economy.'

Spruce set about his task with characteristic thoroughness. He had already spent two years in the Andes of Ecuador, so knew his way around and had acquired several influential friends. He was, according to Wallace, 'tall and dark, with fine features of a somewhat southern type, courteous and dignified in manner, but with a fund of quiet humour which rendered him a most delightful companion.... In all his words and ways he was a true gentleman.' He got on well with most Ecuadorians of every social class.

Although constantly on the move in scores of collecting ventures, his base was Ambato, an agreeable town on the main intramontane road from Quito south towards Guayaquil. In Ambato his host was one Manuel Santander and he had gained the affection of the entire Santander family. In the following decade, after Spruce had returned to England, Santander wrote to his 'never forgotten friend' about botanical matters, but added how one of his daughters saluted him 'most affectionately... with many caresses' and another had named her son 'Ricardo' after him. This good impression on local people stood him in good stead in the quest for cinchona.

Spruce explained to his friend John Teasdale: I wanted 'to make myself acquainted with the different sorts of barks, and to ascertain what facilities existed for procuring their seeds, etc., or, more properly speaking, what difficulties had to be overcome, and I assure you they are not slight ones.' One hazard at this time was a possible war against Peru; and then a civil war erupted between Ecuador's highlanders and those from coastal Guayaquil. The armies in these conflicts press-ganged every available man, commandeered horses and mules and, whenever victorious, ransacked captured towns. As an Englishman, Spruce was theoretically safe, but he had to keep constant watch on his belongings, and supplies became scarce and expensive.

Spruce contemplated seeking red-bark cinchona on the western flank of the mighty snow-capped volcano Chimborazo. But he decided against this, because in August 1859 it would be snowing incessantly with winds so violent that they could hurl both a horse and its rider over a precipice. So he went instead on a two-week ride southwards along the main road to Riobamba and Alausi, and then westwards down slippery trails and up to the flanks of Mount Asuay. He had five horses and mules: one ridden by him and one by his servant, with the others 'laden with my baggage, consisting of drying paper, clothing and bedding, and a copious supply of tea, coffee, and sugar – articles rarely to be met with in a countryside where there are no inns' and where the *campesinos* drink only *aguardiente* brandy and sour *chicha* maize beer. During this journey as on all others over the previous two years, Spruce was constantly botanizing. He described the flora in detail to his mentors Bentham and Hooker. At one point he climbed to the *páramo* (high-altitude moor) of Teocajas. 'Anything more desolate than this paramo I have nowhere seen. It is one great desert of movable sand, in which the distant patches of Cacti, Hedyotis, and a succulent Composita only render its nakedness more apparent.... It may be imagined how cheerless was a slow ride of nearly 20 miles [30 kilometres] over such a waste, rendered all the more gloomy by a leaden sky overhead, and a piercing wind which came laden with mist and fine sand.'

Near his destination, Spruce was 'almost in despair' at the difficulty of getting alfalfa or hay for his animals and food and accommodation for himself. But he had some luck. He descended to a lowland sugar plantation and its owner, Señor José León, immediately invited him to stay in his comfortable hacienda. Spruce was then told to seek a man called Bermeo, who was an experienced cinchona-bark collector. 'I at once secured his services and, as he turned out an honest, active fellow, I took him with me in all my subsequent excursions in the district' throughout the month of August 1859. They had an arduous trip, pushing along trails too overgrown for riding, rigging up a makeshift bridge over a stream called Pumacocha, finding a place to swim their horses across, and building a shelter of palm fronds.

The quest was at first painfully disappointing, because profligate quinine collectors had killed all the cinchona trees in the area. Bark from the lower trunk was considered the most valuable, so they stripped all of this. Spruce plunged deep into forests 'meeting every few minutes with prostrate naked trunks of the cinchona, but with none standing.' Bermeo climbed into the forest canopy, but could never see the large red leaves of a living cinchona. Weary and despairing of ever seeing a living red-bark, they came across 'a prostrate tree, from the root of which a slender shoot, 20 feet [6 metres] high, was growing. My satisfaction may well be conceived....' He learned that the red-bark cinchona grew best on stony slopes where there was humus, at an altitude of 900 to 1,500 metres (3,000 to 5,000 feet), in weather as warm as London in summer, but with almost continuous rain for five months a year from January to May. Bermeo also told him that there was more cinchona in another forest – but their provisions would not allow a visit. In the following weeks, Bermeo did his best to gather seeds and plants, but he was hampered by fear of being kidnapped and press-ganged by soldiers. Spruce studied and reported on all other vegetation of these Pacific flanks of the Andes, collecting fine specimens and adding species new to science; but he found that this region's biodiversity was far less than on the eastern, Amazonian slopes.

Spruce was trapped for some weeks by the passage of another army. He then moved to higher forest, seeking the *serrana* variety of cinchona, and had to stay in a wretched dark, smoky hut too low for him to stand (Spruce was far taller than most Ecuadorians) and with violent wind through chinks in its walls and roof. Here he studied other varieties of cinchona bark: one called *cuchicara* ('pig-skin') because it resembled pork crackling, another known as *pata de gallinazo* ('turkey buzzard's foot'). There are in fact forty species of the evergreen cinchona trees. They are part of the huge family of madders, Rubiaceae, which includes some 10,000 species in 630 genera all over the world. Cinchona trees have small fragrant blooms, either yellow, white or pink, hanging in clusters at the end of branches. Their fruits are small oblong capsules containing many small winged seeds. What matters is of course the bark, which contains four alkaloid chemicals, including quinine. But the wonder drug quinine is found in commercial concentration in only four species of cinchona.

In October 1859 Spruce made his way back north to Riobamba and his base at Ambato. The remainder of that year was occupied in more botanical excursions, interrupted by the defeat of the highland army in the civil war and the arrival of victorious looters from the lowlands. He decided to spend all the year 1860 working 'to fulfil my commission from the Indian Government to procure seed and plants of Red Bark.' The early months were devoted to further intensive research into cinchona. He knew that there were many varieties of the genus, and the bark collectors told him that there were six species of red bark tree (so-called because its colourless sap rapidly turned red when exposed), but 'what we call the *true* Red Bark is now known in Ecuador only by the name "Cascarilla roja"'.

On the western flanks of mount Chimborazo the trees flourished below high-altitude páramo and a dense belt of forests, and above the woodland of the plain of Guayaquil. Much of this region was owned by an ex-president of Ecuador, General Flores (who was embroiled in the civil war); but he had leased three farms with the best cinchona to a Señor Cordovez of Ambato. Two other good areas belonged to the church of Guaranda (a town 55 kilometres (35 miles) southwest of Ambato) which had leased them to a Dr Francisco Neyra. Spruce had to seek permission from these two gentlemen to take the seeds and plants he wanted. 'At first they were unwilling to grant me it at any price.' But Spruce was well known and liked, and when it came to botany he was passionate, patient and persistent. So 'after a good deal of parley, I succeeded in making a treaty with them, whereby, on payment of 400 dollars [in local currency], I was allowed to take as many seeds and plants as I liked, so long as I did not touch the bark.'

Cordovez and Neyra also agreed to give Spruce essential help in procuring workmen and beasts of burden. Neyra in particular did everything he could to assist Spruce's enterprise. He told all his *cascarillos* that when they collected bark for him they should also seek seeds and plants for the English collector. Neyra suggested that a place called El Limón would be the best centre for operations. This had once been home to the finest swathe of red-bark cinchona 'ever seen'. The trees had all been felled years before, but many roots had sprouted shoots that were by now stout little trees, large enough to bear blossoms and fruit. A few local people had learned to husband these trees because their bark was valuable – even though they did not know that it was used for quinine to lower malarial fevers. In addition to his agreement with Cordovez and Neyra, Spruce wrote to Hooker that he was 'also in treaty with the owners of the woods near Loja which produce the *Cinchona condaminea*'. Loja was a fifteen-day journey to the south, and the two species of cinchona blossomed simultaneously; so Spruce had to arrange for others to collect for him in those southern woods.

Spruce pressed ahead with his preparations, gathering a team of bark-collectors, pack animals and supplies, and enlisting a very competent local friend, Dr James Taylor of the town of Riobamba, to accompany him.* Spruce wrote that Taylor was well educated (he remembered much Greek) and that 'he is a very kind-hearted, honourable man, which can't be said of many Englishmen I have met in South America'. Another letter from Clements Markham reached Spruce in May 1860 to say that a skilled Kew gardener

* Taylor, from northern England like Spruce himself, had been in Ecuador for thirty years, first as a lecturer in anatomy at the University of Quito, then medical attendant to President Flores. He had settled in Cuenca and then in Riobamba, and had recently married the widow of one of Simón Bolívar's generals.

called Robert Cross was on his way to assist. This was to prove an inspired move by Markham, because although Spruce was a superb botanist he had no experience of gardening. Markham also had the foresight to leave some Wardian cases (in which plants could grow during transport) for Spruce at Guayaquil. Taylor took the horses down to the lowlands to meet Cross and guide him up to Spruce, but the gardener had fallen ill, and then encountered many obstacles and delays from the civil war, so he did not reach the others until late July.

The civil war was a constant threat and impediment. Once, four raw recruits threatened Spruce with lances. He contemplated drawing his pistol, but decided that a bottle of aguardiente brandy would be more effective – which it was. The marching armies worked their pack animals so hard that many were abandoned. Hundreds of wretched mules and horses died from infected saddle sores, and the stench of their corpses was unbearable. Worse was the abuse of Indians. Spruce once found his men bound ready to be taken off as forced porters, as were their mules – a campesino's most prized possession. Luckily, the Englishmen had done a favour for the colonel, so persuaded him to release the men and mules. On another occasion it was Taylor who saved Spruce's men and animals from press-gangs. Spruce was furious at the abuse of the poor Indians, who were robbed of their livelihood, goods and freedom. Although he got on well with all classes, he particularly identified with the underdogs, and 'it will not be wondered at that I cursed in my heart all revolutions.'

Richard Spruce's health was always precarious. As a young man in Yorkshire he had suffered severe coughs and colds, 'congestion of the brain', and painful gallstones. In South America he often mentioned his frail health and, as we saw, he almost died of malaria on the upper Orinoco in 1854. Yet, between these ailments, Spruce was constantly collecting and plunging out on tough explorations. His simple diet, spartan way of life and regular exercise kept him reasonably fit. But in March 1858, soon after climbing into the Andes, he was 'confined to bed for four days with fever and rheumatic pains from head to foot', and for half of June he was bedridden in Ambato with fever. Worse was to come two years later. During an exploratory venture from Riobamba in early April 1860, Spruce was 'struck deaf in the left ear'. He went to Baños on the Amazonian foothills 'to bathe in hot springs for my deafness'. But on 29 April came 'THE BREAKDOWN. Woke up this morning paralysed in my back and legs. From that day forth, I was never able to sit straight up, or to walk about without great pain and discomfort, soon passing to mortal exhaustion.' He spent the month of May under medical treatment at Ambato, but with very little improvement to his health. In his official report, Spruce wrote that he was disabled by a severe attack of rheumatism: he later described it as 'a rheumatic and nervous affection, almost amounting to paralysis'. This sudden blow – which changed Spruce's life for ever – has never been explained. It cannot have been an insect-transmitted tropical disease, because he was up in healthy Andean valleys. It might have been very rare Guillain-Barré syndrome, a disease that affects the nervous system with ascending paralysis, often triggered by intestinal infection. In desperation, Spruce contemplated leaving the collection of cinchona seeds to Taylor; but the doctor hoped that his friend might recover the use of his limbs down in the warmer woods. So the expedition finally started for the bark forests of Chimborazo on 12 June. Sadly for Spruce, he found that 'fogs and damp' negated the warmth of the woods. But he soldiered on, determined 'to execute to the best of my ability the task I had undertaken, [although] I was too often in that state of prostration when to lie down quietly and die would have seemed a relief.'

The cinchona expedition reached El Limón on 18 June 1860 and immediately set to work. For three months Spruce superintended 'the work of getting plants and seeds of *Cinchona succirubra*. The seeds were all gathered under my eye, and were dried, sorted, and packed by myself. Partly on horseback and partly dragging myself about on foot by the aid of a long staff, I explored pretty thoroughly the neighbourhood...' This was followed by a further sixteen days' collecting in the nearby valley of Las Tablas.

There were many problems in gathering the precious seeds and seedlings. The worst was that in recent years dealers had come to Limón and purchased every red-bark tree for a dollar each, taking the bark and leaving the stripped tree for firewood – which was what local people wanted. 'Every ridge and valley for a great distance around was searched [by the dealers], and the destruction of the Bark went on unsparingly.' Thus, although in the two valleys Spruce saw some 200 red-bark trees standing, only two or three were undisturbed saplings. All the rest were shoots, growing out of old tree stumps 1.2 to 1.5 metres (4 or 5 feet) in circumference. This destruction distressed Spruce because '*Cinchona succirubra* is a very handsome tree, and in looking out over the forest I could never see any other tree at all compared to it for beauty.' The only shoot that had grown into a full tree was 15 metres (50 feet) high. It branched at a third of its height and its crown was a symmetrical paraboloid of dense foliage. Its large oval leaves were shining deep green, mixed with blood-red decaying ones. Another problem for Spruce was that there were plenty of large *Cinchona magnifolia* trees, but these had no quinine and were useless to him. A third problem was that when the local people heard that he wanted to buy seeds, they started stripping these from every panicle before they were ripe. He rapidly went to tell everyone that he would pay only for material collected by him or Taylor.

The English gardener Robert Cross finally arrived at the end of July, pale and thin from his illness, but eager to set to work. He brought fifteen Wardian cases that Markham had left in Guayaquil in the previous December. Cross and Spruce decided to try to propagate their own *Cinchona succirubra* from cuttings. They fenced a piece of ground, dug a pit, filled it with rich soil, and in a few days Cross had planted over a thousand cuttings. 'He afterwards put in a great many more, subjecting them to various modes of treatment; and he went round to all the old stools [stumps] and put in as many layers [shoots] from them as possible.' But Spruce commented that 'only those who have attempted to do anything in the forest' could imagine how difficult this was. They lacked glass to make more Wardian cases, they had no canvas or other materials, myriad insects were a constant threat, the civil war fighting and rough roads made it impossible to get what they wanted from Guayaquil, and whenever there was full sunshine the seedlings had to be heavily watered by 'all the hands' rushing down to a glen and up a steep slope with heavy buckets. After a few weeks the cuttings began to root, but 'they were attacked by caterpillars, which also had to be combated'. Characteristically, Spruce gave full credit to the gardener Cross for surmounting these and other obstacles with 'unremitting watchfulness'.

Meanwhile, by mid-August 1860 Spruce noticed that capsules on some of the cinchona saplings seemed to have ripened. 'An Indian climbed the tree, and breaking the panicles gently off, let them fall on sheets spread on the ground' so that every seed was caught. The capsules were then left to dry for up to ten days, and by early September all the seeds had been gathered and dried. Cross sowed eight of them, and the Englishmen watched delightedly as these began to germinate, push out radicles and develop seed-leaves. One died, but they had 'seven healthy little plants' to dispatch along with the rooted cuttings – and they had proved that properly dried seeds did not lose their vitality, despite their 'excessive delicacy'. By late September Spruce was proud that 'I had now gathered about 2,500 well-grown capsules... namely, 2,000 from ten trees at Limon, and 500 from five trees at San Antonio. Good capsules contain 40 seeds each... so that I calculated I had... at least 100,000 well-ripened and well-dried seeds.' In the civil war, General Flores had won a battle to control Guayaquil. So Spruce was able to take his well-dried seeds down to the port, and on 14 October he packed them on a steamer to Panama, en route to Jamaica and then Kew Gardens in England.

There was still the problem of sending the growing seedlings. He returned, up the rivers and trails, and bargained for a raft and crew to take the plants to the port. All this cost $90. The raft consisted of twelve balsa-wood logs bound with bignonia lianas, over 18 metres (60 feet) long by 3.5 metres (12 feet) wide, on which there was a bamboo deck area, 3 by 11 metres (10 by 36 feet), fenced around and with a thatched roof [PL. 62]. This was exactly what they needed for their Wardian cases, with room for sleeping and cooking. Spruce supervised preparing the cases. The gardener Cross arrived with his young plants, delayed because of difficulty getting oxen and mules, making cylindrical baskets for the plants, and tying each in wet moss and fixing it in the basket – 'delicate operations which Mr. Cross could trust to no hands but his own'. They then had two men gather earth, sand and leaf compost, packed the Wardian cases with this mixture, and planted 637 saplings into them.

After days of heavy rain, they finally set off down the rushing Ventanas river, with three raftsmen to paddle and punt their unwieldy craft. They failed at one sharp turn: 'The raft went dead on, and through a mass of branches and twiners that hung over the middle of the river. The effect was tremendous: the heavy cases were hoisted up and dashed against each other, the roof of our cabin smashed in, and the old pilot was for some moments so completely involved in the branches and the wreck of the roof, that I expected nothing but that he had been carried away; he held on, however, and at last emerged, panting and perspiring, but with no further injury than a smart flogging from the twigs, which indeed none of us escaped.' There were 'many other perils' smashing into the bushy banks. One moonlit night was spent clearing a mess in which 'the leaves of the precious plants were sorely maltreated.' Cross often lent his strong arms to the stern oar, and thanks to his tender care the plants were all in good shape when they reached Guayaquil on 27 December. A carpenter was found to fix glass into the cases and nail them for the ocean voyage. There were problems over steamer sailings, but Spruce finally saw his eighteen months' work sail off 'commodiously arranged on the poop deck' and in the very competent care of Cross. It was a proud moment. In his official report, Spruce recalled how during the cinchona operation 'all engaged in it had run frequent risk of life and limb; but a far greater source of anxiety to me were the contretemps... [that] threatened to bring our work to naught' in a land devoid of roads, peace and plenty.

Richard Spruce was probably the only botanist who could have succeeded in the daunting task of collecting and dispatching seeds and saplings of *Cinchona succirubra*. He declared: 'Most fervently do I desire that the experiment of forming plantations of Cinchonas in the East Indian possessions of the British Crown may be successful.' His wish was realized. He was congratulated by the India Office, in a letter dated 29 January

1862: 'The Secretary of State for India... Sir Charles Cross... expresses his sense of the important and useful service you have rendered; and the zeal and ability you have displayed in collecting so large an amount of scientific information, which will be of most essential service to the cultivators of the plants in India, and which could not have been procured elsewhere.' His friend Wallace later confirmed that Spruce's 'long-continued labour and extreme care were crowned with success. The young plants reached India in good condition, and the seeds germinated and served as the starting-point of extensive plantations.' The cinchona trees were planted in the Nilgiri hills of northern Kerala, at Darjeeling near Sikkim in northern India, and in central Ceylon (Sri Lanka). These locations were at the right altitude, but none had the regular year-round rain or cool mists of Ecuador. Despite this, the Indian plantations did yield enough quinine to alleviate the fever of thousands of malaria victims, and they saved lives.

Having succeeded so brilliantly in sending cinchona to England and India, Spruce could justifiably have ended his eleven years in South America; but he did not. For the first half of 1861 he spent the rainy season at Daule, a village north of Guayaquil, collecting a little during breaks in the weather. Despite 'unfortunately having very little strength left for work of any kind', and severely scalding a foot which confined him to his hammock for eighteen days, he wrote his cinchona report and continued enthusiastic botanizing of the local flora. Then in August 1861, 'being very sick', he went further up the Daule river to spend seven months at Chonana, the farm of a General Illingworth whose son-in-law was his physician. Of course, no one could diagnose Spruce's devastating and mysterious ailment, let alone cure it.

On 11 October 1861 Spruce suffered another disaster. The reputable trading house of Gutierrez & Co. of Guayaquil failed and suspended payments, so that Spruce lost nearly £1,000 lodged with it. 'The blow was so sudden that I had no time to withdraw my property.... Even Gutierrez himself did not comprehend how it had happened.... [My savings] remain to share the fate of the other debts of the firm, and if I ultimately recover a thousand dollars of it I shall think myself well off.' It emerged that the bankruptcy was caused by embezzlement of some $360,000 and the theft of great quantities of cacao and other produce from the company's warehouses. This had been perpetrated by its Ecuadorian cashier and English head bookkeeper, acting in collusion and using a trail of false accounting, forged invoices and bogus shipments. The English bookkeeper, Thomas Clarke, tried to escape by ship: Gutierrez boarded it and at least recovered a wad of stolen cash, but he was unable to have the robber arrested. The cashier, Icaza, was of a prominent local family, so he 'walks about Guayaquil holding his head as high as ever' knowing that should Gutierrez try to charge him he himself risked being stabbed in a dark street one night. Spruce spent the last three months of 1862 in Guayaquil 'trying to redeem some of my lost property'. But it was in vain. He had lost most of his savings from twelve years of superb but punishing botanical collecting, including the removal of the world's greatest plant-based pharmaceutical.

It was suggested that a dry, hot climate might improve Spruce's acute 'rheumatism'. So he spent seven months of 1862 on the arid coast. Most of the Pacific shores of Ecuador are covered in lowland tropical rain forest. But just south of Guayaquil there is an abrupt biogeographical boundary, with dry deserts stretching south along the entire coast of Peru and northern Chile. We now know that this desert is caused by the cold Humboldt Current bringing Antarctic waters to Peru's Pacific shore. This cold ocean means that any rain clouds from Amazonia that cross the Andes are swept out to sea. However, at irregular intervals warm currents slosh across the Pacific Ocean, strike the coast of Ecuador and flow south to cover the cold Humboldt waters. Since this event usually occurs at Christmas, fishermen of northern Peru call it 'El Niño' after the infant Christ. The effect can be devastating, both for fishing (fish stocks plunge with the cold water) and because heavy rains destroy flimsy structures on the usually bone-dry coastal desert. It was Spruce's misfortune to move to the arid coast in an El Niño year, so there were unexpectedly heavy rains – the first in seventeen years. However, although his health worsened, 'with great toil I managed to collect and preserve specimens of everything' of the vegetation that bloomed on the desert and in new ephemeral lakes. It is one of nature's great spectacles when rain causes a bare desert suddenly to be 'clad with a beautiful carpet of grasses, of many different species, over which were scattered abundance of gay flowering plants.'

At the end of 1862 Spruce wrote despairingly that 'after loss of health came wreck of fortune' even though his wants and aspirations were so modest. He asked a friend to consider 'how entirely disabled I am; how rarely I can sit to a table to do anything, but must write, eat, etc., in my hammock; how I cannot walk except for short distances, nor ride on horseback without being in danger of falling from an arm or a leg suddenly turning stiff. I had never calculated on losing the use of my limbs....'

Having failed to recover his savings or his health, Spruce decided that he must get to a genuine desert environment to escape the rains and vapours that he thought were afflicting him. So at the start of 1863 he took a steamer for some 350 kilometres (220 miles) to the port of Paita and nearby town Piura, at the edge of the intensely hot and dry Sechura desert of northern Peru. He spent sixteen months in and around Piura and the low Amotape hills to the north. He was the first major botanist to study the carob *Prosopis* trees, wild manioc, candelabra cacti, willows and endemic vegetation of these unique dry forests – some of which are nowadays protected in the Amotape national park. Spruce wrote an eighty-page report on the people, plants and geography of the rarely visited valleys of Piura and Chira, which the British Foreign Office published as a booklet.

Richard Spruce finally decided to return to England. On 1 May 1864 he boarded the Pacific mail steamer at Paita, reached Panama after five days, crossed the Isthmus, and arrived in Southampton at the end of the month. Before embarking for home, he wrote that although his pains were slightly lessened, for the past few years life had been 'barely tolerable a burthen… [so] I see plainly I can never hope to regain my former activity, or indeed be able to undertake any occupation whatever'. He had been away for fifteen years, the prime of his adult life, from age thirty-one to forty-seven.

12

LATER LIVES: THE AMAZON LEGACY

When Alfred Russel Wallace reached London in October 1852, after the shipwreck and weeks in an open boat, his only clothing was 'a suit of the thinnest calico'. But he was met by his friend and agent, the admirable Samuel Stevens, with the good news that he had had the foresight to insure Wallace's collection. So Stevens took him to buy a warm ready-made suit and be measured for other clothes at Stevens's own tailor, as well as to a haberdashers for 'other necessaries'. (The insurance payment, of £200, came quickly because the shipwreck was well known.) A week staying with Stevens's mother in south London, enjoying her excellent cooking, restored Wallace to his 'usual health and vigour'. He then rented a house, near the Zoo in Regent's Park and not far from the British Museum (which then housed what is now the Natural History Museum collection). He moved his own mother, his sister Fanny and her photographer husband Thomas Sims in with him, and by Christmas they were all comfortably settled in.

Wallace spent the first half of 1853 writing, and starting to know London's scientific establishment. His first two books, *Palm Trees of the Amazon and Their Uses* and *A Narrative of Travels on the Amazon and Rio Negro*, were both written at great speed and published in that year. Neither book had much success. *Palm Trees* was published by John Van Voorst in London with a print-run of only 250 copies. Wallace himself had to meet most of the costs. His largest payment was to the leading botanical artist, Walter Fitch of Kew, to make lithographs of the trees, from pencil drawings that were in the tin box that Wallace grabbed from his cabin in the burning ship. (Wallace was unhappy about one illustration showing people in front of a great Tukano maloca: 'I am sorry to say, [these figures] are as unlike the natives as are the inhabitants of a London slum.') The book's few sales just covered its publishing costs. Sir William Hooker, the magisterial director of the Royal Botanic Gardens at Kew, wrote a critical review. He felt that this small volume was 'more suited to the drawing room table than to the library of a botanist'. He particularly noted that Wallace talked at length about the use of piaçaba bristles in making brooms and brushes, but had failed to collect any fruit or blossoms from this palm during the two years he had lived in its heartland. Wallace had sent a copy of the book to his friend Richard Spruce, who was then far up the Rio Negro. Hooker asked Spruce's opinion; and when this reached him, he rather unfairly published it as if it were a formal review. In his letter, Spruce praised the book's drawings as 'very pretty' and a few as botanically successful. He liked the descriptions of economic uses of palm trees. But other aspects of the book appalled the meticulous expert, particularly the artist's rendering of palms. 'The most striking fault of nearly all the fig[ure]s of the larger species is that the stem is much too thick compared with the length of the fronds, and that the latter has only half as many pinnae as they ought to have. The descriptions are worse than nothing, in many cases not mentioning a single circumstance that a botanist would most desire to know.'

Wallace's *Narrative of Travels* was published equally modestly. He found a small publisher called Reeve and Co. and took no payment beyond a promise of half any future profit. In the event there was none. Only 750 copies of the book were printed, and when Wallace returned from Asia a decade later a mere 500 of these had been sold.

Apart from the print run, there was nothing modest about the 542-page book itself. Its full title was *A Narrative of Travels on the Amazon and Rio Negro, With an Account of the Native Tribes, and Observations on the Climate, Geology, and Natural History of the Amazon Valley*. It had a tinted frontispiece of the chapel at Nazaré in Belém, a map of northern South America, a fold-out chart of vocabularies (in English and eleven other languages), and was enhanced by Wallace's drawings of some rock engravings, Tariana artifacts, granite rock formations and other physical features along his travels.

Darwin and others thought that the book contained too few scientific data. This was unfair, since it was clearly intended for general readers for most of whom it was a first introduction to the exotic world of the Amazon. The main text was a delightful account of Wallace's remarkable adventures, full of colourful information about people, customs and towns, and anecdotes and descriptions of plants, insects, animals and fishes. This culminated in a thrilling account of the ship *Helen* catching fire and sinking. There followed audacious chapters on physical geography and geology, climate, vegetation (particularly palm trees, and other commercial plants including rubber) and zoology (with detailed information about monkeys, bats, jaguars, caimans, turtles and so forth). All this was impressive for someone without formal qualifications in any of those disciplines. A valuable final chapter was 'On the Aborigines of the Amazon'. This was the first anthropological account of the peoples of the Uaupés, with a list of sixty-five articles that they manufactured, followed by the appendix on vocabularies.

There are of course plenty of errors in this book, partly because so much about Amazonia was unknown or wrongly interpreted in the mid-nineteenth century. But *Narrative of Travels* is a lively travel book, a wonderful read and a fine achievement for a young man just turned thirty, writing at speed and having lost most of his notes, journals and collections. It is totally truthful, with no exaggeration, sometimes humorous, and largely devoid of the prejudices one would expect of a young Englishman of that period. It deserved to sell far better than it did.

In his chapter on geography, Wallace correctly identified the four superlatives about the Amazon: that its basin surpasses that of any other river in the world; it has the most abundant rainfall, so that it is 'entirely covered by dense virgin forests'; its 'richness of vegetable productions... is unequalled on the globe'; and its body of fresh water is far greater than that of any other river. (In modern knowledge and parlance, the Amazon and adjacent areas have about 60 per cent of the world's tropical rain forests; these contain the richest and most diverse ecosystem; and it discharges 20 per cent of all the water flowing into the oceans from rivers.) Wallace knew and revered books by the distinguished German scientists who had written about the Amazon – Humboldt who was there in 1800 and Spix and Martius in 1820. But he boldly used his own observations, either to corroborate or disagree with their guesses about such matters as the reasons for black- and white-water rivers, the Amazon's rate of flow, altitudes above sea level and the annual rise and fall of rivers.

The greatest error made by all these observers, and by others up to the mid-twentieth century, was to equate luxuriant tropical vegetation with rich soil. Wallace wrote that the 'richness of vegetable productions and universal fertility of soil... [were] unequalled on the globe... and [made Amazonia] capable of supporting a greater population, and supplying it more completely with the necessaries and luxuries of life'. All visitors imagined that hard-working Europeans or North Americans could turn the region into prosperous farmland, where feckless Portuguese and unambitious Indians had failed. Wallace was convinced that 'an earthly paradise might be created.... I fearlessly assert that here the "primeval" forest can be converted into rich pasture and meadow land, into cultivated fields, gardens and orchards....' He even mused about leaving his beloved England to 'live a life of ease and plenty in the Rio Negro'. Spix and Martius had been even more excited about turning Amazonia into a thriving land with great cities and teeming traffic along its rivers. Two observers sent in 1853 by the United States government, William Lewis Herndon and Lardner Gibbon, reported that 'an active and industrious population' could make the river's banks 'produce all that the earth gives for the maintenance of more people than the earth now holds.' (Interestingly, Herndon met Wallace and Bates in Henrique Antonij's house in Manaus during the brief period that they were together there in 1850; but he forgot their names.) Bates often made the same mistake about Amazonia's potential, in his letters to Stevens and his later book; and Spruce even wrote a pamphlet encouraging Europeans to settle in the Canelos forests of eastern Ecuador. It is extraordinary that so many intelligent observers repeated the same fantasy, when they saw repeatedly how myriad insects and blights destroyed farms and even their own specimens. Various experiments to bring 'industrious' outsiders into Amazonia ended in tragic failures. It was not until a century after Wallace that scientists started to realize that, far from being 'universally fertile', soils under rain forest are impoverished and acidic – because the constantly growing exuberant vegetation recaptures every scrap of nutrient and there is no winter when humus can accumulate.

During the busy year 1853, in addition to writing the two books, Wallace lectured at various learned societies. To the Royal Geographical Society, he spoke about the Rio Negro. His explanation for the contrast between its black water and the 'white' water of other rivers was fairly accurate. He correctly reckoned that the former was caused by decaying vegetation whereas the colour of the latter was because they 'gather much light-coloured sedimentary matter' from the Andes. No one then appreciated that it was tannin in decaying vegetation that produced the dark hue, and Wallace wrongly guessed that granite riverbeds were a factor.

His greatest geographical achievement was to make a remarkably accurate map of the Rio Negro and particularly a larger-scale one of the Uaupés. Both were published in the RGS *Journal*. Surveying was Wallace's only profession, from his years of working with his older brother William. But he made his fine traverse of the Uaupés when 'the only instruments I possessed were a prismatic compass, a pocket sextant, and a watch. With the former I took bearings of every point and island visible on the voyage, with sketches....' He obtained a few latitudes with his sextant, reckoned the speeds of boats and canoes, and, having broken his thermometers, had to guess altitudes by the time taken to boil water – in fact the altitude was only about 100 metres (330 feet) above sea level in that flat expanse of forest. All this was achieved when he was suffering from malaria, dysentery and hunger, had no assistant, was trying to collect saleable specimens, and was constantly battling fearful rapids and waterfalls.

Wallace plunged into geology with characteristic hubris and some success. He made interesting observations about granite in outcrops such as Cucui on the upper Rio Negro, and he wondered why he had not seen a single fossil. He noted that banks of the lower Rio Negro were often sandstone – but he could not know that the table mountains of the Guiana Shield (which he saw in the distance but did not reach) were also largely Precambrian sandstone.*

Wallace wrote that the most striking event on his travels was to meet the rarely contacted indigenous peoples of the Uaupés. He was utterly thrilled by the Tariana, Tukano, Wanana and Cubeo. Seeing their unspoiled beauty was 'as if I had been instantaneously transported to a distant and unknown land'. He was one of the first observers to note the degrading effect of their becoming 'imbued with the prejudices of civilization' and adopting Christian 'forms and ceremonies'. As we have seen, his remarks about these peoples were pioneering and valuable ethnography. It was salutary to have his and Spruce's glimpses of the oppression that Indians suffered on their remote rivers – abuse that would otherwise have gone unrecorded. He and Spruce were also among the first to be interested in indigenous rock art.

Dwarfing these peripheral achievements was the main objective of the journey to Brazil: collection and observation of flora and fauna. Henry Bates wrote that they had gone there in 1848 'for the purpose of exploring the Natural History of [the Amazon's] banks... and to gather facts, as Mr. Wallace expressed it in one of his letters, "towards solving the problem of the origin of species," a subject on which we had conversed and corresponded much together.' They did not solve the mechanism of evolution – as we shall see, Wallace was to do that in 1858 – but they did notice interesting aspects of the geographical distribution of species.

One observation by Wallace was that swarms of some birds and insects crossed the immensely wide Amazon river and its major tributaries, whereas for others these were insurmountable barriers. This was particularly true of primates. 'On going up the Rio Negro the difference in the two sides of the river is very remarkable. In the lower part of the river you will find on the north the *Jacchus bicolor* and the *Brachyurus couxiu*, and on the south the red-whiskered *Pithecia*. Higher up you will find on the north the *Ateles paniscus* [coatá or

* The Canadian Charles Hartt published the first serious study of Brazilian geology in 1870, but it was not until well over a century after Wallace that more was learned about plate tectonics, the uplifting of the Andes and the geological formation of the continent. No one in Wallace's day knew that ancient Precambrian geological shields lay beneath both central Brazil and the Guianas to the north.

spider monkey], and on the south the new black *Jacchus* and the *Lagothrix humboldtii* [brown woolly monkey].'* He boldly challenged the great Spix who, in his work on the monkeys of Brazil, was unaware of this phenomenon even though 'the fact is generally known to the natives'. Wallace observed that, moving up tributary rivers to their narrower sources 'they cease to be a boundary, and most of the species are found on both sides of them.' He appreciated that many aspects of geographical distribution still begged to be answered. Why were groups of the same species of monkey separated by great tracts of forest? What physical features determined boundaries of habitats – was it temperature, as defined by isothermal lines? Why did some rivers and mountain ranges form biological boundaries while others did not? He often observed similar problems of distribution among birds and insects. Why did some butterfly species have a very limited range – could it be related to how recently they had arrived in a location? Wallace imagined that the lower Amazon was recently formed (perhaps accurately: no one knows for sure), and he deduced from this an example of evolution: that butterflies peculiar to the lower Amazon were 'among the youngest species, the latest in the long series of modification which the forms of animal life have undergone.'

Soon after reaching Belém, Wallace worried about a different phenomenon. The accepted wisdom of the day was that birds' beaks and foraging methods were adapted to their habitats and favourite foods. He found, on the contrary, that four distinct families – goatsuckers, swallows, tyrant flycatchers and jacamars – all pursued the same insects in flight; but they used totally different techniques to do so. Similarly the bills of ibises, herons and spoonbills were utterly different and 'peculiarly formed.... Yet they may be seen side by side, picking up the same food from shallow water on the beach.' To confirm this he drew on his experience of preparing his bird specimens, because 'on opening their stomachs, we find the same little crustacea and shellfish in them all.' Equally, pigeons, parrots, toucans and chatterers – 'families as distinct and widely separated as possible' – could all be seen feeding simultaneously on the fruit of the same tree. The untrained observer had successfully refuted the accepted wisdom about birds' beaks and foraging techniques.

The greatest legacy of Wallace's time in the Brazilian Amazon was the collections themselves and, even more so, their preparation. When Wallace sailed in 1848 his experience of collecting, skinning, preserving and mounting specimens was rudimentary; when he returned four years later he was expert in all these procedures. He had learned by

* Wallace's *Jacchus*, *Brachyurus* and *Pithecia* are species of little saki monkeys (*sagüi* in Portuguese) now either of the *Callithrix* genus (17 species including *C. jacchus*), or *Saguinus* (eight species of marmosets including *S. bicolor*, *S. mystax* and *S. imperator*, the last so-named because its sideburn whiskers resemble those of the Austrian emperor Franz Josef), or *Pithecia* (*parauacú* in Portuguese and monk saki in English; four species including *P. pithecia*), or *Chiropotes* (including *C. satanas*, the cuxiú marmoset).

watching and quizzing his native Brazilian helpers and by personal trial and error – months of practice in punishing conditions, and with many mistakes and mishaps. In those years, he and Bates sent back an amazing 14,712 *species*, of which over half had never been seen in Europe. (The number of *specimens* was of course far greater, because they collected many examples of each species.) Whereas Bates was happiest as an entomologist and Spruce as a botanist, Wallace eagerly studied every branch of the natural world.

Their efficient agent Samuel Stevens sold and therefore dispersed much that Wallace sent him in the early years. Tragically the naturalist's precious private collection, his menagerie of live birds and animals, and the fruits of his final year of collecting were all lost in the shipwreck. However, examples that survive from his earlier Amazonian collecting, particularly in the Natural History Museum and the Royal Botanic Gardens, Kew, are expertly mounted and neatly labelled. Wallace's four years of field experience were essential for his subsequent collecting in Southeast Asia.

In addition to his work for the Royal Geographical Society, in the first half of 1853 Wallace spoke to other gatherings about such diverse topics as the umbrella bird, monkeys of the Amazon, fish related to the so-called 'electric eel', insects eaten by indigenous peoples, skipper butterflies and butterfly habits. Every one of these lectures involved geographical *distribution*. From them, and a corpus of later work, Wallace has been credited as a founder of the discipline of biogeography.

All this made Wallace known in London's scientific circles. Samuel Stevens took him to the Entomological Society, of which he was a council member. Everyone had heard of the loss of Wallace's collection and notes in the burning *Helen* and how, in Stevens's vivid account, he had 'narrowly escaped death in an open boat, from which, after long privation and suspense, [he was rescued] in the midst of the Atlantic Ocean.' Edward Newman, president of that Society, reminded its expert members that they owed everything to the actual field collector who 'devotes his time, by night and by day; at all seasons, in all weathers; at home and abroad, to the positive capture and preservation of specimens.' He particularly praised Wallace and Bates (who was still in Amazonia). So, as Wallace's biographer Michael Shermer said, 'during his eighteen months at home Wallace became an intellectual insider – a member of the scientific club'.

With his restless energy and wonderfully eclectic intellectual curiosity, Wallace wanted to return to the field. Various factors steered him towards Southeast Asia. He heard a lecture about the zoology of the Malay peninsula; read Goodrich's new *Universal History*, which said that this region was a 'new world' for undiscovered flora and fauna; and heard that the Dutch had efficient colonies in the Celebes (now Sulawesi) and the

Moluccas (now the Maluku islands). Importantly, he had a chance meeting with Sir James Brooke, who had recently become Rajah of Sarawak in northwest Borneo and who warmly invited Wallace to visit him. So Wallace spent as much time as he could in the Natural History Museum, 'examining the collections, and making notes and sketches of the rarer and more valuable species of birds, butterflies, and beetles of the various Malay islands.' He bought Prince Charles-Lucien Bonaparte's encyclopaedic, 800-page *Conspectus Generum Avium* (1850) and then consulted all the books cited in it about Malayan birds. With characteristic efficiency he 'copied out in abbreviated form such of the characters as I thought would enable me to determine each.' He could not have done this without his years of experience in Amazonia, and that fieldwork proved amazingly helpful when he was identifying birds in those new forests and islands.

The next step was to get a passage to the Malay archipelago. When Wallace lectured to the RGS he met its president, the distinguished geologist Sir Roderick Murchison, and found him 'one of the most accessible and kindly men of science'. In June 1853, Wallace put forward an impressive proposal: to make Singapore his headquarters and then to spend a year each in six groups of islands. The Society noted that 'his chief object is the investigation of the natural history' of these islands, more thoroughly than it had ever been done. Also, to please the Society, he proposed to pay much attention to the geography of the region, making many measurements and readings. He cleverly did not ask for a cash grant, just for influence in getting a free passage. So he was invited to attend the RGS Committee on Expeditions, and made a good impression. It resolved that Murchison should ask the government to take Wallace to Singapore on one of its ships, and also for letters of introduction to the Dutch and Spanish for their East Indian and Philippine colonies.

Having launched his plea, Alfred Wallace went off on holiday to Switzerland with his boyhood friend George Silk, now a financial officer for the Diocese of London. But he was caught out when an offer of free passage came too quickly – he was still walking in the Alps. Back in London, various naval ships were to take him, but there were changes and delays – these reminded Wallace of all the interminable waits for boats in the Amazon. He finally sailed, in March 1854, with a first-class ticket on a P&O liner to Egypt and then on others to Singapore (since there was no Suez Canal), all obtained for him by Murchison.

This book is about Wallace, Bates and Spruce in South America. For Alfred Russel Wallace, his four years in the Amazon were a prelude and preparation for an amazingly full life. But they taught him collecting skills and routines in tropical rain forests, and they had a direct bearing on the two discoveries for which he is most famous: evolution by natural selection, and the 'Wallace Line' in biogeography.

Soon after arriving in the East, Wallace was staying in a house of Rajah Brooke of Sarawak, when he considered the extent to which geographical distribution of animals and plants affected the origin of species. He, Bates and many others agreed that species were constantly changing, some very slowly, others faster. 'Every species has come into existence coincident both in time and space with a pre-existing closely allied species.' This gradual change is also affected by factors such as geographical distribution, geology (the rise and fall of land masses from the oceans), and anatomical organs. He drew on his own collecting of birds and insects to illustrate geographical aspects of the theory.

Wallace sent his paper, 'On the Law which has Regulated the Introduction of New Species', to Stevens, who got it published in the *Annals and Magazine of Natural History* in 1855. (This is now often referred to as the 'Sarawak' paper.) Stevens forwarded the paper to Bates, who was far up the Amazon at Ega (Tefé) at the time. Amazingly, Bates was able to get a letter about it to his friend – even though Wallace was in an equally obscure location on the far side of the world. Bates was delighted by the paper and wholly agreed with it. 'The idea is like truth itself, so simple and obvious that those who read and understand it will be struck by its simplicity; and yet it is perfectly original.' But Bates added a mild counter-claim: 'The theory I quite assent to, and, you know, was conceived by me also, but I profess that I could not have propounded it with so much force and completeness.' Darwin at first dismissed the paper as containing nothing new, but friends such as the geologist Charles Lyell alerted him about its significance, as did a letter from Wallace himself. Darwin replied to this that 'I can plainly see that we have thought (very) much alike & to a certain extent have come to similar conclusions.... I agree to the truth of almost every word of your paper....' Then Darwin delivered what has been described as a delicate no-trespassing notice: 'This summer will make the 20th year (!) since I opened my first note-book on the question how & in what way do species & varieties differ from each other – I am now preparing my work for publication, but I find the subject so very large, that though I have written many chapters, I do not suppose I shall go to press for two years.'

What Darwin feared occurred in February 1858. Wallace was at the main settlement on the small island of Ternate, and then at a village called Dodinga on the nearby larger island of Gilolo.* He was recovering from the fevers and shivering chills of acute malaria, either caught locally or a recurrence of his near-fatal attack on the Uaupés six years previously. As he described it in his later autobiography, he was pondering Malthus's *Principles of Population*, which he had read as a young man. Malthus had written that human population was kept in check by disease, war, famine and other calamities. So animals, which often breed faster than humans, must be controlled by far greater and constant destruction. 'It occurred to me to ask the question, Why do some die and some live? And the answer was clearly, that on the whole the best fitted live. From the effects of disease the most healthy escaped; from enemies, the strongest, the swiftest, or the most cunning; from famine, the best hunters or those with the best digestion; and so on. Then it suddenly flashed upon me that this self-acting process would necessarily *improve the race*, because in every generation the inferior would inevitably be killed off and the superior would remain – that is, *the fittest would survive*.' Changes of environment (land and sea, climate), food supply or enemies, would of course be very gradual, giving plenty of time for animals to adapt or be isolated from one another. 'The more I thought it over the more I became convinced that I had at length found the long-sought-for law of nature that solved the problem of the origin of species.' In another later version he wrote that 'there suddenly flashed upon me the idea of *the survival of the fittest* – that the individuals removed by these checks must be, on the whole, *inferior* to those that survived [his italics]. Then... the whole method of specific modification became clear to me, and in the two hours of my [malarial] fit I had thought the main points of the theory.'

Wallace knew that the Dutch mail boat would call at Ternate on 9 March 1858, so he had no time to lose. During three successive nights, he rapidly wrote the theory as 'On the Tendency of Varieties to Depart Indefinitely from the Original Type' and made a fair copy for dispatch. He also dashed off letters to Bates (in which he made no mention of the new theory), to Bates's brother Frederick, and to Darwin 'in which I said that I hoped the idea would be as new to him as it was to me, and that it would supply the missing factor to explain the origin of species. I asked him, if he thought it sufficiently important, to show it to Sir Charles Lyell....' Decades later Wallace wrote slightly different accounts of his flash of inspiration. He made no mention of it at the time, either in his journal or his later book *The Malay Archipelago* (1869). But there is no question about the brilliance of the paper he sent to Darwin.

* Ternate is in northern Indonesia, off Halmahera Island (then called Gilolo) in the Moluccas between Sulawesi and Papua (New Guinea). When the western half of New Guinea was (disgracefully) given to Indonesia, they initially called it Irian Jaya, but have now changed this to Papua.

When Charles Darwin received the 'Ternate paper' he was shattered. Lyell had warned him to hurry and finish his great book, but he had not yet done so. Now he thought that the younger scientist had stolen his thunder. He wrote to Lyell that 'I never saw a more striking coincidence. If Wallace had my m.s. sketch written out in 1842 he could not have made a better short extract! Even his terms now stand as Heads of my Chapters. So all my originality, whatever it may amount to, will be smashed.' As Michael Shermer notes, Wallace's paper was actually not quite as identical as Darwin thought it was. But Darwin was in an agony of indecision, not wanting to behave 'in a paltry spirit' towards Wallace, but equally in despair at losing primacy of the great idea of evolution by natural selection.

His two closest friends Lyell and Joseph Hooker came up with an honourable compromise solution: Darwin must put Wallace's paper forward to be read to the Linnean Society, but he should also read a sketch outlining these ideas that he had written as long ago as 1844, as well as a letter of September 1857 that added the important principle of divergence of species.* This joint reading took place on 1 July 1858. The meeting was entitled: 'On the Tendency of Species to Form Varieties; and on the Perpetuation of Varieties and Species by Natural Means of Selection'. Neither author was present: Darwin was grief-stricken because his baby son Charles had just died of scarlet fever; and Wallace was on the far side of the world. There was no conspiracy to cheat Wallace. A letter from Lyell and Hooker was read out to the meeting, stressing that 'two indefatigable naturalists, Mr. Charles Darwin and Mr. Alfred Wallace, ...having independently and unknown to one another, conceived the same very ingenious theory to account for the appearance and perpetuation of varieties... on our planet, may both fairly claim the merit of being original thinkers in this important line of inquiry.' The Society's under-secretary then read out the three documents.

The meeting at the Linnean Society was attended by only thirty members, and the minutes about it were brief. Only a few appreciated what a momentous idea was being launched. Hooker later wrote that the subject was too novel for 'the old school' to comprehend; but Lyell's and Hooker's approval 'rather overawed the Fellows, who would otherwise have flown out against the doctrine.' The entomologist Alfred Newton was one of the few to be overcome when he read the paper. He did not know whether to be delighted, or upset that he had never thought of it himself. 'Herein was contained a

* Sir Charles Lyell (1797–1875) was the famous author of the standard work, *Principles of Geology*; Joseph Hooker (six years older than Wallace, the same age as Spruce) was a distinguished botanist. In 1865 when his father Sir William died, Joseph succeeded him as Director of Kew Gardens, which he ran admirably for twenty years.

perfectly simple solution of all the difficulties which had been troubling me for months past… like the direct revelation of a higher power…. [Here] was an end of all the mystery in the simple phrase "Natural Selection".'

There had not been time to consult Wallace about the reading, nor to let him correct his paper for publication. But Wallace's first reaction was to be delighted that eminent men of science such as Lyell and Hooker had thought so highly of it to have the paper read and published. Both Stevens and Darwin sent him copies of the printed paper. In one letter Wallace wrote to thank Hooker profusely for organizing the reading of both papers: 'It would have caused me such pain & regret had Mr. Darwin's excess of generosity led him to make public my paper unaccompanied by his own much earlier & I doubt not much more complete views on the same subject.' He felt that the joint reading 'while strictly just to both parties, is so favourable to myself.' Decades later, Wallace said that both Darwin and Hooker wrote to him 'in the most kind and courteous manner' and gave him 'more honour and credit than I deserved, by putting my sudden intuition – hastily written and immediately sent off… on the same level with the prolonged labours of Darwin….' He urged his friend George Silk to read his paper in the Linnean *Journal*, with 'some complimentary remarks thereon by Sir Charles Lyell and Dr Hooker, which (as I know neither or them) I am *a little* proud of.'

Darwin was spurred into finishing his book, and *On the Origin of Species by Means of Natural Selection* was published by John Murray in November 1859. He sent a copy to Wallace, with an accompanying letter that 'my Book would have been & be a mere flash in the pan, were it not for you, Hooker & a few others'. Wallace, in turn, was bursting with praise for the magisterial book. He wrote to George Silk urging him to read Darwin's *Origin of Species* repeatedly. 'I have read it through five or six times, each time with increasing admiration. It will live as long as the 'Principia' of Newton. It shows that nature is… a study that yields to none in grandeur and immensity…. Mr. Darwin has given the world a *new science*, and his name should, in my opinion, stand above that of every philosopher of ancient or modern times. The force of admiration can no further go!!!' To his friend Henry Bates, who was by now back in England: 'I know not how… to express fully my admiration of Darwin's book. [I myself could] *never have approached* the completeness of his book, its vast accumulation of evidence, its overwhelming argument, and its admirable tone and spirit. I really feel thankful that it has *not* been left to me to give the theory to the world. Mr. Darwin has created a new science and a new philosophy.' Wallace thought that no single man had ever launched a new branch of knowledge so thoroughly, drawing so many facts into one system.

It is not surprising that the relatively unknown collector in the Moluccas was overawed by Darwin. But some modern commentators find it curious that he continued to be so deferential in later life. Wallace reminded people that his own discovery had been completely independent and yet, as one biographer Martin Fichman wrote: 'It is primarily by Wallace's own efforts that the theory of evolution by natural selection is usually known as Darwinism.... When Wallace published his masterly textbook exposition of the theory... in 1889, he titled it, simply, *Darwinism*' – despite the fact that his own book made significant improvements on his mentor's original concept. Other modern writers have sought to show that Darwin could have plagiarized the inspirational idea, but this conspiracy theory has been demolished. So we are left with the attractive concept that both authors behaved as perfect Victorian gentlemen. Darwin put Wallace's paper forward even though he had not yet published the theory himself, and Wallace never claimed priority.

Soon after the publication of their papers in the Linnean *Journal*, Charles Darwin expressed his gratitude in a letter to Bates: 'What a fine philosophical mind your friend Mr. Wallace has, & he has acted in relation to me like a true man with a noble spirit.' Bates then forwarded Wallace's flattering letter about the *Origin of Species* to his new friend Darwin. In reply, Darwin again praised Wallace's modesty and admirable behaviour: 'He rates me much too highly & himself much too lowly.... But what strikes me most about Mr. Wallace is the absence of jealousy towards me; he must have a really good honest & noble disposition, a far higher merit than mere intellect.' Darwin often expressed his thanks to Wallace. When the younger thinker in 1870 published a book called *Contributions to the Theory of Natural Selection*, Darwin wrote: 'My dear Wallace, There has never been passed on me, or indeed on anyone, a higher eulogism than yours. I wish that I fully deserved it. Your modesty and candour are very far from new to me.' It was highly satisfying that 'we have never felt any jealousy towards each other, though in one sense rivals. I believe that I can say this of myself with truth, and I am absolutely sure that it is true of you.'

It is incredible that between 1855 and 1859 Wallace and Bates were able to correspond with one another, when they were on opposite sides of the planet and each in an extremely remote tropical forest. A letter from Bates had to catch a quarterly steamship from Ega down to Manaus, then be transferred to another steamer down the lower Amazon to Belém; it was then put on an infrequent transatlantic ship, which was probably a sailing ship at that time before the Amazon rubber boom; in London the letter went to Samuel Stevens, who then forwarded it to Wallace in Southeast Asia. There was no Suez Canal, so post went to Alexandria (by sea from Southampton for heavier packets, but by trains to Marseille and then ship for letters); then overland to the Red Sea and by other ships to India or Singapore; it was then routed to Batavia (now Jakarta) on Java in the Dutch East Indies (Indonesia); the Dutch had a postal packet that took five weeks to do a circuit of their islands, and this had to find Wallace wherever he happened to be. The reply from Wallace took this cumbersome route in reverse. It is not clear how the naturalists paid for their letters. Britain had created postage stamps in 1840, and Brazil in 1843 was the second country to do so; but these were only for internal postage – the Universal Postal Union governing international mail was not founded until 1874.

Wallace's other great discovery was that animals on Borneo were completely different to those on Celebes (Sulawesi) – only 115 kilometres (70 miles) east across the Makassar Strait. Further south, he noticed the same phenomenon when he collected on Bali and on Lombok Island nearby to the east. 'I was thus enabled to determine the exact boundary between two of the primary zoological regions, the Oriental and the Australian.' This extraordinary biogeographical divide became known as The Wallace Line. In a letter of 1857 he noted that the islands of Bali and Lombok 'though of nearly the same size, of the same soil, aspect, elevation and climate, and within sight of each other, yet... belong to two quite distinct zoological provinces, of which they form the extreme limits.' Then he wrote to Bates in January 1858: 'In this archipelago there are two distinct faunas rigidly circumscribed, which differ as much as do those of Africa and South America, and more than those of Europe and North America; yet there is nothing on the map or on the face of the islands to mark their limits. The boundary line passes between islands closer together than others belonging to the same group.... In mammalia and birds the distinction is marked by genera, families, and even orders confined to one region; in insects by the number of genera, and little groups of peculiar species, the families of insects having generally a very wide or universal distribution.' Then, with amazing prescience, he guessed to his former colleague that one side is 'a separated portion of continental Asia, while the eastern is a fragmentary prolongation of a former west Pacific continent'. Wallace explained this 'line' more fully in a letter in the journal *Ibis* in 1859 and in the Linnean *Journal* in 1860. He correctly supposed that the fauna had evolved differently on separate land masses. So he anticipated by a century the theory of plate tectonics or continental drift. We now know that in ancient geological time, Australia had drifted from the southern land mass Gondwanaland, whereas the Malay archipelago had separated from Asia or Pangaea. Although this discovery was on the far side of the globe, Wallace's earlier noting that species were separated by Amazonian rivers undoubtedly helped him to make this dramatic new observation.

Wallace returned to London in April 1862. During his eight years in Southeast Asia, he had journeyed for 'fourteen thousand miles [22,500 kilometres] within the Archipelago, and made some sixty or seventy separate journeys' during which he collected an amazing 125,660 specimens including many great rarities. He called this 'the central and controlling incident' of his life, and it greatly surpassed his four years in Amazonia. He wrote a masterful account of his eastern adventures, *The Malay Archipelago: The Land of the Orang-utan; and the Bird of Paradise. A Narrative of Travel with Studies of Man and Nature* (1869), which was, deservedly, a huge success. The two great journeys established him as one of the foremost field naturalists of his era.

Back in England, Wallace was still troubled by recurring malaria, but was otherwise lean and fit from his twelve years of tropical collecting. His moustache and sideburns were eventually extended into a beard, which he kept for the rest of his life. (By 1869 the beard had filled out and was starting to turn grey. It later grew larger and became white like his full head of hair. So in old age Wallace looked like a biblical prophet, with his kindly intelligent face peering out from small round spectacles.)

Now aged thirty-nine, he wanted to settle down. He had difficulty in finding a wife. The first lady he wooed, the daughter of a chess-playing friend, turned him down repeatedly; but in 1866 he married Annie Mitten, the twenty-year-old daughter of the pharmacist and moss botanist William Mitten. They spent their honeymoon in north Wales, clambering on hills around Snowdon. Although Wallace was on his honeymoon, he did not devote all his attention to his bride Annie. He noticed curved valleys and argued, in a paper 'Ice-marks in North Wales' (*Quarterly Journal of Science*, January 1867) that these smooth basins were caused by pressure of glaciers rather than water erosion. Wallace's theory is still universally accepted. Despite this honeymoon distraction, and a twenty-three year gap in their ages, it was a happy marriage that lasted until Wallace's death half a century later. They were proud of their two sons and a daughter.

The other element of a settled life – a regular job – eluded Wallace. Various employment applications were unsuccessful. He had earned a lot from his Southeast Asian collections, notably of gorgeous birds of paradise; but that had been spent. So money was always a concern. He lost about £1,000 investing in Welsh slate quarries, and then another sum on lead mines. The only good news came in 1881: an annual pension of £200, obtained for him by Darwin writing personally to the prime minister, Gladstone. Not daunted by lack of funds, Wallace spewed out books, papers, lectures and ideas in every year of his long life.

Wallace's restless, enquiring mind led him into endless branches of learning and social science, some very unorthodox. Although he was a convinced Darwinist, this did not prevent Wallace from querying some aspects of the great book. He criticized Darwin for including domestic animals selectively bred by humans alongside wild ones subject to survival of the fittest. More controversially, Wallace felt that the theory did not work with human beings. Their enlarged brains, far better than those of apes, were identical in both 'primitive' aborigines and sophisticated members of learned societies. So he argued that with mankind 'an Overruling Intelligence has watched over the action of those laws [of natural selection], so... [producing] the indefinite advancement of our mental and moral nature'. Wallace also became passionate about spiritualism, having convinced himself that occult tappings and table movements were not tricks. He wrote and lectured about spiritualism in Europe and the United States, and tried in vain to convert his scientific colleagues to this manifestation of the paranormal. He suffered a long and expensive legal action against a fanatical flat-earther. For a decade he was fervently opposed to vaccination, even though he had no medical qualifications. Throughout his life, Wallace advocated the nationalization of land as a radical social reform that he felt would lead to social progress. He wrote that it was a crime against humanity that one child should be born a millionaire and another a pauper. His visits to the United States impressed him with its geology, natural history and dynamic society. He was fascinated by astronomy and man's place in the universe. He dabbled in reform of the Church of England (even though he was an atheist), imperialism, model farms and modern museums.

Alongside some views that we would regard as blind alleys, he was producing endless scientific and social research and thought. Michael Shermer compiled a list of all Wallace's published works: this included twenty-two books, from *Palm Trees of the Amazon* in 1853 to *The Revolt of Democracy* in 1913, and a staggering 747 papers, articles, reviews and comments. The evolutionary biologist Stephen Jay Gould felt that his masterwork was 'the monumental *Geographical Distribution of Animals* [1876], which essentially established the science of zoogeography and brought evolutionary factors fully into the geographical distribution of animals.' All who come into contact with Alfred Russel Wallace and his writings fall under the spell of his energy, exuberance and enthusiasm, his curiosity, honesty and open-mindedness, his intellectual brilliance, the elegance and clarity of his writing (and, we are told, his lecturing), and his amiability.

Wallace was also a competent artist – better than his friends Bates or Spruce. When the brig *Helen* was on fire, in 1852, Wallace had rushed into his smoke-filled cabin and grabbed a tin box – the only thing he saved from the shipwreck. This contained fifty sketches, mostly of fishes, that he had made on the upper Rio Negro in the previous year. In 1904 he donated these to the Natural History Museum, modestly saying that they might be of interest to people in the Zoology Department. They are wonderfully accurate drawings, precisely to scale. But they are also works of art, treasures of the Museum's archive.

In 1905, when eighty-two, he published his two-volume autobiography, *My Life: A Record of Events and Opinions*. The quirky Wallace was delighted by one of many favourable reviews, which praised his achievements but then read: 'On many subjects Mr. Wallace is an antibody. He is anti-vaccination, anti-state endowment of education, anti-land-laws, and so on. To compensate, he is pro-spiritualism, and pro-phrenology, so that he carries, as cargo, about as large a dead weight of fancies and fallacies as it is possible to float withal.' Wallace loved being called an 'antibody'. There have been many books about this brilliant, engaging, but often eccentric and controversial scientist – notably Peter Raby's fine biography (2001) and impressive studies by Michael Shermer in 2002, Ross Slotten (2004), Martin Fichman (2004), a series of papers edited by Charles Smith and George Beccaloni (2008) and anthologies of Wallace's writings by Andrew Berry (2002) and Sandra Knapp (1999).

During his lifetime Wallace received recognition for his tremendous scientific work. He was elected a Fellow of the Royal Society, and in 1868 that premier scientific body awarded him its Royal Medal, twenty-two years later its Darwin Medal, followed in 1908 by its prestigious Copley Medal. Oxford University gave him a (rare) honorary doctorate. In 1892 there were gold medals from both the Royal Geographical and Linnean societies. 'Isn't it awful?,' he complained to his daughter Violet. 'Two medals to receive – two speeches to make, neatly to return thanks, and tell them, in a polite manner, that I am much obliged, but rather bored!' In 1908, on the fiftieth anniversary of the joint Darwin-Wallace reading, the Linnean Society held a jubilee celebration. Wallace felt he had to attend, even though he dreaded adulation and thought it 'outrageous' that he was being put on a level with the 'mighty genius' Darwin. He made a charming and typically modest acceptance speech for a new Darwin-Wallace Medal. The president of the Linnean said that 'Never was there a more beautiful example of modesty, of unselfish admiration for another's work, of loyal determination that the other should receive the full merit of his independent labours and thoughts' than had been shown by the two great naturalists.

In that same year 1908 came the highest honour of all. King Edward VII included him among the twenty most eminent people in the arts, sciences and academia, in the recently created Order of Merit. Wallace excused himself from attending an investiture in Buckingham Palace. He wrote to a friend that it was 'quite astonishing and unintelligible' that the OM should be given to a red-hot radical, land nationalizer, socialist, anti-militarist, etc., like him. But he wore the Order with pride, and included it after his name on the title page of his last book. He was still working at his desk in his ninetieth year, 1913, when he felt faint, and died soon after. Some friends suggested that Wallace should be buried beside Darwin in Westminster Abbey, but he had told his wife Annie that he wanted a modest grave near their house in Broadstone, Dorset.

Before leaving Brazil in June 1859, the thirty-four-year-old Henry Walter Bates wrote to his family that they would soon see 'an oldish, yellow-faced man in big whiskers (I have a moustache now...) make his appearance in King Street, Leicester.' (He kept his broad moustache, bushy sideburns and a full head of hair above a high forehead for the rest of his life. These were originally brown, but of course turned white in later years.) As a dutiful son, he resumed work in the family hosiery business. He may have hoped that this would please his mother Sarah; but this was not to be, for in January 1860 – a few months after his return – she died of a liver disease, aged only fifty-seven. Her sister moved into the house to look after the Bates men: the 66-year-old widower Henry, for whom the hosiery business was all that mattered, and his sons Henry, Frederick (now also a keen amateur entomologist) and Samuel.

About this time Bates met a nineteen-year-old, Sarah Ann Mason, the daughter of a butcher with a successful stall in the local market. After eleven years of celibacy in the Amazon, the naturalist was obviously delighted by the company of an attractive girl fifteen years younger than him – even though she was of lower social class and shared none of his intellectual interests. She became pregnant in the summer of 1861, and a daughter Alice was born on 2 February 1862. Sarah Ann signed the birth certificate with an 'x', probably because she was illiterate. She and Bates married eleven months later, on 15 January 1863 in a registry office in St Pancras, London. In the parlance of the time, he did the honourable thing and made an honest woman of her. It proved to be a successful union. Henry Walter and Sarah Ann were happily married for the rest of Bates's life, and they had another daughter and three sons. But the marriage did not enhance the scientist's position in society. As Bates wrote to his friend Darwin: 'Mrs. B. is a plain domesticated woman, so there you have it.'

Back in England, Bates's first concern was has personal collection. Samuel Stevens stored it for a while in his premises in Covent Garden, and even engaged temporary staff to help sort and catalogue the specimens as they were unpacked. Bates wrote that in

Amazonia he had collected 14,000 species of insects, 360 of birds, 140 of reptiles, 120 of fishes, 52 of mammals, 35 molluscs and a few zoophytes (animals resembling plants) – a grand total of 14,712 species. He modestly said that, because he had often collected in places unexplored by other naturalists, 'no less than 8000 of the species here enumerated were *new to science* [his italics], and these are now occupying the busy pens of a number of learned men in different parts of Europe to describe them.' As time went by, he admitted that he himself was too poor to maintain his personal collection purely for scientific research – a pity, since it 'would have formed a fair representation of the fauna of a region not likely to be explored again for the same purpose in our time.' To support his growing family, Bates sold his private butterfly material to the taxonomists Frederick Godman and Osbert Salvin, 'of whose unrivalled collection it forms an important part.' Henceforth he studied his beetles; but in 1891 he sold the longicorns to René Oberthür of Rennes, France, keeping only ground beetles.

Over the years, the Natural History Museum had had first pick of Bates's material, so that this national collection had the largest set after his own. One distinguished lepidopterist, William Hewitson, complained to Bates that 'I see nothing of your choice things – they are all picked out by the museum people.' The Museum also had the advantage of frequent journeys by Bates from Leicester to London to supervise his material's 'detailed arrangement'. As a perfectionist he insisted on this unpaid work, because when he and Wallace had visited the national collection twelve years earlier Bates was shocked to find it 'in the utmost confusion; scarcely a genus in proper order and duly named' so that it was almost useless to butterfly collectors trying to identify their catches.

Some entries in the Museum's register recorded how much it paid the Amazonian collector. In one of Bates's earlier consignments, in 1851, 55 of the prettiest butterflies were bought for 3 shillings each (15 pence in modern currency, a hundred times that in value), 70 for 2/6d (12.5p), and 31 smaller ones for 1/6d (7.5p). The Museum acquired less than half the collections he sent back: the rest were sold into private hands in Britain and other parts of Europe, such as to the Austrian collectors Rudolph and Catejan Felder.

Most of the material sent by Bates is assimilated into the Museum's main series, but some panels of Bates's mounted specimens survive. Where a butterfly's thorax has not been damaged by pinning and repinning (by other collectors) it might have been shipped by Bates in a triangular paper envelope or mounted 'low-set in the style of the British saddle-type setting' that was common at that period. Many of the Museum's specimens still have Bates's neatly written labels giving the time and location of each insect's capture. An equally thrilling treasure is two of his notebooks, now in the entomological archive. One, from 1853, has about a hundred plain pages in hard marbelled covers, and its entries deal only with butterflies and moths. The other, from 1856 or later, is twice as long and also includes beetles, ants, termites, flies, dragonflies and other insects. This has numbered entries for over 900 specimens, all extremely detailed, by this expert who could spot anatomical peculiarities and comparisons with other insects. Bates's handwriting is invariably small, very neat and legible, in brown ink in regular straight lines. There are vivid descriptions of every insect he collected. A typical entry is about *Anteros* butterflies (a genus of Riodinidae): 'The very beautiful little Butterflies... distinguished by the spangles of golden & silvery appearance with which the undersurfaces of their wings are furnished & by the general thick clothing of scales & pubescence with which their body & limbs are clothed, are chiefly found in the driest and hottest weather', followed by its preferred habitat, rapid flight, body description and behaviour. To one recent biographer 'these bring alive and render comprehensible the glorious butterfly that would otherwise lie hidden beneath the words of its formal description.' These notes transport readers back to the lonely collector with his net and pin-cushion, running between the mighty trees of the sumptuous Amazon forests. Amid the neat written entries are jewel-like watercolours of butterflies, beetles and other creatures, delicately done, life-sized, amazingly accurate in colour and detail and as fresh as if they had just been painted. Seeing these splendid notebooks, it is difficult to imagine the primitive conditions in which he worked, often damp or humid, cramped, dimly lit, with poor pens and ink and plagued by termites and other voracious pests.

Apart from his magnificent collections, what had Bates achieved during those eleven arduous years? He had grown from a novice of twenty-three to a veteran of thirty-four, with unrivalled field experience. The self-educated hosiery apprentice was now an established scientist, albeit without formal training or higher education and therefore unable fully to enter the scientific elite. His years on the Amazon led to a best-seller, a major corpus of scientific papers, a theory named after him and a job at the heart of British geography.

As we have seen, Charles Darwin published *On the Origin of Species by Means of Natural Selection* in November 1859, a few months after Bates's return to England. Bates immediately espoused this momentous theory. As we also know, Wallace and Bates had originally gone to the Amazon to discover more about the origin of species. Wallace later hit on the idea of survival of the fittest governing change whereas Bates, during his years on the Amazon, was more concerned with geographical distribution causing distinctions between varieties and species. In his notebook in 1855 he recorded an elegant little insect being 'a link in the gradation of affinities' (i.e. evolution) between two species of beetle. We saw how, when Wallace sent him his first ground-breaking paper on laws governing the introduction of new species (the 'Sarawak' paper of 1854), Bates replied that he fully agreed with it – but he had added that 'the theory, ...as you know, was conceived by me also.' Wallace answered that his own and Bates's great collections would 'furnish most valuable material to illustrate and prove the universal applicability of the hypothesis.' But he was referring to geographical distribution, species by species, which 'has never yet been shown as we shall be able to show it.'

Bates in 1860 sent his thoughts about evolution by natural selection to Darwin, together with a paper he had written about intermediate varieties of evolving insects. His letter is lost, but he must have been thrilled by Darwin's flattering reply: 'Your name has for very long been familiar to me, & I have heard of your zealous exertions in the cause of Natural History. But I did not know that you had worked with high philosophical questions before your mind. I have an old belief that a good observer really means a good theorist.... I am delighted to hear that you, with all your large practical knowledge of Nat. History, anticipated me in many respects & concur with me.... I see by your letter that you have grappled with several of the most difficult problems, as it seems to me, in Natural History – such as the distinctions between the different kinds of varieties, representative species etc.' (Of course, Darwin was delighted to have Bates – fresh from eleven years of Amazonian collecting – confirm his still controversial new theory.)

To bolster his credentials as a scientist as well as a naturalist collector, Bates embarked on what was to be a six-year series of lectures and published papers under the general title 'Contributions to the Insect Fauna of the Amazon Valley'. The first was about the common butterfly genus *Papilio*, delivered to the Entomological Society of London in two parts, on 5 March and 24 November 1860, and published in its *Transactions*. In those days it was unusual for a work of taxonomic description to venture into interpretation. Bates disagreed, feeling that his eleven years of meticulous fieldwork *did* qualify him to discuss wider issues. Two contemporary entomologists, William Distant and Edward Clodd, complimented him on this breakthrough. Distant felt that there were three types of entomologist: field collectors, philosophical observers who advanced new theories, and taxonomists concerned with systematic classification of species and genera. 'Bates almost uniquely proved himself a master in each, and with the exception of his old travelling companion Wallace, is approached in that respect by scarcely another living entomologist.' Darwin complimented him on not being one of 'the mob of naturalists without souls'.

In March 1861 Bates proudly sent his *Papilio* paper to Darwin. In a covering letter he said that he had not anticipated the conclusion he reached: that many of the Amazon's insects derived from 'the Guiana region [which] must have been the seat of an ancient &

peculiar fauna transmitted through vast lapses of time.' This was amazingly prescient. Neither Bates nor Darwin knew that the Guiana Shield in the north of South America was in fact a continuation of a Precambrian geological formation in West Africa. This had been separated over millions of years by plate tectonics – a theory that was not established for another century. The great antiquity of insects in the Guiana region was confirmed in 1983 by an expedition onto the Neblina table mountain between Brazil and Venezuela. Its scientists found that the closest relatives of some insects on this massif were still in West Africa. (This author was on that expedition. Neblina is the source of the Baria-Pacimoni river explored by Spruce in 1854.) Bates also argued that the Ice Age had not greatly impacted on Amazonia. In this he disagreed with Darwin, who thought that global glaciation to the north might have caused an invasion of the tropics by cooler fauna – perhaps even causing an extinction of some tropical species. Bates wrote that his *Papilio* butterflies showed plainly that there had always been 'an equatorial fauna rich in endemic species, and extinction cannot have prevailed to any extent within a period so comparatively modern as the Glacial epoch in Geology.'

Darwin immediately wrote back that he had read every word of Bates's papers with extreme interest. 'They seem to me to be far richer in facts on variation, & especially on the distribution of varieties & subspecies, than anything which I have read.... I hope in my future work to profit by them....' Darwin admitted that he had been tempted by the idea of the Ice Age affecting South America, but was convinced to the contrary by Bates's evidence based on his years in those forests: 'I am quite staggered with the blow & do not know what to think.' In a letter to Joseph Hooker, Darwin wrote: 'How well [Bates] argues, and with what crushing force against the glacial doctrine. I cannot wriggle out of it: I am dumbfounded.' Bates was thrilled by this warm response and praise from the great scientist. He replied immediately how gratified he was 'to find that my paper is likely to be useful to you'. He urged Darwin to use insects to illustrate 'the problems you occupy yourself with' because they were small, clearly marked and easy to collect.

Although the Natural History Museum had the cream of Bates's collections, its narrow-minded curators declared themselves unimpressed by his early papers. Bates was disheartened by this. He wrote to Hooker: 'I have had so little countenance from scientific men since my return, as not to feel inclined to publish at all.' Darwin consoled him: 'I can understand that your reception at the B. [Natural History] Museum would damp you; they are a very good sort of men, but not the sort to appreciate your work.' Darwin felt that '*too much*' (his italics) systematic work had blunted the faculties of the Museum's specialists. He invited Bates to stay with him at Down House, just southeast of London. Joseph Hooker was also staying, and he later wrote: 'We there spent several days together, and I can remember none more enjoyable. There was such a fascination in his [Bates's] manner and character, and such a boyish hearty enjoyment of his return to his native country.... Darwin's appreciation of him was whole-hearted and all-round.' It was remarkable that Henry Bates, a hosier's apprentice from a provincial town, who had left school at fourteen and was entirely self-educated, and who had just spent a decade of his adult life deprived of all intellectual conversation or even speaking many words of English, could not only hold his own with two of the finest scientists of his day – both older and socially far grander than him – but actually impress and form genuine friendships with them. With Darwin this affection was to last to the end of his life, and Bates corresponded regularly with both men.

Writing to Darwin in March 1861, Bates said that there was one aspect of natural selection that entomology could help to illustrate, 'viz. that of mimetic analogies.... Some of these resemblances are perfectly staggering – to me they are a source of constant wonder & thrilling delight. It seems to me as though I obtain a glimpse of an *intelligent motive* pervading nature, as well as of the mighty never-resting wonder working laws that regulate all things.' Mimicry was to be the exciting theme of his next paper, which he read to the Linnean Society of London on 21 November 1861 (published in that Society's *Transactions* in the following year). This was another in his series on 'Insect Fauna of the Amazon', this time about heliconid butterflies (in Bates's time a mixture of two current subfamilies of Nymphalidae: Heliconinae and Ithomiinae). It was probably the most important piece he ever wrote, because of what he said about mimicry. He discussed remarkably exact counterfeiting among butterflies and moths, then wicked 'cuckoo' bees and flies which 'mimic in dress various industrious or nest-building bees' in order to get into their hives and become parasites. He then described species that imitate others of totally different genera, by being almost identical in colours and markings. These had understandably confused many lepidopterists, including Bates himself: 'Although I had daily practice in insect-collecting for many years, and was always on my guard, I was constantly being deceived by them when in the woods.... These imitative resemblances, of which hundreds of instances could be cited, are full of interest, and fill us with the greater astonishment the closer we investigate them; for some show a minute and palpably intentional likeness which is perfectly staggering.'

Mimicry should not be confused with protective camouflage, which occurs throughout the natural world. Bates gave examples of camouflage. 'Many caterpillars of moths... have a most deceptive likeness to dry twigs and other objects. Moths themselves very frequently resemble the bark on which they are found, or have wings coloured and veined like the fallen leaves on which they lie motionless... or are deceptively like the excrement of birds on leaves.' Among beetles, one species looked just like caterpillar dung, another resembled glittering drops of dew on leaf tips, a third had precisely the colour and shape of the bark of its favourite tree. There was a *Scaphura* katydid (of the family Tettigoniidae) that wonderfully imitated its main enemy: large sand wasps that were always hunting crickets in order to feed their eggs. And another 'pretty cricket' looked like tiger beetles (of the family Carabidae, subfamily Cicindelinae) and lived on trees that these beetles frequented. Predators could also use such tricks. Hunting spiders, for instance, might look like insects or inanimate objects among which they moved in order to fool their prey.

Mimicry was when a defenceless insect mimicked the shape and colouring of a dangerous or inedible one. Bates was fascinated by this discovery. One species of moth that frequented flowers in daytime 'wore the appearance' of a wasp, evidently to deceive 'insectivorous animals which persecute the moth but avoid the wasp.' Another extraordinary protective imitation was by a very large caterpillar, which startled him by stretching itself out from a tree's foliage in order to look like a poisonous viper. It dilated the three segments behind its head, with large pupillated spots on either side to resemble the reptile's eyes; and when it threw itself backwards its recumbent feet imitated the keeled scales on the snake's crown.

All species must have some means of 'holding their own in the battle of life'. Bates pondered why the *Heliconidae* group of butterflies was so flourishing, and concluded that it must be because a nasty smell secreted by glands near the anus made them unpalatable.* The juices that produced this powerful scent stained a collector's hands yellow, and it took repeated washings to remove the odour; Edward Clodd called them the 'skunks of the insect world'. Conspicuous patterns and colours of these butterflies – and of other insects – were a warning to predators that there must be a catch. If a bird did eat one and suffered its disagreeable taste, it would be a case of once bitten, twice shy. 'I never saw the flocks of slow-flying *Heliconidae* in the woods persecuted by birds or dragon-flies, to which they would have been easy prey.' And when they settled on leaves, they were not molested by lizards or predatory flies that often pounced on butterflies of other families. But even more fascinating was the less numerous family of Leptalidae (now the subfamily Dismorphiinae of the family Pieridae). These did not make the disagreeable secretion, so they were temptingly edible. Their answer was to

* The distinguished entomologist William Overal explains that Bates worked on a colourful group of Amazonian butterflies that he spelled *Heliconidae* (his italics). These he divided into two groups. One was the *Danaoid Heliconidae*, now known as *Ithomiini* (clear-wing butterflies) and closely related to milkweed butterflies (subfamily Danainae). The other group was, for Bates, the *Acraeoid Heliconidae*, now recognized as the subfamily Heliconiinae (long-wing or postman butterflies) based on the genus *Heliconius*. Both groups are within the family Nymphalidae and both consume plant poisons that confer protection from predators. Ithomiine butterflies obtain alcaloid poisons from the floral nectar they ingest as adults, while heliconiine caterpillars feed on poisonous *Passiflora* vines (the genus of passion-flowers). Ithomiine and lookalike heliconiine butterflies exhibit co-evolved mimicry complexes throughout the Amazon. Bates was not the last butterfly collector to be deceived in the field.

imitate the heliconids 'in their dress, and thus share their immunity'. In a flash of inspiration, Bates realized that for some inoffensive species mimicry is 'apparently its only means of escaping extermination by insectivorous animals.' This precarious survival strategy became known as Batesian Mimicry. It is the young entomologist's most enduring legacy.

Bates acknowledged that the *process* by which creatures came to mimic others was a mystery. But the *reason* for it was 'quite clear on the theory of natural selection recently expounded by Mr. Darwin in the 'Origin of Species.' For instance, edible *Leptalis* (now *Dismorphia*) butterflies gradually came to imitate disagreeable Ithomiini ones [PL. 76]. 'The principle can be no other than natural selection, the selecting agents being insectivorous animals, which gradually destroy those... varieties that are not sufficiently like *Ithomidae* to deceive them.'* Of course the topic is hugely complicated, with many varying types of imitation – some of them between different species that are both inedible.† At other times mimicry is not very faithful. But an insect's enemies 'who seek the imitator but avoid the imitated' would 'generation after generation' eliminate the poor mimics so that 'only the others [would be] left to propagate their kind.' This was survival of the fittest.

The officers of the Linnean Society were lukewarm about Bates's paper on mimicry; but this was compensated by fulsome praise from Darwin. He wrote that 'In my opinion it is one of the most remarkable and admirable papers I ever read in my life. The mimetic cases are truly marvellous, and you connect excellently a host of analogous facts.' Darwin modestly said that he was pleased that he had written little about mimicry in the *Origin of Species* since 'you have most clearly stated and solved a wonderful problem. No doubt, with most people, this will be the cream of the paper.... I cordially congratulate you on your first great work.' Reviewing the paper in a learned journal, Darwin wrote that Bates had very clearly described the markings and mimicry of the various species of butterfly. But he had then 'given these facts (of adaptive resemblance) the requisite touch of genius, and hit upon the final cause of all this mimicry.' Bates's discovery helped Darwin to elaborate a hypothesis he had advanced in *Origin of Species* into a theory in his next two books: *The Variation in Animals and Plants under Domestication* (1868) and *The Descent of Man, and Selection in Relation to Sex* (1871).

* William Overal comments that 'the clincher is that *Dismorphia* is a six-footed butterfly while the Ithomiini are four-footed.'

† A contemporary Brazilian lepidopterist, the German Fritz Müller, explained the problem of two inedible species mimicking one another by arguing that each new bird had to learn which butterfly was unpalatable by trying to eat one. Thus if several disagreeable species of butterfly or other insect looked alike, it would mean that fewer were killed as part of the predators' learning process. This complementary concept is known as Müllerian Mimicry.

When Bates returned to England he was 'much depressed in health and spirits' after his eleven years of collecting in remote forests close to the equator. Darwin wrote how sorry he was to hear that 'your health is shattered.... I can sympathise with you fully on this score, for I have had bad health for many years & fear I shall ever remain a confirmed invalid.' Because of Bates's illness, his work in the hosiery business in Leicester, his marriage, and his lectures and learned papers, he saw little prospect of ever writing a popular book about his remarkable travels. After two years, he had 'almost abandoned the intention of doing so. At that date I became acquainted with Mr. Darwin, who, having formed a flattering opinion of my ability for the task, strongly urged me to write a book.' Darwin kept encouraging Bates, with detailed advice on striking out every unnecessary word, keeping his 'style transparently clear & throwing eloquence to the dogs', and writing plenty about ants – since 'more notice has been taken on slave ants in the Origin than of any other passage'. Of course, Darwin's support was not wholly disinterested: he needed all the allies he could muster to confirm the theory of evolution by natural selection, and nobody could match the proof from Bates's years of fieldwork. Darwin urged the younger naturalist to use his own publisher, John Murray; and he offered to write praising the 'force of intellect & knowledge & style' in Bates's letters. Bates gratefully accepted this offer. By February 1862 he had passed the first five chapters to Murray, who was impressed by them. Darwin was amazed when he heard the financial arrangement: 'I never heard of such terms being offered for first work. You may depend he thinks very highly of your book.' But the author confessed that he found writing a tedious chore. Darwin replied: 'How gets on your book? Keep your spirits up. A book is no light labour.' Finally, Bates proclaimed that 'under this encouragement the arduous task is at length accomplished.'

Bates's *The Naturalist on the River Amazons* was published in two volumes in January 1863. It 'leapt at once into the front rank amongst books of travel and original observation.' Charles Darwin gave it a six-page 'Appreciation' in the *Natural History Review*, and immediately wrote an encomium to Bates: 'You have written a truly admirable work, with capital original remarks, *first rate* descriptions, and the whole in a style which could not be improved.... Every evening it was a real treat to me to have my half-hour in the grand Amazonian forest, and picture to myself your vivid descriptions.' Another tribute came from the famous ornithological artist John Gould, who had never been to South America. When Gould met Bates he exclaimed: 'Bates, I have read your book: I have seen the Amazons!' The entomologist David Sharp also praised this magnum opus: 'Its keynote is a profound love of nature, its mode of expression, simple truthfulness; that it should be permanently popular is a credit to our nation.' Another scientist, William

Distant, later wrote that the book was 'universally read, no naturalist's library is without it, and it will go down to posterity between Darwin's "Naturalist's Voyage", and Wallace's "Malay Archipelago".'

The book was such a success that the 1,250 print-run of the first edition sold out in a few months. John Murray hurried to publish a second edition in the following year, but (to Darwin's astonishment) asked a reluctant Bates to make a massive abridgement. Thus, the two-volume original's 774 pages were reduced to 466 in the single-volume second edition by omitting much scientific content – a process that we would now call dumbing down. Important passages were dropped. These included: how indigenous people were affected by having no domesticable animals; how insects like the sand wasp have an acute sense of locality; 'the origin of species as illustrated by certain butterflies of the genus *Heliconius*'; a full investigation of termites and their colonies; equally interesting observations about thirty-eight species of monkey; and scattered references to mimicry – which led to Bates's celebrated theory. The full first edition was later republished, but only in 1892 just after the author's death.

Among all the praise, some jealous rivals queried the magnitude of Bates's discovery of species new to science. He wrote to Hooker that 'On Monday morning I fell into a nest of hornets at the British [Natural History] Museum, in the shape of a knot of the leading curators (Dr. [John] Gray at the head) criticising fiercely my statement [in the book's Preface] of having found 8,000 new species out of 14,700.' Bates gave Hooker a detailed breakdown to support his claim, to which the distinguished botanist replied that he had never for a moment doubted the number of new species.

Hooker wrote that Gray was notorious for abusing everybody. Bates should turn a blind eye, ignore criticism from these small-minded men, and leave them in command of their molehill while he scaled a mountain with his book. 'Remember that entomologists are a poor set, and it behoves you to remember this in dealing with them. It is their misfortune not their fault.' But he warned Bates that these museum professionals instinctively regarded him as an interloper. 'It is extremely difficult to *establish a footing* in London scientific society.... There is no use blinking this fact, that to establish your position will take several years of good, hard, unremunerative scientific writing.' Hooker's warning was sadly correct. In 1862 a position became vacant as Entomological Assistant to the Curator of Zoology in the Natural History Museum. Bates applied eagerly, and he was supported by the Entomological Society. But he was rejected on technicalities of his age, thirty-seven, and lack of formal education – perhaps also because he was seen as too Darwinian. The job went instead to a member of the Museum's staff, an obscure poet called Arthur

O'Shaughnessy who had done no scientific research.* A year later, Bates hoped that the Museum would pay him to make a new catalogue of its butterflies; but he was again rejected, from academic and social snobbery – and jealousy of his unrivalled fieldwork.

In 1863, Bates married Sarah Ann and moved his young family to Haverstock Hill in London. His annual income was only £123. Of this, £100 came from his two brothers, as a compensation or reward for leaving their Leicester hosiery business; the rest was dividends from a few shares resulting from the £800 he had saved from his eleven years of collecting.

Early in 1864 the Royal Geographical Society needed to appoint a salaried chief executive, with the title Assistant Secretary. The Society, founded in 1830, had for three decades largely been run by unpaid honorary officers – and it was in an administrative and financial mess as a result. Francis Galton, a distinguished member of the Society's council, recalled that they were desperate to find an efficient administrator. John Murray strongly recommended Bates. The RGS councillors knew Bates only 'as an enterprising traveller, as a naturalist of high distinction, and as a charming writer.' But when Murray assured them of his 'methodical and orderly ways and his business-like habits' they eagerly offered him the job. (It is very rare for a publisher to call an author 'business-like' – Bates must have negotiated his book contract masterfully.) Charles Darwin also endorsed Bates as the ideal candidate. Although Bates himself would have preferred to work in zoology or entomology, 'his services had been rejected by the officials or the rules of the [Natural History] Museum'. Bates had a young family to support and needed an income, so he accepted. There was one other person who would have liked the post: Alfred Russel Wallace, returned from Southeast Asia in 1862. But when Bates was chosen, he good-naturedly conceded that his friend was 'much better qualified'.

Running the Royal Geographical Society was no sinecure. Most presidents served for short terms, so the Assistant Secretary was the Society's permanent 'directing mind... superintending the immense correspondence with travellers and geographers all over the world; the arrangement of evening meetings; often the revision, sometimes re-composition, of the papers read; the editing of the *Proceedings*; and the mass of work incidental to the departure and return of travellers, one and all of whom... went straight to Bates for counsel and help, and never came empty away.' In the event, Bates was to run the RGS for twenty-eight years. These decades were a golden age for British exploration and geography, and for

* It was said that O'Shaughnessy was the nephew of a mistress of a trustee of the British Museum, that he could not tell a moth from a butterfly, that he was too short-sighted to distinguish insects clearly and too clumsy to handle them delicately, and that he knew more about etymology (the derivation of words) than entomology.

the RGS – which grew from the amateurish club of 1864 to be the world's premier geographical society.

Bates was considerably helped by having for twenty-five years the dynamic Clements Markham as the Society's Honorary Secretary. Markham, five years younger than Bates, was working for the India Office. He was a notable traveller and writer and, as we have seen, the instigator of Spruce's removal of cinchona trees from Ecuador.* Looking back at Bates's achievements for the RGS, Markham recalled how he had reorganized its office, library and map room; improved accounting methods; made the presentation of papers easier for authors; and in 1870 oversaw the 'difficult and arduous work' of moving to larger premises at 1 Savile Row.† In 1873 Bates organized the lying in state of the great explorer David Livingstone in the Society's new seat, with thousands of admirers filing past the body before it was removed for burial in Westminster Abbey.

Markham, like all Bates's obituarists, stressed his invariable courtesy, charm and tact, and noted the constant information and advice he gave to travellers, explorers and geographers from all parts of the world. The Assistant Secretary worked extremely hard, not least in tactfully editing (often rewriting) papers for the *Journal* and *Proceedings*. With Francis Galton, he wrote much of the Society's famous *Hints to Travellers* manual; and he wrote, revised or contributed to a dozen other books. Bates's eleven years of fieldwork in the Amazon gave him unassailable authority in all this. As Markham wrote: 'His own rich stores of information were invaluable to all who needed help in their work, and over and over again they enabled him to supply a missing clue in some difficult inquiry, or to elucidate and piece together isolated facts.... Colleagues, ...geographers and travellers have always been... impressed by his ability and knowledge, ...the soundness of his judgement, [and the] sympathising and kindhearted way of giving his opinion or advice....' Galton echoed this. For him, Bates 'was always a frank and helpful adviser, kindly natured in taking the best view of things, and perfectly upright and trustworthy.... [His colleagues] in the Society regarded him with unreserved admiration, affection, and respect.'

During Bates's tenure, the RGS organized or heavily supported eighteen expeditions. These started with David Livingstone's Congo explorations, and the various Livingstone Relief Expeditions prior to Henry Stanley's meeting in 1871 (which was not

* Markham was later to orchestrate, with Hooker, the removal of rubber trees from the Amazon to Malaya. This removal of *Hevea brasiliensis* rubber seedlings eventually led to the great Southeast Asian rubber plantations and the collapse of the Amazon rubber boom that had been based on tapping wild trees. After Bates's death in 1892, Markham served for thirteen years as one of the Society's most distinguished presidents.

† Forty-one years later in 1911, this building on Savile Row became the tailor Gieves & Hawkes when the Society moved to its current premises, Lowther Lodge opposite Kensington Gardens. Both enterprises are still in these locations.

arranged by the RGS); Stanley's disastrous Emin Pasha Relief Expedition (1887–1888); there were many ventures in Central and East Asia, including those led by George Hayward, Ney Elias, Henry Trotter, Douglas Freshfield and Martin Conway; and the start of fascination with polar regions in George Nares's 'Farthest North' expedition of 1876. Each of these, and others, would have involved the Assistant Secretary in a great deal of hard but very rewarding work. (This author knows about organizing big research projects, having led or helped to launch eleven during my twenty-one years as the Society's fourth chief executive (now called Director). Throughout that time, I was inspired by having a portrait of Bates above my desk.) Bates's decades in office were the apogee of the British Empire, and the RGS was seen as an unofficial extension of the Foreign and Colonial Offices. Bates encouraged explorers to report on the natural resources and potential of places they visited – which could have been a prelude to colonization. But he also stressed scientific discovery, which has always been the Society's main concern. Bates was also keen on getting geography recognized as an academic subject. The RGS fostered prizes for geography in schools, with Bates as an examiner, and in 1877–1879 organized a series of lectures on the physical geography of oceans, land masses and plant distribution. There was a move to improve geographical education, and Bates having 'pondered long, as was his wont... came to an affirmative conclusion, and took the matter up with characteristic energy.' So the Society was instrumental in persuading both Oxford and Cambridge universities to appoint Readers in Geography.

All this work for the RGS was demanding on Bates's time and energy. Wallace regretted that his friends' duties there hampered his work in natural history. This is true. But running the Society for so many years and building it into a world-class institution was a great achievement. The Amazon entomologist must have enjoyed meeting so many of the world's explorers, from Richard Burton and Henry Stanley to the brilliant Russian anarchist Prince Peter Kropotkin. And, after his eleven years on the Amazon, he could hardly have been expected to undertake another great research expedition.

Bates continued to publish papers on butterflies and beetles. But he ceased to ponder evolution and the distribution of species, or to theorize about nature. Instead, his work was systematic – the 'naming, arranging, labelling, etc.' for which he had once scorned the curators of the Natural History Museum. He defended this in an address when he was President of the Entomological Society: he now argued that the insect world was so vast that classification was a priority. David Sharp, a successor as president of that Society, agreed: 'Some have expressed regret that, since his paper on Mimicry, he has favoured us with no further wide generalisations or ingenious suggestions. The reason of this is

[that]... descriptive entomologists... know how immense is the work to be accomplished....' Even within systematics, Bates was innovative: he reclassified various families of butterfly, and his 'epoch-making' changes were 'universally followed' by lepidopterists. Another leading entomologist, Frederick Godman, recalled how Bates worked with characteristic energy on large consignments of beetles for a survey of them in Central America. He worked on longicorns (of the family Cerambycidae) between 1879 and 1886, on Geodephaga from 1881 to 1884, and on lamellicorns (of the family Scarabaeidae and allies) from 1886 to 1891 – all, of course, in spare time from his job at the RGS.

As we saw in his Amazonian travels, Bates was by nature cautious, conservative and pragmatic. He gradually sold his personal collections to pay for his two houses (one in London and a cottage in Folkestone) and his family. He educated and then established two sons as sheep farmers in New Zealand and another as an electrical engineer, and he married off his daughters. In the end, he was left with only his collection of Carabidae, ground beetles, on which he became a world authority.

Bates received some recognition: fellowships of the Zoological Society, the Entomological Society (of which he was twice President), in 1871 the more prestigious Linnean Society, and in 1881 – after a vigorous campaign by Darwin – the most illustrious Royal Society. In 1872 the Brazilian Emperor Pedro II, on a state visit to Britain, invested Bates (and also Clements Markham) as a knight of the Order of the Rose. But the Royal Geographical Society, which in 1892 gave a gold medal to Wallace, failed to make any award to its own loyal servant and exceptional explorer. Bates loved going to meetings of the various societies, as well as all those in his own RGS (which he had to attend), and he was a keen member of the Kosmos dining club. He died in February 1892, of influenza and bronchitis, aged sixty-seven and still actively running the Society.

The Brazilians were pleased by Bates's eleven years' work in their forests – even though he did not leave any collections in their country, as modern scientists do. The book was published in Portuguese in 1876, has remained in print in Brazil ever since, and is often cited by modern Brazilian authors. Even though Bates criticized some Brazilians and their morals, his love of the country and its nature shone through. A governor of the state of Amazonas wrote in 1892 that Bates had been the first to predict a splendid future for the region 'and to describe its bewildering splendours.... Many an inhabitant of our regions still retains a vivid remembrance of the English naturalist... and his bold and perilous excursions in our forests.'

The final word on Bates should go to the writer Grant Allen: 'Henry Walter Bates had, in my humble judgement, one of the profoundest scientific intellects I have ever known.' Allen felt that Wallace and Bates were 'Darwinians before Darwin… and few were more learned, none more self-effacing, more modest, more retiring than "Bates of the Amazons".' Allen recalled (with some hyperbole) an evening in the drawing room of Edward Clodd, Bates's neighbour in north London. There were about ten guests, all distinguished explorers or writers. 'Bates broke his wonted reserve in a rare fit of communicativeness, and poured out to a small and sympathetic company the whole story of his hardships…. With marvellous pathos and child-like simplicity he told us, in his pure and exquisite English, a tale of single-hearted devotion and strange labour…. He had the finest forehead I ever beheld upon a human face; and as he talked… we all listened, open-mouthed…. Bates told us with hushed breath how on that expedition he had at times almost starved to death; how he had worked with slaves like a slave for his daily rations of coarse food; how he had faced perils more appalling than death; and how he had risked, and sometimes lost, everything he possessed on earth with a devotion that brought tears into the eyes of grown men who heard him.'

Richard Spruce was the last of the three naturalists to return from South America, landing in Southampton on 28 May 1864. He had been away longest, for fifteen years, and at forty-seven he was older than the others. He returned with his health shattered, and most of his modest savings gone with the crash of the highly reputable trading company in Guayaquil. There was no one to welcome him: his father had died in 1851 and his impoverished mother Mary in the year before his return.

However, Spruce was pleased to be back. He spent a while in London visiting his friends: Alfred Wallace, George Bentham at Hurstpierpoint and Sir William Hooker and other botanists at Kew. He met Clements Markham, organizer of the cinchona quest, who 'had the great pleasure of receiving him in my house' in London. He also met another correspondent, Daniel Hanbury. Hanbury's family had a thriving pharmaceutical business in London, and the young Daniel had urged George Bentham to seek cinchona trees for their quinine. Bentham had put him in touch with Spruce, and they corresponded for several years. Hanbury and Spruce 'at once established an intimacy which quickly ripened into a close friendship'. The modest Spruce took lodgings in Kew Green, in order to work quietly on his plant collections. He explained to Hanbury that he preferred this solitary existence. People had urged him to go to meetings of learned societies and 'in short [to] make a lion of myself in every possible way…. However, as you know, I have nothing of the lion in my constitution.' He did not mind dangers of exploration, 'but to stand before an assemblage of men and talk to them about myself & my doings, demands a degree of pluck to which I have no pretention.'

In 1865 one of Spruce's sponsors, the Earl of Carlisle, arranged for him to return to his roots in Yorkshire. He was settled rent-free in the village of Welburn, which belonged to the Earl's Castle Howard estate, 20 kilometres (12 miles) northeast of the city of York. He lived in Welburn for eleven years – it was where his father had taught and he himself was schooled – and then for seventeen in Coneysthorpe, another of the estate's villages just outside its walled park. Wallace described Spruce's stone house in the latter village as 'a humble dwelling... sitting-room about 12 feet [3.5 metres] square, with a bedroom of equally limited proportions'. It was (and still is) a sturdy two-storey, semi-detached cottage, with a tiny garden giving onto the village's broad grassy main street. It is now called 'Spruce Cottage' and bears a plaque about him [PL. 66].

When Spruce returned to Yorkshire, his friend Daniel Hanbury acted as his informal agent in London, and sent him any medicines, books or 'special delicacies' he wanted to buy in the capital. In return, Spruce gave Hanbury a mass of botanical information for his pharmaceutical research. For the next eleven years, until Hanbury's death in 1875, the friends wrote to one another almost daily.

Spruce's health continued to be appalling. He could scarcely move. Eminent doctors in and near London were baffled. They sought in vain for a digestive cause but 'found it easier to hide their ignorance under the accusation of hypochondria, and to prescribe brandy-and-water every three hours.' A Yorkshire doctor thought that the pain was due to a 'stricture of the rectum': his simple treatment of 'enemas and gentle opiates' improved Spruce's condition to the extent that he could work at his microscope for short periods and even take brief walks. But by 1867 Spruce was able to write only reclining in his easy-chair with a book across his knees as a table. Attempts to sit up and work at his microscope 'brought on bleeding of the intestines'. During the final twenty years of his life, from 1873 onwards, his friend Wallace wrote that 'he rarely went far outside his small cottage, alternating only from chair to couch, with an occasional walk round the room or in the very small patch of garden. What was especially trying to him was that, for months or even for years together, he was unable to sit up at a table to write or use a microscope, and could never do so for more than a few minutes at a time with intervals of rest on a couch.'

Richard Spruce never married. But he was so amiable that he always had devoted carers. For the last two decades of his life 'he was carefully looked after and nursed by a kind housekeeper and a little girl attendant, who were also his friends and companions'.

After the death of Daniel Hanbury, Spruce found another bryologist (as liverwort specialists were known) to be his professional helper, factotum and confidant. This was Matthew Slater, who lived at Malton, the nearest town to Coneysthorpe. Slater described the invalid as 'courteous and dignified in manner but with a fund of quiet humour which rendered him a most delightful companion. He possessed in a marked degree the faculty of order, which manifested itself in the unvarying neatness of his dress, his beautifully regular handwriting and the orderly arrangement of his surroundings.' Both in remote parts of the Amazon and Andes, and 'in his little cottage in Yorkshire, his writing-material, his books, his microscope, his dried plants, his stores of food and clothing – all had their proper place, where his hand could be laid upon them in a moment.' Spruce wrote to Slater 'I do not know what I sh'd do without your aid', and he made him his sole botanical executor. Another lifelong friend was George Stabler, 'who, both as schoolmaster, invalid, and botanist, was in complete sympathy with him'. His friends mentioned that Spruce was always in good humour, and was something of a musician (he played the fiddle and composed a hymn for his local church) and a keen chess player. Like both Wallace and Bates, Spruce was a tremendous letter-writer, answering queries from botanists in Britain and other countries; and a few enthusiasts, including the Duke of Argyll, found their way to the humble cottage to meet the collector in person.

Unknown to the Richard Spruce, his friends – particularly the Earl of Carlisle and his sister – 'had the greatest difficulty in obtaining for him, first [from Lord Palmerston's government] the small pension of £50 a year in 1865, and in 1877, through the long-continued and earnest representations of... Clements Markham, a further pension of £50 from the Indian Government'. These grudging pensions were, of course, recognition of Spruce's labours to get red-bark cinchona to India – where the plants saved thousands of people from malarial suffering, and had also earned considerable income for the nation.

Often prostrate in his chair, or painfully propped at the table, Spruce devoted all his time to writing-up and classifying his magnificent collections. He had sent back from South America over 7,000 different species of flowering plants. Many were conveyed to George Bentham and other leading botanists for further study and classification. Mosses went to one William Mitten: Spruce was angered by the way in which this expert had neglected the collection, 'taken the best specimens of everything for himself', garbled Spruce's attributions, and then pillaged Spruce's catalogue of them for his own *Musci Austro-americani* ('South American mosses'), published by the Linnean Society in 1869. (Mitten became Alfred Wallace's father-in-law, when his daughter Annie in 1866 married Spruce's friend.)

With palms the situation was far better. The great Bavarian botanist Martius, who had been in Brazil and the Amazon in 1818–1820, published a volume on Amazonian palms in his splendid three-volume *Historia naturalis palmarum* (1823). Spruce had always sought to collect palm novelties not already recorded by Martius. The German had heard about Spruce and wrote a series of letters begging him to write about a natural order of plants for his proposed Flora of Brazil. His letters began 'My dear Spruce' and ended 'Your affectionate and admiring friend'. Spruce sadly had to decline, because of his health and other commitments. When Alfred Wallace returned from Amazonia he published in 1853 the little book *Palm Trees of the Amazon and Their Uses*; but as we have seen, much though he liked Wallace as a friend, Spruce was very critical of the book's amateurish botanical descriptions and drawings. In 1865 Joseph Hooker of Kew Gardens sent Spruce his collections of Amazonian palms for him to study in his tiny cottage. The result of this work was the 118-page *Palmae Amazonicae*, published in 1869 as a special issue of the Linnean Society's *Journal (Botany)*. Andrew Henderson of the New York Botanical Garden still considers this 'one of the most important papers on neotropical palm taxonomy.' From five localities in the Amazon region, Spruce 'described 47 new species from at least 62 collections. Of these, 10 are currently accepted at either specific or varietal rank.' More important than his attempt to find new species of palm tree were his acute observations of flowers, fruits and fibres of individual species, his brilliant essay relating distribution of palms to underlying geology in different parts of South America, and many other insights and guesses by such a fine botanist.

As we know, Spruce's favourite class of plants was liverworts. Back in Yorkshire, his greatest enjoyment came from hours of study of minute hepatics under his microscope. Curiously, his first publication on this subject was not until 1876: a paper about a new genus named *Anomoclada*. During those twelve years after his return, he produced a dozen articles on such diverse subjects as cotton-growing in northern Peru, narcotics, poisonous snakes and insects, volcanic tufa, grass fertilization and the great work on palm trees. But although he wrote these other pieces, Spruce devoted most of his time to liverworts, which Wallace described as 'the joy of his early manhood and the consolation of his declining years'. During the previous century, a handful of botanists had written about hepatics of the Americas; but their studies had been based on herbarium material rather than years of observation in the field. So Spruce's greatest work was *The Hepaticae of the Amazon and the Andes of Peru and Ecuador*, published as a book in 1885 (having just appeared in Latin as a separate volume of *Transactions and Proceedings of the Botanical Society of Edinburgh*). As the Dutch expert on hepatics Rob Gradstein recently wrote, 'The fact that Spruce knew all his species in the field, together with his outstanding skills as a taxonomist, his eye for detail and his passion for the hepatics, must be reasons why his book is so good.' Another modern botanist, Raymond Stotler, said that Spruce's journey is 'unparalleled in its consequences for bryology. He not only provided future generations of botanists with a wealth of information about these plants in his published works, but also left a legacy in his distribution of specimen duplicates [in various herbaria].... Access to these collections allows one to share, at least in part, some of the fascination that these plants held for Spruce, as he discovered each and every one.' When it came to systematics, 'Spruce, with the comparatively limited facilities of the day, could make astute observations and interpretations and align seemingly unrelated taxa into natural groupings, or, conversely, cleave a large group of species within a single genus and sequester taxa into workable groups (subgenera).'

In the 600-page book, Spruce recorded 560 species of hepatics, of which almost 400 were new. The novelties included 3 genera, 39 subgenera, 374 species and 132 varieties. (Spruce was cautious about claiming that a genus was new: subsequently most of his subgenera have been raised to genera.) The majority occurred in the Andes, with 38 per cent in the lowland Amazon, and a few species common to both environments. As a lasting tribute, two genera of liverworts have been named after their discoverer: *Sprucella* (by Stephani in 1886), and *Spruceanthus* (by Verdoorn in 1934). Another great tribute was that the book was reprinted by the New York Botanical Garden roughly a century after it first appeared, because 'this work still remains the standard reference' for these plants in South America.

During these years of study and writing, the cinchona trees that Spruce had collected and propagated were thriving in India. As we have seen, the Kew gardener Robert Cross brought them successfully back to England and then took them out to India. He crossed the desert between the Mediterranean and Red Sea in the cooler month of March 1861 – unlike Markham who had taken his consignment of 'yellow-bark' from Calisaya in Peru out in midsummer of the previous year, so that many died of heat. Spruce's red-bark trees thrived marvellously on the Nilgiri hills of southern India, because the local horticulturalists planted them – as Spruce had recommended – on an exposed hillside with plenty of sun and rain. After only a few years, Markham reported gleefully that 'The old jungle had disappeared, and in its place were the rows of Peruvian bark trees with their graceful and beautiful foliage.' By 1866 there were 244,000 trees planted out. Twelve years later, there were half a million. There were eventually cinchona plantations all over India, and in Ceylon (Sri Lanka) there were over a million trees. Clements Markham had envisaged quinine alleviating malaria throughout the subcontinent's bazaars; but his philanthropic ideal was thwarted because most of the bark was sent to England to be manufactured into quinine. Markham was angry that the masses of India were deprived of the febrifuge so that the government could 'enter upon a speculation which will no doubt pay well, but which is entirely foreign to its functions.' This was, however, what happened. Anti-malarial quinine became available only to the British elite and particularly the Army. Expatriates and soldiers disliked the bitter taste of the alkaloid, so they took to mixing their 'Indian tonic water' with gin. Thus Richard Spruce inadvertently helped to invent the cocktail gin and tonic.*

Unlike the two younger naturalists, Spruce received few honours in his lifetime. Soon after his return, Markham got the Royal Geographical Society to make him an Honorary Fellow, in recognition of his maps and explorations in the upper Rio Negro and, of course, his successful hunt for cinchona trees. The University of Dresden awarded him an honorary doctorate for his botanical achievements. Late in life he became an Associate of the Linnean Society and an honorary life member of the Yorkshire Naturalists' Union. Far more enduring tributes were the numerous plants that bear his name.

Spruce died of influenza at Coneysthorpe at the end of 1893, aged seventy-six – a respectable age for someone who had endured fifteen years of tough exploration and thirty years of chronic disability. All the learned journals carried eloquent obituaries that stressed what a fine and delightful man he had been and the magnitude of his botanical achievements. His friend and admirer Alfred Wallace gave him a magnificent posthumous present. He laboriously assembled thousands of letters, papers, journals, manuscripts and publications and wove them into the narrative of *Notes of a Botanist on the Amazon and Andes* – two volumes containing 1,055 pages plus Wallace's own 32-page Biographical Introduction – published by Macmillan in 1908. But after that Spruce's fame faded.

The man who revived awareness of Spruce's greatness was Richard Evans Schultes of Harvard University. Schultes was himself the founder of the discipline of ethnobotany (the study of indigenous peoples' use of plants), a pioneer in scientific study of hallucinogenic plants famed as discoverer of the 'magic mushroom' peyote, an extraordinary explorer of Amazon forests in search of blight-resistant rubber trees during the Second World War, and hailed as the twentieth-century's outstanding botanist of tropical South America. He was unqualified in his admiration of Richard Spruce. The Harvard professor's student

* In later years, the Dutch in the hills of Java won the contest for producing quinine. This came from good luck and good management. The luck was to buy some *C. calisaya* seeds from a British alpaca dealer Charles Ledger and his brilliant and brave Bolivian Aymara helper Manuel Mamani. By 1874 these few seeds had grown into 12,000 trees. The Dutch managers then developed a cinchona hybrid that yielded twice as much quinine as British plantation trees, as well as new methods of husbandry. By the early twentieth century the Dutch had a virtual monopoly, supplying 97 per cent of the world's quinine.

and biographer Wade Davis wrote that 'Schultes's love of Spruce, a raw atavistic association bordering at times on obsession, became his strength, allowing him to endure, encouraging him always to achieve more [in his own amazing explorations], and providing his closest experience of spiritual certainty. Asked... whether it might be possible that he had, subconsciously or unconsciously, modelled his life and career on that of Spruce, Schultes replied, "Neither. It was conscious."' The professor described Spruce's 'epoch-making phytogeographic studies and collections in the Amazon Valley and the northern Andes' as one of the 'most extraordinary feats of botanical exploration'. He found it amazing that so few people had heard of Spruce, 'undoubtedly one of the greatest explorers of all times'.

From the 1950s onwards, Schultes published a string of papers and books that frequently praised Spruce and his collections. In 1970 he and others launched an appeal for a plaque on Spruce's cottage in Coneysthorpe. This was rapidly subscribed, and in September 1971 George Howard of Castle Howard presided over a ceremony of unveiling. In 1993, on the centenary of Spruce's death, the Linnean Society held an annual regional meeting in Yorkshire, in honour of one of that county's most distinguished sons. The proceedings of the conference's twenty-seven papers were edited by Mark Seaward and Silvia FitzGerald and published as *Richard Spruce (1817–1893): Botanist and Explorer*, by the Royal Botanic Gardens, Kew. Kew's Director, Sir Ghillean Prance, wrote that the conference had been a remarkable experience. 'The diversity of papers presented and continuing research reflect not only the spirit and genius of Richard Spruce, courageous explorer of the Amazon and meticulous observer of plants and people, but also the rich diversity of the Amazon itself.'

Several tributes to Spruce quote the same eloquent passage in one of his letters to Daniel Hanbury: 'I look on plants as sentient beings, which live and enjoy their lives – which beautify the earth during life and after death may adorn my herbarium.... [Even if they have no medicinal or commercial value to man] they are infinitely useful where God has placed them.... They are at the least useful to and beautiful in themselves – surely the primary motive for every individual existence.'

The lives of the three naturalists were transformed by their time in South America. They gained much from those often tough years of fieldwork. But what did they give back to the host countries? They got on well with almost everyone they met, brought knowledge of the outside world to people who had often met no foreigners, and made some close friendships – notably Bates with Antonio Cardozo in Ega, Wallace with João Antonio Lima on the Rio Negro, and Spruce with the Santander family in Ambato. Bates's *The Naturalist on the River Amazons* was the first comprehensive and sympathetic picture in English of that part of Brazil. It sold well and had a powerful impact, which was why the Emperor of Brazil knighted the author. (Wallace's book was also good, but its sales were minimal; and Spruce's *Notes of a Botanist* was not compiled – by his friend Wallace – until after his death.)

The naturalists' main gift to the host countries was in scientific knowledge about them. Wallace, and to a lesser extent Spruce and Bates, made pioneering anthropological studies of indigenous peoples, including their rock art and use of hallucinogens. This was disinterested research that yielded nothing to the collectors' meagre earnings. Far more important were their tremendous discoveries in natural history, notably Bates's 8,000 species new to science and all the plants identified by Spruce. They removed all this material and lodged none in Brazilian or Andean institutions, as modern scientists do: but they could not do so, because such museums and research institutes did not exist in their day. Their contribution to greater understanding of the Amazonian and Andean environments was a lasting achievement, of which they could be proud.

67

68

69

70

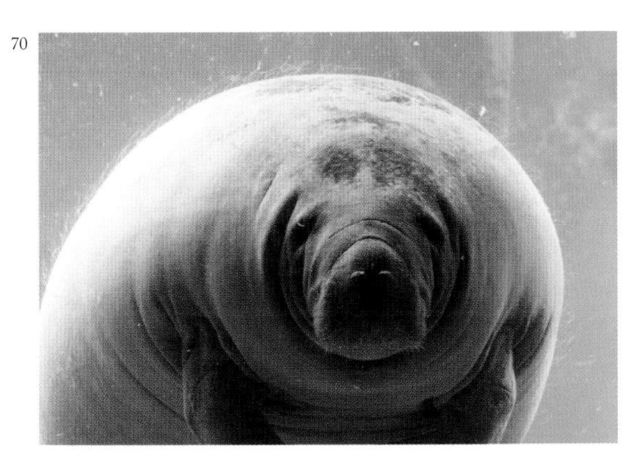

71

72

67 A hyacinth macaw, eagerly sought by Wallace and Bates on the Tocantins river.

68 Wallace collected brilliantly coloured cocks-of-the-rock (*Rupicola rupicola*) after an arduous expedition up the Cubaté tributary of the upper Rio Negro.

69 The naturalists collected many species of toucan, and Bates and Wallace also kept them as pets.

70 Manatees (*Manatus inunguis*) are freshwater mammals, sought-after and endangered because of their meat and fat, but difficult to catch. Bates and Wallace each studied and ate these fascinating animals.

71 Camouflage is common throughout Amazonia. Here a long-horned grasshopper (*Tettigoniid*) perfectly imitates the leaves and twigs of its favourite plant.

72 The cutia (*Dasiprocta aguti*) is a small, shy and largely nocturnal rodent. Bates once embarked on a nighttime hunt for cutia and slightly larger paca rodents.

73

74

73 Travellers sail for 1,000 km (600 miles) up the mighty Rio Negro before encountering its first rapids, at São Gabriel da Cachoeira. Wallace's and Spruce's indigenous boatmen spent days battling up these ferocious falls, before reaching the small fort and village of São Gabriel. This illustration shows three common types of boat and different methods of paddling.

74 Spruce bought a gaudy hammock fringe, of hummingbird, cock-of-the-rock, toucan and other feathers, and sent it to his patron the Countess of Carlisle, for display in Castle Howard – where it is still on view. It took a girl in Tomo four months to make.

75 When Spruce reached Esmeralda in 1853, he marvelled that its location was the most magnificent he had seen in South America. It lay beside the Orinoco, was overlooked by Cerro Duida, and was visited by attractive Maquiritare indigenous people. But a plague of biting gnats made it almost uninhabitable.

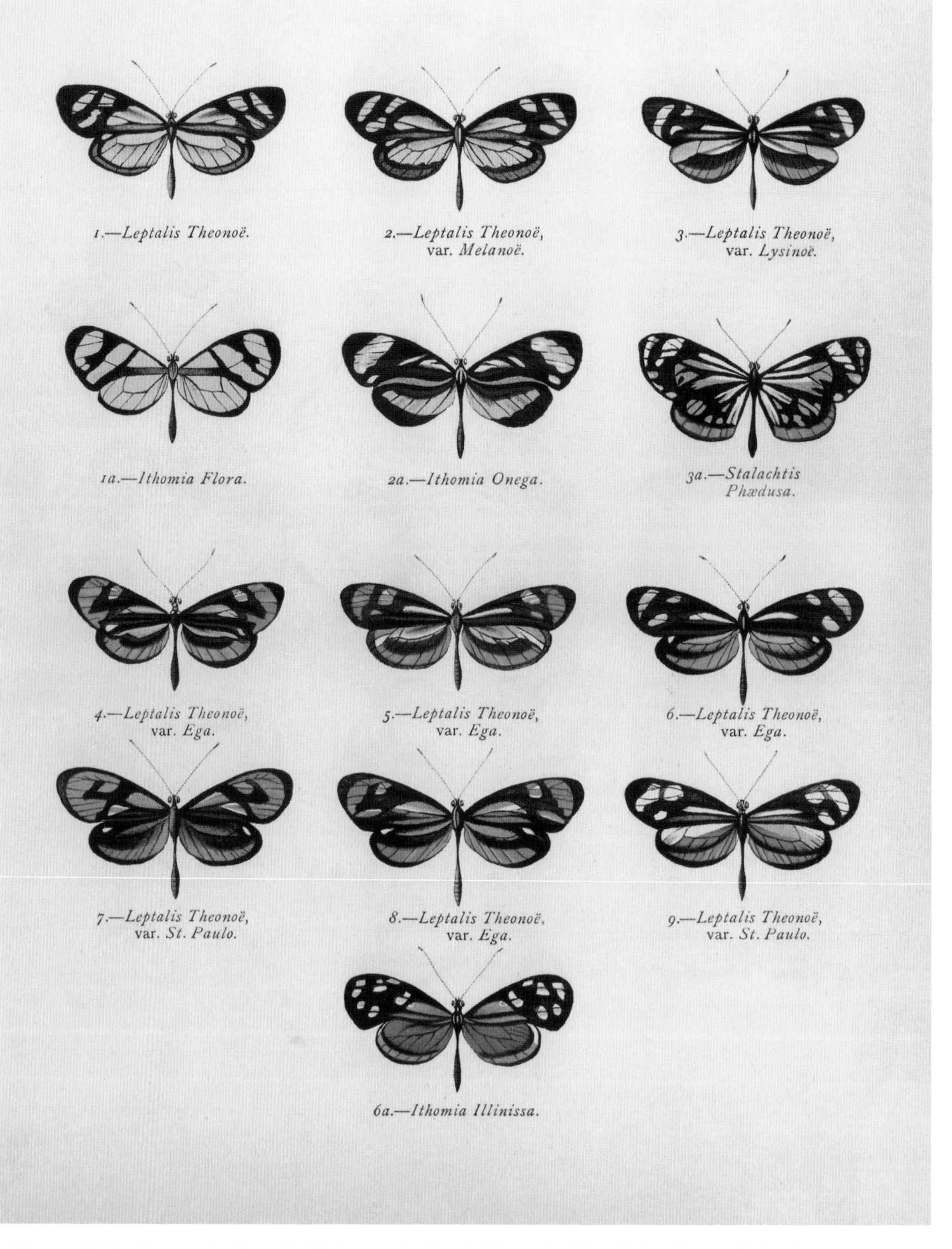

76 'Batesian Mimicry' is when a harmless and edible insect survives by mimicking an inedible or foul-smelling one. In this illustration to his 1862 paper for the Linnean Society, Bates showed edible *Leptalis* (now *Dismorphia*) butterflies exactly imitating the wing patterns of unpleasant-tasting Ithomiini of the second and bottom rows – in the hope that birds would hesitate to eat them.

TIMELINE

	Alfred Russel Wallace	Henry Walter Bates	Richard Spruce
	1823 8 January, born in **Usk**, Wales 1828 Wallace family moves to **Hertford** 1828–1836 Primary schooling, Hertford Grammar School 1837 In **London** with apprentice brother John 1837–1843 Land-surveying with older brother William 1844–1846 Teaching in primary school in **Leicester** 1844 Meets Bates, starts collecting insects 1846–1847 In **Neath**, Wales, with brother John	1825 8 February, born in **Leicester** 1838 Ceased schooling and becomes hosiery apprentice. 1839–1843 Self-taught at Mechanics' Institute 1843 First published article, on beetles 1844 Bates and Wallace meet in **Leicester**	1817 10 September, born in **Ganthorpe**, Yorkshire Primary school in **Ganthorpe** 1840–1844 School teacher near **York** 1845–1846 Botanical collecting in the **Pyrenees** 1846–1848 Botanical work, teaching in Yorkshire
1848	In **London**, plans Amazon trip with Bates 26 April – 28 May Sails from **Liverpool** to **Belém** 26 August – 30 September On **Tocantins** with Bates October, Wallace and Bates agree to go separate ways 3 November – January 1849 To **Mexiana, Marajó** islands	26 April – 28 May, sails from **Liverpool** to **Belém** 26 August – 30 September, on **Tocantins** with Wallace October, Wallace and Bates agree to go separate ways October – 13 February 1849, at **Caripí** near Belém	
1849	In Belém May – July, to **Guamá** and **Capim** rivers, for tidal bore 12 July, brother Herbert Wallace arrives, with Spruce 22 August – 19 September, sails up to **Santarém** September – October, to **Monte Alegre** for rock art November – 31 December, **Santarém–Óbidos–Parintins–Manaus**	February – June, in **Belém** 8 June – 19 July, returns to **Cametá** on **Tocantins** 5 September – 11 October, sails to **Santarém, Óbidos** 11 October – 19 November, in **Óbidos** 19 November – 18 January, sails to **Manaus** via **Parintins**	7 June – 12 July, sails from **Liverpool** to **Belém** 10–27 October, sails from **Belém** to **Santarém** 19–28 November, to **Óbidos** 17 December, to **Trombetas/ Erepecuru** rivers
1850	In **Manaus**, excursion up **Rio Negro** in January May – July, visit to **Manaquiri** on **Solimões** 31 August – 19 October, up **Rio Negro** to **São Gabriel** rapids, **Guia** November – December, **Cubaté** river and hills for cocks-of-the-rocks	18 January – 26 March, two months in **Manaus** 26 March – 1 May, sails to **Ega (Tefé)** May 1850 – March 1851, ten months in **Ega** September – October, excursions for turtles and turtle eggs	January – October, back in **Santarém** for nine months 8 October – 10 December, to **Vila Nova–Lagos channels–Manaus**

	Alfred Russel Wallace	Henry Walter Bates	Richard Spruce
1851	27 January – 31 March, from **Guia** up **Rio Negro** into **Venezuela**, portages to and from **Javita** April, returns from upper **Rio Negro** to **Guia** June – August, on **Uaupés**, sees little-contacted Tariana 1 – 15 September, sails to **Manaus** November, sails back up **Rio Negro** to lower **Uaupés** December – January 1852, severe malaria at **São Joaquim** on lower **Uaupés**, sees Spruce in **São Gabriel**	March – April, sails from **Ega** down to **Belém** 9 June, Herbert Wallace dies of yellow fever November, **Belém** to **Santarém**	In and around **Manaus**, including to **Lages** June, excursion to **Manaquiri** on **Solimões** 14 November – 15 January 1852, sails up **Rio Negro** to **São Gabriel**
1852	16 February – April, up **Uaupés** beyond **Iauareté** to **Colombia** May, descends **Rio Negro** 17 May – 10 June, in **Manaus** 10 June – 2 July, down **Amazon** to **Belém** 12 July, sails to **England** on *Helen*; 6 August, shipwreck fire; 15 August, rescued from open boat; 1 October, reaches England	In Santarém six months until: 8 June – 21 September, up **Tapajós** and **Cupari** rivers September 1852 – August 1853, eleven months in **Santarém**	January – August, in and around **São Gabriel da Cachoeira** 21 August, leaves **São Gabriel** for **Uaupés**, ascends to **Iauareté** September, five months at **Ipanoré** on **Uaupés**
1853	October, *Palm Trees of the Amazon* published December, publication of *A Narrative of Travels on the Amazon and Rio Negro*	In Santarém until: 26 August – 30 November, **Alter do Chão** on lower **Tapajós**	March – 11 April, from **Uaupés** to **San Carlos de Río Negro** 24 June, fear of Indian uprising 19 July, climbs part of **Cucui** outcrop 27 November – 24 December, up **Casiquiare** to **Esmeralda** on **Orinoco** 28 December – 6 January 1854, up **Cunucunuma** to Maquiritare people
1854	March, sails to **Singapore**. 1854–1862, Travels and collecting in **Malaya**, **Sarawak** and **Dutch East Indies**	In **Santarém** for nine months until: September – December, to **Vila Nova (Parintins)** on **Amazon**	27 January – 24 February, explores **Pacimoni** tributary of **Casiquiare** 24 February – 26 May, in **San Carlos** 26 May – late June, up **Guainía**, watershed to **Atabapo**, down **Orinoco** to **Maipurés** July – August, returns to **San Fernando de Atabapo**, acute malaria 13–28 August, returns to **San Carlos de Río Negro** 23 November, leaves **San Carlos**, foils murder attempt by boatmen 23 November – 22 December, descends **Rio Negro** to **São Gabriel**, **Manaus**
1855	'Sarawak' paper, on distribution of species	**Santarém** for six months until: 11–19 June, steamer up to **Ega**	14 March – 2 April, steamer from **Manaus** to **Nauta**, Peru 16 April – June, up **Huallaga** from **Nauta** to **Tarapoto**, Peru
1856		In **Ega** sixteen months until: 7 September – 25 January 1857, 4½ months to **Tonantins**, **Fonte Boa** on **Solimões**	Twenty-one months in **Tarapoto**, with excursions to nearby mountains and forests

	Alfred Russel Wallace	Henry Walter Bates	Richard Spruce
1857	Notices division of fauna by 'The Wallace Line'	In **Ega** seven months until: 5 September – 2 February 1858, five months at or near **São Paulo de Olivença** on **Solimões**	23 March – 1 July, from **Tarapoto** down **Huallaga**, up **Pastasa** and **Bobonaza** rivers to **Baños**, Ecuador Excursions around **Baños**
1858	February, 'Ternate paper' on evolution by natural selection 1 July, read to Linnean Society with papers by Darwin	February, ill with malaria, returns to **Ega** Twelve months in **Ega**, in poor health	Botanical collecting from **Ambato**, **Riobamba** and **Quito**
1859	Darwin publishes *On the Origin of Species by Means of Natural Selection*	3 February – 17 March, steamers **Ega** to **Manaus** and **Belém** 2 June, leave **Belém** for **New York** and **London** 1859–1862, in **Leicester** for the family hosiery business, but often in London to supervise his collections	July, receives commission to collect cinchona fever-bark trees July – October, seeks cinchona at **Alausi** October – March 1860, cinchona research, **Ambato** and **Riobamba**
1860		21 November, paper on butterflies to Linnean Society, that led to theory of Batesian Mimicry	29 April, medical breakdown, paralysis 12 June – December, collects cinchona seeds, propagates saplings in forests below **Mount Chimborazo** 24–27 December, raft to **Guayaquil**, specimens shipped
1861		1861–1873, lectures and papers on insects of the Amazon Corresponds with and meets Darwin, stays at Down House	**Daule** and **Chonana** near Guayaquil, for health and collecting 11 October, loss of savings, with bankruptcy of Guayaquil trader
1862	April, returns to **London**	2 February, Sarah Ann Mason has their daughter	Seven months on dry coast of **Equador**, for health In **Guayaquil** trying to recover funds
1863		15 January, moves to **London** and marries Sarah Ann January, publishes *The Naturalist on the River Amazons*	January, sails to **Paita**, Peru Fifteen months in dry **Piura** and **Amotape**, Peru
1864		Second, abridged edition of *The Naturalist* Appointed Assistant Secretary, Royal Geographical Society	1–28 May, sails **Paita–Panama–England**
	1866 Marries Annie Mitten 1869 Publication of *The Malay Archipelago* 1892 Fellow of The Royal Society, medals from the RS, Royal Geographical and Linnean Societies 1905 Publication of *My Life* 1908 Awarded The Order of Merit 1913 7 November, dies in his house in **Dorset**, aged seventy-five	1881 Elected Fellow of the Royal Society; previously of the Zoological, Entomological and Linnean Societies 1892 February, dies in **London** aged sixty-seven, while still running the RGS	1865–1876 Given a cottage in **Welburn**, Castle Howard, Yorkshire 1869 *Palmae Amazonicae* published by the Linnean Society 1876–1893 Moves to another cottage in nearby **Coneysthorpe** 1885 *The Hepaticae of the Amazon and the Andes* published 1893 Dies of influenza, aged seventy-six

REFERENCES

CHAPTER 1

p.10 *expectations*', Bates, *The Naturalist on the River Amazons* (1864 ed.), 7.
p.10 *foliage*', Bates to Stevens, Pará, 1 June 1849, *Zoologist* 8 (1850), 2665.
p.10 *expectation*', Wallace, *My Life*, 269–70.
p.11 *trees*', Bates to Stevens, Pará, 16 March 1849, *Zoologist* 8 (1850), 2664.
p.11 *blue sky*', *height*', Bates, *Naturalist* (vol.1), 27, 27–28.
p.11 *flashing about*', Bates to Stevens, Pará, 16 March 1849, *Zoologist* 8 (1850), 2665.
p.11 *attention*', Bates, *Naturalist* (vol.1), 26.
p.11 *earth*', Wallace, *'Tropical Nature' and Other Essays* (1878), 29.
p.12 *lianas*', *traverse it*', Spruce, *Notes of a Botanist on the Amazon and Andes* (vol.1), 7, 256.
p.12 *alders*', Spruce, *Notes* (vol.1), 19.
p.12 *parasites, &c.*, Bates to Frederick Bates, Ega, 30 May 1856, *Zoologist* 10 (1856), 5658–9.
p.12 *Paradise*', Bates, *Naturalist* (1863 ed.; vol.2) 416; (1864 ed.) 460.

CHAPTER 2

p.13 *health*', Clodd, 'Memoir' in Bates, *Naturalist* (1892 ed.), xvii.
p.13 *pursuits*', Advertisement in *Leicester Journal*, Jan. 1819, in Crawforth, *The Butterfly Hunter*, 28.
p.14 *deeds*', *became*', Note by Frederick Bates in Markham's 'Obituary of Henry Walter Bates', *Proceedings of the Royal Geographical Society* 14 (1892), 245, 246.
p.14 *knowledge*', Allen, 'Bates of the Amazons', *Fortnightly Review* 58 (1892), 800.
p.14 *institution*', *fifteen*', Clodd, 'Memoir', xvii, xviii.
pp.15–16 *injured him*', *abounded*', *difficulties*', Frederick Bates, in Markham's 'Obituary', 247.
p.15 *insects*', Exhibition advertisement, 8 May 1840, in Crawforth, *The Butterfly Collector*, 31.
p.16 *specialist*', Allen, 'Bates of the Amazons', 801.
p.18 *hypocrite*', O'Hara, 'Dead men's embers', *Saturday Night Press* (2006), 75.
pp.18–19 *Reason*', Wallace, *My Life; a record of events and opinions* (1908, condensed ed.), 45.
p.19 *privacy*', Wallace, 'A British convert / A distinguished convert', *The New Nation* 1(50) (1891), 135; in Claeys, 'Wallace and Owenism', in Smith & Beccaloni, *Natural Selection & Beyond*, 245.
pp.20–21 *species*', *about it*', *C-of-E way*', *returned*', *skeptics*', Wallace, *My Life* (vol.1), 232, 227, 17, 78, 87.
p.21 *butterflies*', *already done*', Wallace, *My Life* (vol.1), 237.
p.22 *species*', Wallace to Bates, 11 April 1846, Wallace Family Archive, in Raby, *Alfred Russel Wallace: A Life* (2001), 28.
p.23 *against it*', Slotten, *The Heretic in Darwin's Court*, 31.
p.23 *banks*', *together*', Bates, *Naturalist*, 'Preface' to first ed. (1863; vol.1), iii.
p.23 *look over*', *species*', Wallace, *My Life* (vol.1), 256, 257.
p.24 *thought*', Wallace to Bates, Neath, 11 April 1846, in *My Life* (vol.1), 256.
pp.25–26 *delightful*', Edwards, *A Voyage up the River Amazon*, 201.
p.26 *collections*', Wallace, *My Life* (vol.1), 264.
p.26 *interior*', Wallace and Bates to Sir William Hooker, 30 March 1848, in the Kew archive, in Raby, *Alfred Russel Wallace*, 31.
pp.26–27 *merchants*', *interest*', *journey*', *scantiest*', Wallace, *My Life* (vol.1), 264, 266, 267, 267.
p.28 *own profession*', *year*', Wallace 'Biographical Introduction' to Spruce, *Notes* (vol.1), xxii, xxii.
p.28 *life*', quote by George Stabler of Ganthorpe, in Wallace, *idem*, xxiv.
pp.29–30 *century*', *botanists*', *species*', *valley*', *collected*', *thirty*', Wallace, *idem*, xxiii, xxiv, xxviii, xxxiii.
p.30 *assistant*', Spruce, *Notes* (vol.1), 1.

CHAPTER 3

p.31 *natural history*', Editor's comment, *Annals and Magazine of Natural History* 2(3) (1849), 74.
p.31 *landwards*', *port*', Bates, *Naturalist* (vol.1), 2, 3.
p.31 *city*', Wallace, *A Narrative of Travels on the Amazon and Rio Negro*, 6.
p.32 *foliage*', Bates to his friend Edwin Brown in Burton-on-Trent, 17 June 1848, in *Zoologist* 8 (1850), 2837.
p.37 *general*', Edwards, *A Voyage up the River Amazon...* (1847), 10.
p.37 *natives*', Prince Adalbert, *Aus meinem Reisetagebuch* (1847), translated by Schomburgk &

Taylor, *Travels of His Royal Highness Prince Adalbert of Prussia* (1849), 154–55.
p.37 *boxes', deshabillé',* Wallace, *Narrative*, 4, 31.
p.37 *scampering about',* Bates to Edwin Brown, 17 June 1848, *Zoologist* 8 (1850), 2837.
p.38 *shade',* Wallace, *Narrative*, 8.
p.38 *appetites',* Bates to brother Frederick, 1848, *Zoologist* 8 (1850), 2715.
p.38 *available',* Bates to Brown, *Zoologist* 8 (1850), 2837.
p.38 *position', population',* Bates, *Naturalist*, 24.
p.38 *skins',* Wallace, *Narrative*, 8.
p.38 *women',* Bates, *Naturalist*, 3
p.39 *passed', England',* Wallace, *Narrative*, 6, 47.
p.39 *spontaneously',* Bates to Brown, 17 June 1848, *Zoologist* 8 (1850), 2837.
p.39 *oxen', place',* Bates, *Naturalist* (vol.1), 6, 3.
pp.39–40 *good-humoured', pleasure for me', work in',* Bates to Edwin Brown, 17 June 1848, *Zoologist* 8 (1850), 2887, 2338–9, 2838.
p.40 *England',* Bates, *Naturalist* (vol.1), 55.
p.40 *populace',* Wallace, *Narrative*, 19.
p.40 *colours', homes',* Bates, *Naturalist*, 54.
p.40 *night',* Wallace, *Narrative*, 19.
p.41 *active',* Bates, *Naturalist* (vol.1), 32.
p.41 *handsome',* Wallace, *Narrative*, 21.
pp.41–42 *talk', motions', tree', all three', forest',* Bates, *Naturalist* (vol.1), 33–34, 34, 35, 31.
p.43 *tenacious', bark',* Bates, *Naturalist*, 37; Wallace, *Narrative*, 28.
p.43 *learn', foliage',* Wallace, *Narrative*, 32, 33.
p.43 *species',* Bates to Stevens, Pará, 16 Aug. 1848, *Zoologist* 8 (1850), 2716
p.44 *England',* Bates to Edwin Brown, Burton-on-Trent, 19 Oct. 1848, Zoologist 8 (1850), 2839.
p.44 *observed',* Wallace, *Narrative*, 51.
pp.44–45 *fellow', himself to us', gloriously',* Bates, *Naturalist*, 66, 67.
p.45 *beautiful',* Wallace, *Narrative*, 53.
p.45 *away',* Bates, *Naturalist*, 68; Wallace, *Narrative*, 54.
p.45 *temple',* Bates, *Naturalist*, 69.
p.46 *humour', country house', orders',* Wallace, *Narrative*, 55, 56, 56.
p.46 *family',* Bates, *Naturalist*, 71.
p.46 *daughters',* Wallace, *Narrative*, 56–57.
p.47 *indolent', habits', horn', signs', river', blackberries',* Bates, *Naturalist*, 72, 73, 74.
p.47 *cream',* Wallace, *Narrative*, 79.
p.47 *strong',* Smith, *Amazon, Sweet Sea* (2002), 93.
p.48 *productions',* Wallace, *Narrative*, 59.
p.48 *reside',* Bates, *Naturalist*, 75.
p.48 *skins',* Wallace, *Narrative*, 61.
pp.48–49 *Tocantins', roof', defeated',* Bates, *Naturalist*, 75, 77, 76.
p.49 *vicinity', resemble',* Wallace, *Narrative*, 75, 74.
pp.50–51 *people', happily away', characters', bottom', course',* Bates, *Naturalist*, 79, 79, 80, 44, 82.
p.51 *fall itself',* Wallace, *Narrative*, 72.
pp.51–53 *Indians', downwards', so forth', converse', story-teller', place', shells',* Bates, *Naturalist*, 82, 82–83, 84, 85, 85, 88, 86.
p.53 *scarce',* Bates to Stevens, 23 Oct. 1848, *Annals and Magazine of Natural History* 2(3) (1849), 74–75.
pp.54–55 *upset', all over', away', miserable', feeding-place', bird',* Wallace, *Narrative*, 68, 68, 77, 82, 65, 76.
p.55 *long ago',* Spruce to Hooker, Pará (Belém), 3 Aug. 1849, 'Letters from Spruce (1842–1890)' in the Archives of the Royal Botanic Gardens, Kew.
p.56 *companion',* Bates to Stevens, Pará, 30 Aug. 1849, *Zoologist* 8 (1850), 2667.
p.56 *see',* Wallace, *Narrative*, 385.
p.56 *competitive',* Raby, *Alfred Russel Wallace*, 45.
p.56 *mania',* Vladimir Nabokov, *Speak Memory, An Autobiography Revisited* (1967), 99, quoted in Crawforth, *The Butterfly Hunter. The Life of Henry Walter Bates* (2009), 89.
p.56 *Marajó',* Bates to Stevens, Pará, 24 Feb. 1849, *Zoologist* 8 (1850), 2717.

CHAPTER 4

p.57 *table',* Wallace, *Narrative*, 82.
pp.57–58 *pieces', jaws', birds', them all', handsome',* Wallace, *idem*, 83, 84, 85, 88.
pp.59–61 *generally do', manner', forest', useless', fish', despised', very much', once',* Wallace, *idem*, 92, 94, 95–96, 100, 98–99, 99, 107, 108.
p.61 *found there',* Bates, *Naturalist*, 102.
p.61 *Portuguese',* Bates to Stevens, Belém, 24 Feb. 1849, *Zoologist* 8 (1850), 2717.
pp.61–62 *fellows', perceptible', vertical sun',* Bates, *Naturalist*, 104, 105.
p.62 *beetles',* Bates to Stevens, 24 Feb. 1849, *Zoologist* 8 (1850), 2718.
p.62 *met with',* Bates, *Naturalist*, 107.
p.62 *Ingá, &c.'* Bates to Stevens, 24 Feb. 1849, *Zoologist* 8 (1850), 2718.
pp.62–63 *hair', pleasant', round', wall', unpleasant',* Bates, *Naturalist*, 128, 128, 108, 108, 108.
p.63 *dinner',* Bates to Stevens, Belém, 24 Feb. 1849, *Zoologist* 8 (1850), 2718.
pp.63–65 *flavour', rapidity', subject', head', height', farinha', strangers', guitar',* Bates, *Naturalist*, 109, 113, 115, 117, 117–18, 118, 118, 118.

p.65–66 *host', rice', called', alone',* Bates, *Naturalist,* 121, 121, 122, 122.

p.66 *moonlight',* Bates to Stevens, 2 March 1849, *Zoologist* 8 (1850), 2664.

pp.66–67 *truly is!' river', again', water',* Bates, *Naturalist,* 124, 125, 126, 126.

p.67 *all along',* Bates to Stevens, Pará, 2 March 1849, *Zoologist* 8 (1850), 2664.

pp.67–70 *call them', fate', erected now', have seen', remarkably well', idleness', aspect',* Wallace, *Narrative,* 113, 114, 117, 117, 119, 119, 121.

p.70 *forest', right', annoyance',* Wallace, *Narrative,* 126, 128, 132.

p.71 *haunts', yellow', other',* Bates to Frederick Bates, Belém, 24 Feb. 1849, *Zoologist* 8 (1850), 2717, 2718, 2719.

p.71 *little thing',* Bates, *Naturalist,* 99.

pp.71–72 *barbarism', sailer', versifying', delight',* Bates to Stevens, 31 Aug. 1849, *Zoologist* 8 (1850), 2791.

p.72 *fun', utmost', life', races', blood',* Bates, *Naturalist,* 90, 92, 92, 93.

p.72 *Leicester',* Bates to Stevens, 31 Aug. 1849, *Zoologist* 8 (1850), 2792–3.

pp.72–73 *obtained', Amazonia', servant for me',* Bates, *Naturalist,* 93, 94, 95.

p.73 *description',* Bates to Stevens, 31 Aug. 1849, *Zoologist* 8 (1850), 2792.

p.74 *traders', hold',* Bates, *Naturalist,* 182, 101.

p.75 *cabin',* Bates to Stevens, 31 Aug. 1849, *Zoologist* 8 (1850), 2793.

p.75 *wreck', port',* Bates, *Naturalist,* 101.

p.75 *assistant',* Spruce, *Notes* (vol.1), 1.

pp.75–76 *shot one', fastidious',* Wallace, *Narrative,* 133, 135.

pp.76–78 *completely', side by side', trees for me', flowers', trees', beautiful', spikelets', Hepaticae', valley', machete', anatomy',* Spruce, *Notes* (vol.1), 2, 37, 2, 3, 4, 7, 9, 10, 13.

p.79 *outline', palms', margins', beauty',* Spruce, *idem,* 34, 37, 40.

pp.79–80 *appearance', dots', tree', see them', rivers',* Spruce, *idem,* 38, 39, 41.

p.80 *flowers',* Spruce, *idem,* 3.

p.81 *branches of trees', fist', purple', weaker',* Spruce, *idem,* 17-18, 28, 28–29, 29.

pp.81–82 *upon them', both', veins', harmless', drunk', beneath',* Spruce, *idem,* 29, 30, 31, 32.

p.83 *forest',* Spruce, *idem,* 19.

p.83 *campaign, traveller',* Spruce, *idem,* 54, 55.

p.84 *shore', Luco',* Bates, *Naturalist,* 131, 136.

p.84 *expensive', enjoyment', to be had',* Bates to Stevens, 30 Aug. 1849, *Zoologist* 8 (1850), 2667.

CHAPTER 5

p.85 *possible',* Bates, *Naturalist,* 139.

p.86 *awe',* Wallace, *Narrative,* 137.

p.86 *heat', other',* Bates, *Naturalist,* 143, 147.

p.86 *business',* Wallace, *Narrative,* 138.

p.86 *classes',* Bates, *Naturalist,* 148.

p.95 *service', animals',* Wallace, *Narrative,* 141, 140.

p.95 *Scotsman',* Spruce, *Notes* (vol.1), 62.

p.95 *hour or so',* Wallace, *Narrative,* 141.

p.95 *complexion',* Spruce, *Notes,* 63.

p.95 *ruinous', vegetation',* Wallace, *Narrative,* 139, 141.

p.96 *thither',* Spruce to Hooker, Santarém, 3 Aug. 1849, Archives of the Royal Botanic Gardens, Kew, in Crawforth, *The Butterfly Hunter,* 107.

pp.96–99 *exist there', cattle', comfortable', seeds', against them', company', branches', to see', forms', fruits', health',* Wallace, *Narrative,* 146, 143, 143, 147, 148, 150, 151, 152, 157.

p.99 *conversation',* Wallace, *My Life* (vol.1), 276.

p.100 *morals', fortune',* Bates, *Naturalist,* 151, 150.

p.100 *suffered from it',* Bernardino de Souza, *Lembranças e curiosidades do valle do Amazonas* (Pará, 1873) 113–14.

pp.100–2 *objects', numerals', moulds', crocuses', sunset', insect', won', days', life',* Bates, *Naturalist,* 154, 155, 155, 155, 156, 157, 158, 152.

pp.103–6 *young man', pelting-rain', quarterdeck', guides', knees', upon us', voice', part', prickly', ants', lost', flowers', snuff-box', instrument', broke off',* Spruce, *Notes* (vol. 1), 81, 84, 84, 91, 91, 92, 92, 93, 94, 95, 96, 98, 101, 102, 102.

pp.107–8 *Amazons', interruption', hospitable', people', fertility', countenance', fellow',* Bates, *Naturalist,* 164, 165, 161, 162, 187, 188, 169.

p.109 *together', gentlemanly man',* Wallace, *Narrative,* 159, 160.

pp.110–11 *morning', hovels', sides', ball', melancholy', cruel', days', blood', shower',* Bates, *Naturalist,* 179, 182, 183, 184, 192, 192, 194, 197, 180.

p.111 *moment',* Wallace, *Narrative,* 161.

p.112 *lancets', animal',* Bates, *Naturalist,* 180, 173.

p.112 *profusion',* Bates to Stevens, Ega, 14 June 1850, *Zoologist* 8 (1850), 2942.

p.112 *travelling',* Bates, Journal for 10 Jan. 1850, in Clodd, 'Memoir', xxv.

CHAPTER 6

pp.113–14 *shutters', occupation', community',* Wallace, *Narrative,* 164, 165, 165.

p.114 *Wallace',* Bates, *Naturalist,* 201.

p.114 *every day',* Bates, Journal for 12 March 1850, in Clodd, 'Memoir', xxv. The idea that Wallace and Bates discussed evolution is in Ferreira, *Bates, Darwin, Wallace e a teoria da evolução,* (Brasília, 1990), 28.

p.114 *forgotten',* Bates, *Naturalist,* 201.

p.115 *Barra', type', virtues',* Spruce, *Notes* (vol.1), 201, 201, 202.

p.115 *eating',* Spruce to John Teasdale, 3 Jan. 1851, Spruce, *idem,* 239.

p.115 *Antonij',* Bentham, 'On Henriquezia verticillata, Spruce: a new genus of *Bignoniaceae* from the Rio Negro in North Brazil', *Hooker's Journal of Botany* 6 (1853/54), 337.

p.115 *flowers',* Spruce, *Notes* (vol.1), 202.

pp.115–16 *floor', bread', himself', nothing',* Wallace, *Narrative,* 167, 171, 171, 172.

p.116 *known',* Wallace, 'On the umbrella bird...', *Proceedings of the Zoological Society of London* 18 (1850), 206.

p.116 *plume',* Wallace, *Narrative,* 169.

p.116 *note',* Bates, *Naturalist,* 375. He collected an umbrella bird near Ega (Tefé) on the Solimões, and included a fine drawing of it in his book.

p.116–17 *wounded', cleaned away', unserviceable',* Wallace, *Narrative,* 169, 169, 173.

p.117 *explorations',* Bates, *Naturalist,* 203; also Wallace, *Narrative,* 174.

p.117 *Negro to him',* Bates to Stevens, 22 March 1850, *Zoologist* 8 (1850), 2940; also Bates, *Naturalist,* 204.

p.117 *together',* Bates to Stevens, 14 June 1850, *Zoologist* 8 (1850), 2941.

p.117 *board',* Bates, *Naturalist,* 203.

p.117 *collecting', miserably',* Bates to Stevens, Manaus, 22 March 1850, *Zoologist* 8 (1850), 2940.

p.118 *tree',* Bates, *Naturalist,* 293.

p.118 *terrible', salt ditto',* Bates to Stevens, Ega, 14 June 1850, *Zoologist* 8 (1850), 2942.

p.118 *ground',* Bates, 'Some Account of the Country of the River Solimoens, or Upper Amazon', *Zoologist* 10 (1852), 3591.

p.119 *food',* Bates to Stevens, Ega, 14 June 1850, *Zoologist* 8 (1850), 2943.

pp.119–20 *friends', countries', man', Europe', country', children', village',* Bates, *Naturalist,* 303, 304, 304, 306, 306, 306, 305.

p.120 *nose',* Bates, *Naturalist* (1863 ed.; vol. 2), 205.

pp.120–21 *butterflies', beautiful', heliconia', Europe',* Bates to Stevens, Ega, 15 June 1850, *Zoologist* 8 (1850), 2965, 2965, 2965–6, 2966.

p.121 *province', mention',* Bates to Stevens, 30 April 1851, *Zoologist* 9 (1851), 3231, 3232.

p.121 *Cassidae',* Bates to Stevens, Ega, 28 July 1856, *Zoologist* 15 (1857), 5557.

p.121 *anywhere else',* Bates to Stevens, Ega, 15 June 1850, *Zoologist* 8 (1850), 2965.

pp.121–22 *best things',* Bates to Stevens, Pará, 3 June 1851, *Zoologist* 10 (1852), 3321.

pp.122–23 *weight', imaginable', troublesome', expression',* Bates, *Naturalist,* 406, 408, 408, 408.

p.123 *furies',* Bates, *idem,* 411–12. His picture of this adventure is opposite p.408.

pp.124–25 *crowds', full', huntsmen', branches', well', dish', gathering', labours',* Bates, *Naturalist,* 348, 348, 352, 354, 355, 356, 363, 364.

p.126 *appearance', praias', despise',* Bates, *Naturalist* (1863 ed.; vol. 2), 274, 275, 277.

pp.127–28 *decreasing', shouting', instantly', insects',* Bates, *Naturalist,* 365, 358, 358, 306.

p.127 *as dogs do',* Bates, *Naturalist* (1863 ed.; vol. 2), 278.

p.128 *mind',* Bates, *Naturalist,* 308; also Clodd, 'Memoir', xxv.

p.128 *money',* Bates, *Naturalist,* 308.

p.128 *low ebb',* Bates, Journal for 5 Sept. 1850, in Clodd, 'Memoir', xxv.

p.128 *England',* Bates, Journal for 2 Jan. 1851, in Clodd, *idem,* xxv.

pp.129–30 *high', water', alighted', manner', life', kind', needlework',* Wallace, *Narrative,* 176, 177, 177, 179, 187, 188, 180.

pp.130–37 *ridges', weather', nodes', water!', plant', penetrate', escape', season', shores', undescribed', white with it', Alismas', drakes', below', currants', campo', equal to it', determination', urbane', deal with', voyage', sitios', flowers', people', bed', streams', joy',* Spruce, *Notes* (vol.1), 107, 108, 110, 113, 115, 115, 117, 119, 147, 147, 148, 148, 149, 150, 162, 145–46, 161, 160, 124, 125, 119, 164, 169, 172, 190, 192, 192.

p.137–38 *pitied us', limbs, etc.'* Spruce to Bentham, Barra (Manaus), 1 Jan. 1851, Spruce, *idem,* 208.

p.138 *table', vegetation',* Spruce, *Notes* (vol.1), 201, 207.

p.138 *from it', previous one',* Spruce to Bentham, Barra do Rio Negro (Manaus), 1 April 1850, Spruce, *idem,* 209.

pp.138–39 *species',* Wallace's introduction to this part of Spruce, *idem,* 203.

p.139 *arrow!' tattooed', new to me',* Spruce to Sir William Hooker, Barra do Rio Negro (Manaus), 1 April 1851, Spruce, *idem,* 215, 213, 214.

p.140 *admiration',* Spruce to John Teasdale, Barra do Rio Negro, 17 Aug. 1851, Spruce, *idem,* 244.

pp.140–41 *met with', season',* Spruce to Bentham, Barra do Rio Negro, 7 Nov. 1851, Spruce, *idem,* 228, 231.

p.141 *traverse it'*, Spruce to Matthew Slater, Manaus, Oct. 1851, Spruce, *idem*, 256.

p.141 *preserving them'*, Spruce to Dr Semann, Manaus, 25 April 1851, Spruce, *idem*, 225.

p.142 *clear-headed'*, *Henrique's'*, Spruce, *idem*, 197, 198.

pp.142–43 *system!'*, *abundance'*, *night'*, *hopping'*, *embrace'*, *every one'*, *breasts'*, Spruce to John Teasdale, Manaus, 17 Aug. 1851, Spruce, *idem*, 241, 248, 250, 251, 252, 253, 242–43.

pp.143–44 *task'*, *amusement'*, Spruce to Bentham, Manaus, 7 Nov. 1851, Spruce, *idem*, 231, 232.

p.144 *country'*, Spruce to Bentham, 1 April 1851, Spruce, *idem*, 210.

p.144 *voluntarily'*, Spruce to Bentham, Manaus, 1 Jan. 1851, Spruce, *idem*, 208.

p.145 *alternative'*, *inconvenience'*, Spruce to Bentham, Manaus, 7 Nov. 1851, Spruce, *idem*, 227.

CHAPTER 7

pp.146–52 *year back'*, *voyage'*, *within it'*, *grizzly man'*, *Negro'*, *around us'*, *mud-huts'*, *thickets'*, *country'*, *afford'*, *Rio Negro'*, *island'*, *fishes'*, *comrades'*, *cross again'*, *enjoyment for me'*, *doors'*, *countenance'*, *shirt-sleeve'*, *house'*, Wallace, *Narrative*, 192, 193, 194, 195, 195, 195, 199, 202, 202, 202, 205, 205, 206, 206, 206–7, 207, 210, 216, 217, 221.

pp.153–60 *feathers'*, *disappearing from us'*, *under foot'*, *ant-thrushes'*, *Indians'*, *high'*, *birds'*, *skill'*, *leather'*, *affected'*, *past'*, *boxes, etc.' continent'*, *payment'*, *two days'*, *dances'*, *frame'*, *gold!' wet'*, *forest'*, *inflamed'*, *maggots'*, *alone'*, *piaçaba'*, *ceremony'*, *about them'*, *day'*, *companions'*, Wallace, *idem*, 222, 224, 226, 227, 251, 232, 238, 238, 240, 240, 247, 247, 241–42, 250, 254, 259, 256, 261, 251, 251, 253, 271, 228, 229, 229, 229, 274.

pp.161–63 *peculiarities'*, *human form'*, *bodies'*, *admire'*, *theatre'*, Wallace, *idem*, 476, 478, 492, 490, 276.

p.163 *have to fall'*, Nimuendajú, 'Reconhecimento dos rios Içana, Ayari e Uaupés.' *Journal de la Société des Américanistes de Paris* 39 (1950), 190.

p.163 *harmony'*, *weapons'*, *understand'*, Wallace, *Narrative*, 276, 279, 282, 283, 286.

p.164 *Dabucuri'*, *intoxicant'*, Spruce, *Notes* (vol. 1), 324, 325.

pp.164–65 *breasts'*, *men'*, *reptile'*, *good'*, Wallace, *Narrative*, 293, 296, 297, 297.

pp.165–66 *six hours'*, Wallace, *idem*, 496.

p.166 *bassoons' pain of death'*, Wallace, *idem*, 501.

p.166 *country'*, *carrying them off'*, Spruce to Hooker, San Carlos de Río Negro, 27 June 1853, *Hooker's Journal of Botany* 6 (1853), 33. Spruce mentioned sending these and other indigenous artifacts to Hooker for his Museum. Wallace also made a fine collection of artifacts, but all were lost in his shipwreck. In a note to the 1911 ed. of his *Narrative* (p.352), he wrote that 21 of the 65 objects he listed had also been sent 'by my friend R. Spruce, Esq.' for the Museum at Kew.

p.166 *boat'*, Spruce, *Notes* (vol.2), 416 footnote.

p.175 *night'*, Spruce, Journal for 3 Jan. 1853, *Notes* (vol.1), 331.

p.175 *odour'*, *drinkers'*, *exchange'*, *beads'*, Wallace, *Narrative*, 498, 498, 356, 354.

p.176 *grandfather?'* Spruce, *Notes* (vol.1), 332.

p.177 *language'*, Wallace, *Narrative*, 480. Wallace's publisher (Reeve & Co., London) included a fold-out chart of the vocabularies after p.520 of his *Narrative*, followed by an intelligent appendix describing how he tackled phonetics (pp.521–25) followed by another appendix by an authority on South American dialects called Dr R. G. Latham (pp.525–41). Other studies of indigenous languages were by the Portuguese naturalist Alexandre Rodrigues Ferreira, who visited peoples of the Uaupés in 1785, and by Carl Friedrich Philip von Martius (who did not visit the upper Rio Negro), *Beiträge zur Ethnographie und Sprachenkunde America's, zumal Brasiliens* (2 vols, Erlangen 1863 & Leipzig 1867).

pp.178–82 *killed'*, *found there'*, *disagreeable'*, *animals'*, *orchid-house'*, *spot'*, *inflamed'*, *life'*, *giddy'*, *disposition'*, *Rio Negro'*, *pleasures'*, *fever'*, *swam'*, *diet'*, Wallace, *Narrative*, 300, 306, 306, 307, 308, 308–9, 310, 313, 317, 321, 322, 324, 327, 327, 187.

p.182 *flavour'*, Bates, *Naturalist* (1863; vol.1), 165.

p.182 *milk'*, Wallace, *Narrative*, 186.

p.182 *resembling a fish'*, Bates, *Naturalist* (vol.1), 165.

pp.182–83 *peixe-boi'*, *tubercled'*, *compass'*, Wallace, *Narrative*, 331, 329, 332.

p.183 *hammock'*, Wallace, *idem*, 340; also *My Life* (vol.1), 312.

p.183 *sticks'*, Wallace, *Narrative*, 340.

p.184 *feed himself'*, Spruce to John Smith, Uanauacá, 28 Dec. 1851, Spruce, *Notes* (vol.1), 268.

CHAPTER 8

p.185 *liked'*, *fish'*, *waters'*, Spruce to Teasdale, São Gabriel, 24 June 1851, Spruce, *Notes* (vol.1), 269, 274, 274.

p.186 *botanist'*, Wallace editor's comment, in Spruce, *idem*, 260–1.

p.186 *90 feet high'*, Spruce to John Smith, Uanauacá sitio, Rio Negro, 28 Dec. 1851, Spruce, *idem*, 266.

p.186 *earth'*, Spruce to John Smith, Uanauacá 28 Dec.1851, Spruce, *idem*, 268.

pp.187–89 *task', doing it', requested', couple of days', undertaking', this time', possible way', face alone'*, Spruce, *idem*, 282, 287, 287, 272, 273, 294, 292, 365.

p.189 *heroic ferocity'*, Spruce, *Notes* (vol.2), 69.

p.189 *garrison', banana'*, Spruce to Bentham, São Gabriel, 15 April 1862, *Notes* (vol.1), 293, 294.

pp.189–90 *accompaniment'*, Spruce, *idem*, 307.

p.190 *civilisation'*, Spruce to Bentham, São Gabriel, 18 Aug. 1852, Spruce, *idem*, 299.

p.190 *organic beings'*, Spruce to Wallace, 21 Nov. 1863, in the Wallace family archive held by his grandsons Alfred and Richard Wallace, in Raby, *Alfred Russel Wallace*, 78.

p.190 *without a break'*, Spruce, 'Notes on the Theory of Evolution' in a letter to W. Wilson, 28 May 1870, in the archive of the Royal Botanic Gardens, Kew, in Raby, *idem*, 78.

p.190 *of the boat', take with me'*, Wallace, *Narrative*, 341.

p.191 *perish'*, Spruce, Journal for Sept. 1852, Spruce, *Notes* (vol.1), 321.

p.191 *informed me', furiously', difficulty'*, Wallace, *Narrative*, 343, 345, 347.

pp.192–93 *visited', perpendicular falls', pigeon', happiness', during the rain'*, Wallace, *idem*, 359, 356, 372, 370, 371.

p.193 *alligators, etc.', Antonij'*, Wallace, *idem*, 373.

pp.193–94 *trade with them', bathe'*, Wallace, *idem*, 438, 364.

p.194 *commerce'*, Annual *Relatório* of President A. B. Albuquerque de Lacerda, 1864, p.135, in Carlos de Araújo Moreira Neto, *Os índios e a ordem imperial* (2005), 85.

p.195 *defence'*, Report by Director Jesuino Cordeiro to President of Amazonas, 1853, in Bento de Tenreiro Aranha, ed., *Arquivo do Amazonas* 1(3) (Manaus, 1907), 59. Wright, *História indígena e do indigenismo no Alto Rio Negro* (ISA, São Paulo, 2005), 115. The Brazilian Indian Decree of 1845 was also known as the Regimento das Missões.

p.195 *cachoeiras', Jesuino'*, Wallace, *Narrative*, 365, 366.

p.196 *punishment'*, Wallace, editor's footnote, in Spruce, *Notes* (vol.1), 330.

p.196 *rapid'*, Spruce, *Notes* (vol.2), 482.

p.196 *met with'*, Spruce, *Notes* (vol.1), 323–24.

p.196 *patterns'*, Spruce, *Notes* (vol.2), 483.

p.197 *draughtsman'*, Wallace editor's comment, in Spruce, *Notes* (vol.1), 317.

p.197 *European'*, Spruce, *Notes* (vol.1), 325. The drawings all appeared in Spruce's book (vol.1, 325–28), and their originals are in the archive of The Royal Society in London. He made it clear that his sketches of the Pira-Tapuya woman and the Carapanã man were done at Iauaretê, not in their homes far up the Papuri river.

p.197 *around them'*, Spruce, *idem*, 344–45. The drawings of the girls' faces are in profile on p.345.

pp.197–98 *desert air'*, Spruce to Bentham, Ambato, 22 June 1858, Spruce, *Notes* (vol.2), 208–9.

p.198 *eatables'*, Spruce to Bentham, San Carlos, 2 July 1853, Spruce, *idem*, 329.

p.198 *dog'*, Spruce, *Notes* (vol.1), 340.

p.198 *fungi'*, Spruce to Bentham, San Carlos, 25 June 1853, Spruce, *idem*, 334–35.

pp.198–99 *day', Caesalpinae', Voyrieae', undescribed'*, Spruce to Sir William Hooker, San Carlos de Río Negro, 27 June 1853, Spruce, *idem*, 336.

p.200 *attacked', attack', attack us'*, Spruce to John Teasdale, San Carlos, 1 July 1853, Spruce, *idem*, 351, 352.

p.200 *once a day'*, Spruce to Sir William Hooker, San Carlos, 27 June 1853, Spruce, *idem*, 357.

p.200 *game'*, Spruce to John Teasdale, San Carlos, 20 Nov. 1853, Spruce, *idem*, 374.

p.200 *ought to be'*, Spruce to Sir William Hooker, San Carlos, 27 June 1853, Spruce, *idem*, 357.

p.200 *many'*, Spruce to Sir William Hooker, San Carlos, 17 Sept. 1853, Spruce, *idem*, 382.

p.201 *I had made'*, Spruce, Journal for 15 Aug. 1853, Spruce, *idem*, 363.

pp.201–2 *thousands', bottom'*, Spruce to John Teasdale, San Carlos, 20 Nov. 1853, Spruce, *idem*, 374, 375.

p.202 *rock', task', pay'*, Spruce, Journal for 27 Nov. 1853, Spruce, *idem*, 386, 387.

pp.203–4 *beehive', stung'*, Spruce, Journal for 16 Dec. 1853, Spruce, *idem*, 396.

p.203 *river'*, Spruce, *Notes* (vol.2), 70.

p.204 *new'*, Spruce, Journal for 18 Dec. 1853, Spruce, *Notes* (vol.1), 399.

p.204 *lianas', bushes', hibernation', frontage', shore'*, Spruce, *Notes* (vol.2), 355, 354–55, 354, 356, 356.

p.204 *water'*, Pires & Prance, 'The vegetation types of the Brazilian Amazon', in Prance & Lovejoy, *Key Environments: Amazonia*, 119. This is an excellent modern summary of the different types of forest.

p.205 *South America', equals', pests', subsistence'*, Spruce to John Teasdale, San Carlos, 22 May 1854, Spruce, *Notes* (vol.1), 403, 404–7, 404. *Notes of a Botanist* contains Spruce's charming pencil drawing of Cerro Duida seen from Esmeralda, of which the original is held by The Royal Society, London. A similar view in colour, with Indians in the foreground, was done by Charles Bentley from sketches by Schomburgk, to illustrate Schomburgk, *A Description of British Guiana* (1840) and *Twelve Views of the Interior of Guiana* (1841).

p.205 *respect'*, Schomburgk, 'Journey from San Joaquim to Esmeralda', *Journal of the Royal Geographical Society* 10 (1840), 247; also entry for 22 Feb. 1839, in Rivière, *The Guiana Travels of Robert Schomburgk 1835–1844* (vol.1), 366. Humboldt, *Personal Narrative*, ch. 24.

p.206 *gathered', haws'*, Spruce to Hooker, San Carlos, 19 March 1854, Spruce, *Notes* (vol.1), 441.

p.206 *own'*, Schomburgk, *Journal of the Royal Geographical Society* 10 (1840), 238; or in Rivière, *Guiana Travels*, (vol.1), 356.

p.206–7 *return', indispensibles'*, Spruce, Journal for 2 Jan. 1854, Spruce, *Notes* (vol.1), 410, 412.

p.207 *extreme'*, Spruce, Journal for 21–25 Jan. 1854, Spruce, *idem*, 423.

p.208 *way up'*, Spruce, Journal for 27 Jan. 1854, Spruce, *idem*, 423.

p.209 *life', Custodio'*, Spruce to Sir William Hooker, San Carlos de Río Negro, 19 March 1854, Spruce, *idem*, 446, 447.

p.209 *tassels'*, Spruce, Journal for 31 Jan. 1854, Spruce, *idem*, 424.

p.209 *work'*, Spruce, Journal for 10 Feb. 1854, Spruce, *idem*, 429.

p.210 *heath'*, Spruce, Journal for 14 Feb. 1854, Spruce, *idem*, 432.

p.210 *name'*, Spruce to Hooker, San Carlos, 19 March 1854, Spruce, *idem*, 440.

p.210 *times', wetting', zancudos', flower'*, Spruce, *idem*, 430, 431.

pp.210–11 *time', much more'*, Spruce to Hooker, San Carlos, 19 March 1854, Spruce, *idem*, 443.

CHAPTER 9

p.212 *tree-trunk', cat'*, Bates, 'Some account of the Douroucouli Monkey (Aötes trivirgatus, *Humb.*)', *Zoologist* 10 (1852), 3324, 3326.

p.213 *healthy'*, Bates, *Naturalist*, 205

p.213 *fearfully'*, Bates to Mary Wallace, Pará, 13 June 1851, in the 'Wallace Collection' of the Natural History Museum, London, quoted in Anthony Crawford, *The Butterfly Hunter*, 130.

p.213 *body'*, Bates, *Naturalist*, 206.

p.213 *brain-fever'*, Spruce to Bentham, Manaus, 7 Nov. 1851, Spruce, *Notes* (vol.1), 227.

p.213 *than ever'*, Bates, Journal for 19 April 1851, in Clodd, 'Memoir', xxvi.

p.214 *paid me'*, Bates to Stevens, Pará, 30 April 1851, *Zoologist* 9 (1851), 3230; Stevens had written to him on 3 Oct. 1850 and 8 Jan. 1851.

p.214 *science'*, Clodd, 'Memoir', xxvi.

p.214 *fine things'*, Bates to Stevens, Pará, 3 June 1851, *Zoologist* 10 (1852), 3321.

p.214 *leaves'*, Bates to Stevens, Pará, 8 Oct. 1851, *Zoologist* 10 (1852), 3324.

p.214 *Ouramimeu &c.', charms of'*, Bates to Stevens, Pará, 3 June 1851, *Zoologist* 10 (1852), 3321.

pp.214–15 *fine', specimens'*, Bates to Stevens, Pará, 8 Oct. 1851, *Zoologist* 10 (1852), 3322.

p.215 *country', subsequently made', others'*, Bates, *Naturalist*, 209, 209, 210.

p.215 *long', so many'*, Bates to Stevens, Santarém, 8 Jan. 1852, *Zoologist* 10 (1852), 3450.

pp.215–16 *chrysalides'*, Bates to Stevens, Santarém, 12 April 1852, *Zoologist* 11 (1853), 3726.

p.216 *about it'*, Bates to Stevens, Santarém, 17 May 1852, *Zoologist* 11 (1853), 3728.

p.216 *severe here'*, Bates to Stevens, Santarém, 12 April 1852, *Zoologist* 11 (1853), 3726.

p.216 *notes'*, Bates to Stevens, Santarém, 17 May 1852, *Zoologist* 11 (1853), 3728.

p.216 *shells &c.', beauty'*, Bates to Stevens, Santarém, 4 June 1852, *Zoologist* 11 (1853), 3728, 3729.

p.216 *animals'*, Bates to Stevens, Santarém, 12 April 1852, *Zoologist* 11 (1853), 3727.

p.216 *so forth'*, Bates, *Naturalist*, 238–39.

pp.216–17 *myself', roar', showers'*, Bates to Stevens, Santarém, 4 June 1852, *Zoologist* 11 (1853), 3727.

p.217 *own way'*, Bates to Stevens, Aveyros, 1 Aug. 1852, *Zoologist* 11 (1853), 3801.

pp.217–18 *jib-sail', collected', bed-time', might', village', Europe'*, Bates, *Naturalist*, 240, 250, 243, 252, 250, 253.

p.218 *Erycinidae'*, Bates to Stevens, Aveyros, 1 Aug. 1852, *Zoologist* 11 (1853), 3803.

pp.218–22 *free itself', aviaries', step', valley', cannibalism as this', whites', expression', demeanour', culture', Alter do Chão', lamp', sight', movement', spot'*, Bates, *Naturalist*, 256, 260, 266, 266–67, 273, 270, 274, 274, 276, 278, 279, 279, 278–79.

p.222 *new to me', people'*, Bates to Stevens, Santarém, 18 Oct. 1852, *Zoologist* 11 (1853), 3842, 3843.

p.222 *health'*, Bates, *Naturalist*, 281.

p.222 *wrecked'*, Bates to Stevens, Santarém, 18 Oct. 1852, *Zoologist* 11 (1853), 3842.

pp.222–23 *mammals &c.', new things', natives'*, Bates to Stevens, Santarém, 22 Nov. 1852, *Zoologist* 11 (1853), 3897, 3898.

pp.223–24 *room', society', exertion', Tapajós', of old', think of it'*, Wallace, *Narrative*, 373, 375, 390, 385, 390, 391.

p.224 *actual fire', voyage'*, Wallace, *My Life* (vol.1), 311.
p.224 *saving'*, Wallace, *Narrative*, 393.
p.224 *sovereigns'*, Wallace to Spruce, on the brig *Jordeson*, 19 Sept. 1852, in Wallace, *My Life* (vol.1), 303.
pp.225–26 *time', ocean', tremendously', ground', beheld!'* Wallace, *Narrative*, 394, 396, 402, 403, 400–1.
p.226 *Europe'*, Wallace, *My Life* (vol.1), 306–7.
p.226 *irreparable'*, Bates to Stevens, Santarém, 10 March 1853, *Zoologist* 12 (1854), 4114.
p.226 *Lima'*, Wallace to Spruce, London, 5 Oct. 1852, Wallace, *My Life* (vol.1), 309.
p.226 *heart'*, João Antonio de Lima, São Joaquim, 7 June 1853 (answering a letter from Spruce of 26 April 1853), Wallace, *My Life* (vol.1), 312.
p.227 *holds good'*, Sir John Scott-Keltie, Secretary of the Royal Geographical Society, to Wallace's son William Wallace, 21 May 1915, in Marchant, *Alfred Russel Wallace Letters and Reminiscences* (1916), 29. Scott-Keltie had been told that Wallace's map was still valid – despite errors over longitude – by a former President of Amazonas, Barão F. J. de Santa-Ana Néry.
p.228 *butter', place', structure'*, Spruce to Hooker, San Carlos de Río Negro, 19 March 1854, in *Hooker's Journal of Botany* 6 (1853/54), 334, 335, 335.
p.229 *demoralised'*, Spruce, *Notes* (vol.1), 451. The messianic movement of the 1850s is described in Hemming, *Amazon Frontier* (1995 ed.), 307–13.
pp.229–31 *Venezuela', bitten', troubles', vigilance', fruit', entrance', fever in me', death', alone'*, Spruce, *Notes* (vol.1), 452, 435, 455, 462, 461, 462, 463.
p.231 *prostrated'*, Spruce's Journal, Royal Botanic Gardens, Kew – Wallace omitted some of these details when he compiled *Notes of a Botanist* after Spruce's death. Honigsbaum, *The Fever Trail*, 22.
pp.231–32 *forget', demoniacal', effects', apathy', dollars!'* Spruce, *Notes* (vol.1), 463, 464, 465, 465.
p.233 *art there'*, Spruce to Countess of Carlisle, Tomo, Aug. 1854, in the Castle Howard archive. The letter and hammock were enclosed with one to Hooker of Kew, who forwarded them to Yorkshire.
pp.233–34 *intoxicated', did so', attention', be made', gloomier time'*, Spruce, *Notes* (vol.1), 488, 490, 491, 492.

CHAPTER 10

pp.235–38 *vegetables', interior', strangers', door', Brazilians', days', season', place', school-master', country'*, Bates, *Naturalist*, 215, 210, 210, 211, 211, 211, 212, 213, 214, 211.
p.238 *impossibility'*, Bates to Stevens, Santarém, 10 March 1853, *Zoologist* 12 (1854), 4114.
p.238 *roças'*, Spruce, *Notes* (vol.1), 507.
p.239 *steel blue'*, Bates, *Naturalist*, 220.
pp.239–40 *swamps', elytra'*, Bates to Stevens, Santarém, 10 March 1853, *Zoologist* 12 (1854), 4113. The letter in which Stevens urged him to collect small coleoptera was dated 7 Oct. 1852.
p.240 *Clerii'*, Bates to Stevens, Santarém, 18 Aug. 1853, *Zoologist* 12 (1854), 4319.
p.240 *mine'*, Bates, *Naturalist* (1863 ed.; vol.2), 33; (1864 ed.), 221.
pp.240–41 *poor', wonderful', proceedings', work', progeny', uproar', wonderful'*, Bates, *Naturalist*, 224, 225, 226, 226, 227, 227.
pp.241–42 *nostrils', charm for me', so forth', leaves', come forth'*, Bates, *Naturalist*, 229, 229–30, 231, 233, 236.
p.242 *Diurnes', neuration &c.'* Bates to Stevens, Santarém, 17 Dec. 1853, *Zoologist* 12 (1854), 4320, 4321.
p.243 *arguments'*, Distant, contribution to the Obituary of Henry Walter Bates, *Proceedings of the Royal Geographical Society* 14 (1899), 250.
pp.243–44 *stone', creatures', comers', civilisation', hold'*, Bates, *Naturalist* (1863 ed.; vol.2), 59, 60, 61, 62, 64. Bates called termites 'white ants' as was common at that time; but he knew that they were technically not ants.
p.244 *materials'*, Bates to Stevens, Santarém, 27 March 1854, *Zoologist* 13 (1855), 4551.
pp.244–45 *clothing', sisters', lamps', astonishing'*, Bates, *Naturalist* (1863 ed.; vol.2), 66–67, 68, 68, 69.
p.245 *naturalists'*, Bates to Stevens, Santarém, 27 April 1854, *Zoologist* 13 (1855), 4552.
p.245 *such things'*, Bates to Stevens, Santarém, 18 Aug. 1853, *Zoologist* 12 (1854), 4319.
p.245 *letters'*, Bates to Stevens, Santarém, 27 March 1854, *Zoologist* 13 (1855), 4550.
p.245 *occupation'*, Bates to Stevens, Santarém, 27 April 1854, *Zoologist* 13 (1855), 4552.
p.245 *effect'*, Bates to Stevens, Santarém, 27 March 1854, *Zoologist* 13 (1855), 4550.
p.246 *locality'*, Bates to Stevens, Vila Nova, 15 Dec. 1854, *Zoologist* 14 (1856), 5014.
p.246 *country', nicely'*, Bates to Stevens, Santarém, 30 April 1855, *Zoologist* 14 (1856), 5016.
p.246 *forests'*, Bates to Stevens, Ega, 16 Aug. 1855, *Zoologist* 14 (1856), 5017.
pp.246–47 *forest', colours', species', world', department'*, Bates to Stevens, Ega, 1 Sept. 1855, *Zoologist* 14 (1856), 5018, 5018, 5019, 5019.
p.247 *shoulder', discovery'*, Bates to Stevens, Ega, 28 July 1856, *Zoologist* 15 (1857), 5558.
p.247 *attached to them'*, Bates, *Naturalist*, 388.

p.248 *dozen altogether'*, Bates to Adam White, Ega, 2 May 1857, *Transactions of the Entomological Society of London* 5 (1859), 224; Crawford, *The Butterfly Hunter*, 247–48.

p.248 *magazines'*, Bates to Stevens, Ega, 1 Sept. 1855, *Zoologist* 14 (1856), 5018.

pp.248–49 *steamer'*, Bates, 'Excursion to St. Paulo, Upper Amazons', *Zoologist* 16 (1858), 6160. Curiously, in his book *The Naturalist on the River Amazons*, published five years later, Bates gave the departure date as 7 Nov.: (1863 ed.; vol.2) 367; (1864 ed.) 427.

pp.249–50 *possible', Yankee-looking', reeking', colour', mosquitoes', legs', birds', arborescent'*, Bates, *Naturalist*, 427, 428, 430, 431, 437, 437, 437, 438.

p.250 *travels'*, Bates, 'Excursion to St. Paul, Upper Amazon', *Zoologist* 16 (1858), 6162.

pp.250–51 *Indians', degree', settlement', beams'*, Bates, *Naturalist*, 445, 446, 446, 447.

p.251 *forests'*, Bates, *Naturalist*, 448; also 'Excursion to St Paul, Upper Amazon', *Zoologist* 16 (1858), 6162. Bates gave its ornithological name as *Cyphorhinus cantans*, but this has now changed to *C. arada*.

pp.251–52 *wind', memories', head', spring'*, Bates, *Naturalist*, 448, 448, 379, 379.

p.253 *Lepidoptera'*, Bates, 'Excursion to St. Paul, Upper Amazon', *Zoologist* 16 (1858), 6169.

p.253 *dissipation'*, Bates, *Naturalist*, 450. Bates spelled the tribe's name Tucúna, as was common at that time; but Ticuna (or Tikuna) is how they now spell their name – this also avoids confusion with the totally unrelated Tukano of the Uaupés, the people visited by both Wallace and Spruce.

pp.254–63 *Jurupari', powder', eyes', huts', nose', people', Indians', so forth', whites', women', good ones', anywhere'*, Bates, *idem*, 450, 453, 432, 432, 434, 292, 292, 292, 313, 342, 293, 293.

p.264 *enthusiasm', food', spleen', brave it out'*, Bates, *idem*, 455, 455, 456, 456.

pp.265–66 *so forth', ague', place', woods', Paradise', nourishment'*, Bates, *idem*, 314, 315, 456, 458, 459–60, 460.

CHAPTER 11

pp.268–69 *uninvited', spruceana', forest', stream'*, Spruce to John Teasdale, Tarapoto, Peru, July 1855, Spruce, *Notes* (vol.2), 27, 28, 26–27, 40; also Spruce, 'Notes on some Insect and other Migrations observed in Equatorial America', *Journal of the Linnean Society* 9 (1867), and *Notes* (vol.2), 360.

p.268 *difficulty'*, Spruce, Journal for 28 April 1855, *Notes* (vol.2), 10.

p.269 *trees', Indian'*, Spruce to John Teasdale, Tarapoto, 23 March 1856, Spruce, *Notes* (vol.2), 39, 47.

p.270 *recrossed'*, Spruce, article in *Ocean Highways* (1856), quoted in *Notes* (vol.2), 61.

p.270 *vegetation'*, Spruce, *Notes* (vol.2), 91.

p.270 *world'*, Spruce's notes on 'Collecting near Tarapoto', Spruce, *idem*, 92.

p.270 *tree-ferns'*, Spruce, 'Précis d'un Voyage', *Revue Bryologique* (1886). Bentham must have submitted this account to the French moss review, soon after receiving it. Wallace translated it back into English in Spruce, *Notes* (vol.2), 99.

p.270 *botanists', genera'*, Spruce, *Notes* (vol.2), 101.

p.270 *deserts'*, Spruce, quoted in Schultes, 'Richard Spruce and the potential for European settlement of the Amazon…', *Botanical Journal of the Linnean Society* 77 (1978), 131.

p.271 *Acrostichum', existence', interest for me'*, Spruce to Daniel Hanbury, Welburn, Yorkshire, 10 Feb. 1873, in Wallace, 'Biographical Introduction', Spruce, *Notes* (vol.1), xxxviii, xxxix, xxxix; Schultes, *idem*.

p.271 *feet off'*, Spruce, Journal for 25 April 1855, *Notes* (vol.2), 10.

p.272 *excoriation', find', imminent'*, Spruce, *idem*, 72.

p.272 *boards', month'*, Spruce to Bentham, Tarapoto, Dec. 1855, Spruce, *idem*, 73.

p.272 *collecting'*, Spruce to Bentham, 7 April 1856, Spruce, *idem*, 74.

p.273 *Maynas'*, Spruce to Bentham, Baños, Ecuador, June 1857, Spruce, *idem*, 104.

p.274 *kind'*, Spruce, *idem*, 109.

pp.274–75 *each other', position', palms'*, Spruce, Journal for 21 May 1857, Spruce, *idem*, 121, 122, 123.

p.275 *fruit'*, Spruce, Journal for 17 June 1857, Spruce, *idem*, 140.

pp.275–76 *Litobrochia, etc.'*, Wallace editor's comments, Spruce, *idem*, 166.

p.276 *across'*, Spruce to Bentham, Baños, Ecuador, 1 Sept. 1857. Spruce, *idem*, 176.

p.276 *money', high', death'*, Spruce, Journal for 28 June 1857, Spruce, *idem*, 157, 157, 158.

p.277 *spot'*, Spruce, 'Of some remarkable narcotics of the Amazon Valley and Orinoco', *Ocean Highways: The Geographical Review* 9(55) (1873), 184–93; also in Spruce, *Notes* (vol.2), 414.

pp.277–78 *Malpighiaceae', disagreeable'*, Spruce, *Notes* (vol.2), 421, 416.

p.278 *quantities'*, Spruce, Field notebook, after the passage about his avoiding vomiting at the Dabucuri festival; but Wallace chose to omit this in his magnificent editing of his friend's papers. In Schultes, 'Some impacts of Spruce's Amazon explorations on modern phytochemical research', presented at the III International Pharmacological Congress, São Paulo, July 1966, and published in *Rhodora* 70(783) (1968; pp.313–39), 325. *Rhodora* is the quarterly peer-reviewed journal, founded in 1899, of the New England Botanical Club.

p.278 *exhausted', tobacco', horrible', always are',* Spruce, *Notes* (vol.2), 419, 420, 421. The Persian tales of *A Thousand and One Nights* had been available in English since 1706 and also in Portuguese translation.

p.279 *ground', patient',* Spruce, *idem*, 425, 444. Spruce noted that the Bavarian botanist Carl Friedrich Philip von Martius made the same observation, *Systema Materiae Medicae Vegetabilis Brasiliensis* (Munich, 1843), xvii.

pp.279–80 *bravery', 1852', observations',* Schultes, 'Some impacts of Spruce's Amazon explorations…', *Rhodora* 70(783) (1968), 317–18. Schultes wrote about *Banisteriopsis caapi* in his book, with Hofmann, *The Botany and Chemistry of Hallucinogens* (1973, rev. 1980) and many other papers, including *Journal of Ethnobiology* 6(2) (winter 1986); and his encyclopaedic description of hundreds of hallucinogenic and toxic plants, *De plantis toxicariis e Mundo Novo tropical commentationes*, that appeared in a score of Harvard University *Botanical Museum Leaflets* throughout the 1970s and 1980s; also *Ethnobotany* 5 (1993), 32–35. Wade Davis's *One River* (1996), a homage to his tutor and mentor Schultes, has many thrilling descriptions of *yagé* trances.

p.280 *lasting', tired!'* Spruce, 'On some remarkable narcotics…', *Notes* (vol.2), 428. Spruce had previously bought from a trader a Y-shaped inhaler made by Katauixi people of the Purus (now extinct, absorbed into the Katukina); a similar device was illustrated by the late-18th-century Portuguese scientist Alexandre Rodrigues Ferreira.

pp.280–81 *properties',* Wade Davis, *One River*, 476. The paper in which Schultes described this analysis was: '*De plantis toxicariis e Mundo Novo tropical commentationes II.* The vegetal ingredients of the myristicaceous snuffs of the north-west Amazon', *Rhodora* 70 (1968), 113–60. Milliken & Albert, *Yanomami, A forest people* (1999), 37.

p.281 *alert',* Spruce, *Notes* (vol.2), 429.

p.281 *souls',* Spruce to John Teasdale, 14 Sept. 1857, Spruce, *idem*, 183.

p.281 *time',* Wallace editor's comment, in Spruce, *idem*, 171.

p.282 *tickets',* Joseph Hooker letter to Wallace, quoted in Wallace's 'Biographical Introduction', Spruce, *Notes* (vol.1), xliv.

p.282 *contend',* Daniel Oliver to Wallace, 'Biographical Introduction', Spruce, *idem*, xliv.

p.282 *preserved',* Bentham, 'Obituary Notice of Richard Spruce', *Proceedings*, Botanical Society of Edinburgh (Feb. 1894), in Wallace's 'Biographical Introduction', Spruce, *idem*, xlvi.

p.282 *performance',* Spruce, *Report on the Expedition to Procure Seeds and Plants of the* Cinchona succirubra *or Red Bark tree* (1862), 3; also in Spruce, *Notes* (vol.1), 261–93.

p.283 *devil',* Oliver Cromwell, quoted in Musgrave & Musgrave, *An Empire of Plants: People and Plants that Changed the World* (2000), 146.

p.283 *bark',* Sir John Evelyn, Diary, entry for 6 Aug. 1685. Rocco, *The Miraculous Fever-Tree* (2003), 109.

p.284 *suffering',* Markham, *Peruvian Bark: A Popular Account of the Introduction of Chinchona Cultivation into British India 1860–1880* (1880), 88; Honigsbaum, *The Fever Trail* (2001), 101.

p.285 *economy',* India Office to Spruce, Nov. 1859, Manchester City Library, Spruce Correspondence 5, in Pearson, *Richard Spruce, Naturalist and Explorer* (2004), 59.

p.285 *gentleman',* Wallace, Obituary of Spruce, *Nature*, 1 Feb. 1894; 'Biographical Introduction', Spruce, *Notes* (vol.1), xliii.

p.285 *caresses',* Manuel Santander to Ricardo Spruce, Ambato, 30 June 1867 (translated by Wallace), Spruce, *Notes* (vol.2), 347.

p.285 *slight ones',* Spruce to John Teasdale, Ambato, 14 April 1860, Spruce, *idem*, 228.

pp.286–87 *inns', sand', district', standing', conceived',* Spruce to Sir William Hooker, Ambato, 20 Oct. 1859, Spruce, *idem*, 229, 232, 237–38, 242, 242. The páramo of Teocajas had been the site of the largest pitched battle between Pizarro's conquistadors and an Inca army, in May 1534.

p.287 *Red Bark',* Spruce, 'List of Botanical Excursions', Spruce, *idem*, 259.

pp.287–88 *roja', touch the bark',* Spruce, *Report of the Expedition* (1862), 5, 7; also Spruce, *Notes* (vol.2), 262.

p.288 *condaminea',* Spruce to Sir William Hooker, Ambato, 12 March 1860, Spruce, *Notes* (vol.2), 261.

p.288 *South America',* Spruce to John Teasdale, Ambato, 13 March 1858, Spruce, *idem*, 191.

p.289 *revolutions',* Spruce, *idem*, 297.

p.289 *foot',* Spruce to Bentham, Ambato, 16 March 1858, Spruce, *idem*, 203.

p.289–90 *exhaustion'*, Spruce, 'List of excursions during the year 1860', Spruce, *idem*, 259.

p.290 *paralysis'*, Honigsbaum, *The Fever Trail*, 125. Guillain-Barré syndrome was identified by two French doctors, of those names, in 1916.

p.290 *relief'*, Spruce, *Report of the Expedition*, 10.

p.290 *neighbourhood'*, Spruce, 'List of excursions during the year 1860', Spruce, *Notes* (vol.2), 260.

p.290 *unsparingly', beauty'*, Spruce, *Report of the Expedition*, 24, 30.

p.291 *possible', forest', combated', watchfulness', ground'*, Spruce, *Notes* (vol.2), 294, 294, 295–96, 298.

pp.291–92 *seeds', his own', escaped', maltreated', naught', successful'*, Spruce, *Report of the Expedition*, 71–72, 78, 81, 82, 82–83, 83.

p.293 *elsewhere'*, India Office to Spruce, 29 Jan. 1862, Manchester City Library, Spruce Correspondence 22, in Pearson, *Richard Spruce*, 63.

p.293 *plantations'*, Wallace editor's comment, in Spruce, *Notes* (vol.2), 310.

p.293 *kind'*, Spruce to Bentham, Daule near Guayaquil, 9 March 1861, Spruce, *idem*, 316.

p.293 *very sick'*, Spruce, 'Botanical excursions, 1861 to 1864', Spruce, *idem*, 313.

pp.293–94 *well off', high as ever'*, Spruce to John Teasdale, Chandúy near Guayaquil, 14 May 1862, Spruce, *idem*, 320, 321.

p.294 *property', everything'*, Spruce, 'Botanical Excursions, 1861 to 1864', Spruce, *idem*, 314, 313.

p.294 *flowering plants'*, Spruce, *Notes on the Valleys of Piura and Chira, in Northern Peru* (1864); Spruce, *Notes* (vol.2), 340.

p.294 *limbs'*, Spruce to Daniel Hanbury, Guayaquil, 29 Nov. 1862, Spruce, *Notes* (vol.2), 321.

p.295 *whatever'*, Spruce to Bentham, Amotape, near Paita, Peru, 13 April 1864, Spruce, *idem*, 341.

CHAPTER 12

p.296 *calico', slum'*, Wallace, *My Life* (vol.1), 313, 321.

p.297 *botanist'*, William Hooker, 'Review of Wallace's *Palm Trees of the Amazon*', *Hooker's Journal of Botany* 6 (1854), 62.

p.297 *know'*, Spruce to Sir William Hooker, San Carlos de Río Negro, 19 March 1854, in *Hooker's Journal of Botany* 7 (1855), 213.

pp.298–99 *forests', life', orchards', Negro'*, Wallace, *Narrative*, 405, 404, 336, 337.

p.299 *holds'*, Herndon & Gibbon, *Exploration of the Valley of the Amazon* (1854); Hemming, *Amazon Frontier* (1995), 233–34.

p.299 *matter', sketches'*, Wallace, 'On the Rio Negro', *Journal of the Royal Geographical Society* 23 (1853), 213, 217.

p.300 *unknown land', ceremonies'*, Wallace, *Narrative*, 477.

p.300 *together'*, Bates, Introduction to first ed. (1863) of *Naturalist* (vol.1), iii; Wallace, *My Life* (vol.1), 264.

pp.300–1 *humboldtii', sides of them'*, Wallace, 'On the Monkeys of the Amazon', *Proceedings of the Zoological Society of London* 20 (1852), 110.

p.301 *undergone'*, Wallace, 'On the Habits of the Butterflies of the Amazon Valley', *Transactions of the Entomological Society of London* 2(8) (April 1854; pp.253–64), 257–58.

p.301 *in them all', possible'*, Wallace, *Narrative*, 85.

p.302 *Ocean'*, Stevens, introducing Wallace's paper on Hesperidae, March 1853, in Raby, *Alfred Russel Wallace*, 84.

p.302 *specimens'*, Edward Newman, Presidential address, *Transactions of the Entomological Society of London* 2 (1853), 142; Raby, *idem*, 85.

p.302 *club'*, Shermer, *In Darwin's Shadow* (2002), 77.

p.303 *islands', each', science'*, Wallace, *My Life* (vol.1), 327. Murchison had been President of the RGS in 1843–1845, was now president in 1851–1853, and would serve again in 1856–1859 and for nine more years in 1862–1871.

p.303 *natural history'*, Wallace's proposal to the RGS, June 1853, in its archive; Raby, *Alfred Russel Wallace*, 88. The Expedition Committee's favourable decision was on 22 July.

p.304 *species'*, Wallace, 'On the Law which has Regulated the Introduction of New Species', *Annals and Magazine of Natural History* 16 (1855), 184–96; also in Berry, *Infinite Tropics*, 36–49.

p.304 *completeness'*, Bates to Wallace, Ega, 19 Nov. 1856, in the Wallace Family Archive; Woodcock, *Henry Walter Bates, Naturalist of the Amazons*, 217; Raby, *Alfred Russel Wallace*, 127.

p.304 *years'*, Darwin to Wallace, 1 May 1857, Burkhardt, Smith, et al., *The Correspondence of Charles Darwin* (1985–) (vol.6), 387; Raby, *Alfred Russel Wallace*, 125–26; the remark about a trespass notice is by Desmond & Moore, *Darwin* (1991), 455.

p.305 *survive', species'*, Wallace, *My Life* (1908, single-volume ed.), 191–92.

p.305 *theory'*, Wallace, *The Wonderful Century* (1898), 140; Shermer, *In Darwin's Shadow*, 113; Fichman, *An Elusive Victorian* (2004), 97–98.

p.305 *Lyell'*, Wallace, *My Life*, 191. Ross Slotten analysed Wallace's different accounts of his flash of inspiration, in a note on p.511 of *The Heretic in Darwin's Court*. The first mention was in a letter of 1869 to a German naturalist, then in one of 1887 to Alfred Newton. Wallace then wrote it in three of his own books, from 1891 onwards.

p.306 *smashed'*, Darwin to Lyell, Down, 18 June 1858, in Francis Darwin, *The Life and Letters of Charles Darwin* (1887–1892) (vol.2), 116. It was Darwin's custom to put only the day and not the month on his letters: thus, this one has only '18'. John Langdon Brooks argued that it could have been 18 *May* not June. It was just possible for Wallace's paper to have reached him by then, so that Darwin could have doctored his writings that were read at the Linnean meeting to incorporate the younger scientist's idea. But this conspiracy theory is invalidated by Darwin writing in his private journal for 18 June: 'interrupted by letter from A. R. Wallace'.

p.306 *inquiry'*, Darwin and Wallace meeting of 1 July 1858, *Journal of the Linnean Society of London (Zoology)* 3(9), 53. The letter of the previous September that was read out was addressed to the American botanist Asa Gray.

p.306 *doctrine'*, Joseph Hooker, account written for Francis Darwin's edited *The Life and Letters of Charles Darwin* (vol.2), 126; Raby, *Alfred Russel Wallace*, 139.

pp.306–7 *Selection'*, Alfred Newton, in Wollaston, *Life of Alfred Newton* (1921), 112–13; Raby, *Alfred Russel Wallace*, 140.

p.307 *subject'*, Wallace to Joseph Hooker, Ternate, 6 Oct. 1858, in the Huntington Library, California; Shermer, *In Darwin's Shadow* (2002), 139; Wallace, *My Life* (1908 ed.), 193–94.

p.307 *proud of'*, Wallace to George Silk, Batchian, Nov. 1858, *My Life* (vol.1), 366. The 'Ternate' paper of Feb. 1858 has often been published, including in Berry, *Infinite Tropics*, 52–62.

p.307 *others'*, Darwin to Wallace, 18 May 1860, Burkhardt, Smith, et al., *The Correspondence of Charles Darwin* (vol.8), 218.

p.307 *further go!'*, Wallace to George Silk, Bessir, 1 Sept. 1860 (actually posted from Ternate in October), Wallace, *My Life* (vol.1), 372–73.

p.307 *philosophy'*, Wallace to Bates, Ternate, 24 Dec. 1860, Wallace, *idem*, 374.

p.308 *Darwinism'*, Fichman, *An Elusive Victorian*, 100–1.

p.308 *spirit'*, Charles Darwin to Bates, Down House, Bromley, Kent, 22 Nov. 1860, Stecher, 'The Darwin-Bates Letters', *Annals of Science* 25(1) (March 1969), 6. Dr Stecher had bought all Darwin's letters to Bates from the New York publisher Charles Scribner's Sons in 1933.

p.308 *intellect'*, Darwin to Bates, Dec. 1861, Stecher, *idem*, 20.

p.308 *true of you'*, Darwin to Wallace, Down, Beckenham, 20 April 1870, in Marchant, *Alfred Russel Wallace*, 252.

p.309 *Australian'*, Wallace, *My Life* (vol.1), 356.

p.309 *limits'*, Wallace letter, *Zoologist* 15 (1857), 5415; Michaux, 'Alfred Russel Wallace, Biogeographer', in Smith & Beccaloni, *Natural Selection & Beyond* (2008, pp.166–85), 179.

p.309 *distribution', continent'*, Wallace to Bates, Amboyna (Ambon), 4 Jan. 1858, *My Life* (vol.1), 358–59. Also, Wallace letter in *Ibis* 1 (1859), 111–13; Wallace, 'On the zoological geography of the Malay Archipelago', *Journal of Proceedings of The Linnean Society: Zoology* 4 (1860), 172–84.

p.310 *journeys'*, Wallace, *The Malay Archipelago* (1869), xii.

p.311 *moral nature'*, Wallace, Review of new editions of Charles Lyell, *Principles of Geology* and *Elements of Geology*, in *Quarterly Review* 126 (1869), 394; *My Life* (vol.1), 406; Shermer, *In Darwin's Shadow*, 159.

p.311 *animals'*, Gould, 'A Biographical Sketch', in Berry, *Infinite Tropics*, 18.

p.312 *withal'*, notice in *Reviews*, which Wallace included in a letter to his friend Raphael Meldola, 4 Nov. 1905, Hope Library, Oxford Museum of Natural History, quoted in Raby, *Alfred Russel Wallace*, 274.

p.312 *bored!'* Wallace to Violet Wallace, 20 May 1892, in Wallace Family Archive, in Raby, *Alfred Russel Wallace*, 259–60.

p.312 *thoughts'*, Sir Ray Lankester, President of the Linnean Society, address to the 1908 commemorative meeting, in Marchant, *Alfred Russel Wallace*, 121. Marchant's book gives a full account of the meeting by Sir Joseph Hooker, and the various speeches at it.

p.312 *unintelligible'*, Wallace to Frederick Birch, 30 Dec. 1908, Wallace Family Archive, in Raby, *Alfred Russel Wallace*, 278.

p.313 *Leicester'*, Bates to his mother, Belém, 29 May 1859, in Clodd, 'Memoir', xxx.

p.313 *there you have it'*, Bates to Darwin, 1863, in Woodcock, *Henry Walter Bates*, 254.

p.314 *describe them'*, Bates, Preface to *Naturalist* (vol.1), v. This preface was in the first ed. of 1863, but was omitted in the somewhat abridged single-volume second ed. of 1864.

p.314 *part'*, Sharp, 'Henry Walter Bates, F.R.S.', obituary, *The Entomologist* 25(347) (April 1892), 78.

p.314 *people*', Hewitson, in Goodger & Ackery, 'Bates, and the Beauty of Butterflies', *The Linnean* 18(1) (Jan. 2002), 24.
p.314 *named*', Clodd, 'Memoir', xxx.
p.314 *setting*', Goodger & Ackery, 'Bates... Butterflies', 25.
p.315 *weather*', Bates notebook, in Museum of Natural History, Entomology Library.
p.315 *description*', Moon, *Henry Walter Bates, F.R.S.* (1976), in Goodger & Ackery, 'Bates... Butterflies', 22.
p.315 *affinities*', Bates, notebook, 1855, quoted in Anon, 'Bates and natural selection', *The Linnean* 11(1) (Jan. 1995), 6.
p.316 *show it*', Wallace to Bates, Amboyna, 4 Jan. 1858, *idem*, 66. Wallace wrote further letters to 'My dear Bates', from Ternate on 25 Jan. and 2 March 1858, from Ceram on 25 Nov. 1859, and from Ternate on 24 Dec. 1860: all are in Marchant, *Alfred Russel Wallace*.
p.316 *species etc.*', Charles Darwin to Bates, Down House, Bromley, Kent, 22 Nov. 1860, Stecher, 'The Darwin-Bates Letters', 5–6; also in Darwin, *More Letters of Charles Darwin* (1903), 176–77.
p.316 *entomologist*', Distant, contribution to 'Obituary, Henry Walter Bates, F.R.S.'. *Proceedings of the Royal Geographical Society* 14 (1899), 250; Clodd, 'Memoir', xxx–xxxi.
p.316 *souls*', Darwin to Bates, 20 Nov. 1862, Clodd, 'Memoir', xiv.
pp.316–17 *time*', Bates to Darwin, King St., Leicester, 18 March 1861, in Stecher, 'The Darwin-Bates Letters', 8.
p.317 *Geology*', Bates, 'Contributions to an Insect Fauna of the Amazon Valley. Part I: Diurnal Lepidoptera' *The Transactions of the Entomological Society of London* 5 (1861), 353.
p.317 *profit by them*', *to think*', Darwin to Bates, Down, Bromley, Kent, 26 March 1861, Stecher, 'The Darwin-Bates Letters', 9, 10.
p.317 *dumbfounded*', Darwin to Hooker, in Woodcock, *Henry Walter Bates*, 246.
p.317 *useful to you*', Bates to Darwin, Leicester, 28 March 1861, Stecher, 'The Darwin-Bates Letters', 11.
p.317 *publish at all*', Bates to Dr. J. D. Hooker, King Street, Leicester, 19 March 1861, in Clodd, 'Memoir', xxxiii.
p.317 *work*', Darwin to Bates, Down, Bromley, Kent, 3 Dec. 1861, in Stecher, 'The Darwin-Bates Letters', 20.
p.318 *all-round*', Sir Joseph Hooker, contribution to 'Obituary, Henry Walter Bates, F.R.S.', *Proceedings of the Royal Geographical Society* 14 (1892), 249–50.
p.318 *all things*', Bates to Darwin, Leicester, 28 March 1861, in Stecher, 'The Darwin-Bates Letters', 14.
pp.318–19 *bees*', *staggering*', *leaves*', *wasp*', *life*', Bates, 'Contributions to an Insect Fauna of the Amazon Valley. LEPIDOPTERA: HELICONIDAE', *Transactions of the Linnaean Society of London* 23 (1862; pp.495–566), 506, 507, 508–9, 508, 510.
p.319 *world*', Clodd, 'Memoir', xliii; Wallace, *Darwinism: An exposition of the Theory of Natural Selection* (1889), 234.
pp.319–20 *prey*', *animals*', *Species*', *kind*', Bates, 'Contributions to an Insect Fauna...', 510, 511, 511, 512.
p.320 *work*', Darwin to Bates, Down, Bromley, 20 Nov. 1862, in Clodd, 'Memoir', xiv.
p.320 *mimicry*', Charles Darwin, 'An Appreciation', *Natural History Review*, 3 April 1863, vii; Clodd, 'Memoir', xlii.
p.321 *spirits*', Bates, Preface to *Naturalist* (1863 ed.; vol. 1), iv.
p.321 *invalid*', Darwin to Bates, Down, Bromley, 22 Nov. 1860, in Stecher, 'The Darwin-Bates Letters', 5–6.
p.321 *book*', Bates, 'Preface', *The Naturalist* (1863 ed.; vol. 1), iv.
p.321 *dogs*', Darwin to Bates, Nov. 1861, Stecher, 'The Darwin-Bates Letters', 18.
p.321 *book*', Darwin to Bates, in Woodcock, *Henry Walter Bates*, 250.
p.321 *labour*', Darwin to Bates, 20 Nov. 1862, in Clodd, 'Memoir', xiv.
p.321 *accomplished*', Bates, Preface to *Naturalist* (1863 ed.; vol. 1), iv.
p.321 *observation*', Clodd, 'Memoir', lxi.
p.321 *descriptions*', Darwin to Bates, Hartfield, Tunbridge Wells, 30 April 1863, Clodd, 'Memoir', lxii; Charles Darwin, 'An Appreciation', *Natural History Review* 3 (1863), vii–xiii..
p.321 *Amazons*', Clodd, 'Memoir', lxii.
p.321 *nation*', Sharp, 'Henry Walter Bates, F.R.S.', *The Entomologist* 25(847) (April 1892), 79.
p.322 *Archipelago*', Distant, contribution to 'Obituary. Henry Walter Bates, F.R.S.', *Proceedings of the Royal Geographical Society* 14 (1892), 251.
p.322 *Heliconius*', Clodd, 'Memoir', lxxi.
p.322 *14,700*', Bates to Joseph Hooker, 12 May 1863, in Clodd, 'Memoir', lxv.
p.322 *fault*', *writing*', Hooker to Bates, 13 May 1863, in the Archive of Kew Gardens, in Clodd, 'Memoir', lxvi; also Goodger & Ackery, 'Bates, and the Beauty of Butterflies', *The Linnean* 18 (2002), 24.
p.323 *habits*', Galton, contribution to 'Obituary, Henry Walter Bates, F.R.S.', *Proceedings of the*

Royal Geographical Society 14 (1892), 256. The first paid head of the RGS was actually Norton Shaw, but he resigned in 1863; the ubiquitous Clements Markham had lunch with Bates and suggested that he take the job, but Bates still hoped for a museum job; so the RGS appointed a Mr Greenfield, but he had died prematurely after only a few weeks in the post.

p.323 *Museum',* Sharp, 'Henry Walter Bates, F.R.S.', *The Entomologist* 25(347) (April 1892), 79. In a short-lived diary, Bates recorded in his entry for 15 Feb. 1864 that he went to discuss the job offer with Mr Crawfurd, a vice-president of the RGS: Clodd, 'Memoir', lxxiv.

p.323 *away',* Clodd, 'Memoir', lxxv–lxxvi.

p.324 *work', advice',* Markham, 'Obituary. Henry Walter Bates, F.R.S.', *Proceedings of the Royal Geographical Society* 14 (1892), 255.

p.324 *respect',* Galton, in 'Obituary. Henry Walter Bates, F.R.S.', *Proceedings of the Royal Geographical Society* 14 (1892), 256.

p.325 *energy',* Grant Duff, Annual Address on the progress of geography', *Proceedings of the RGS* 14 (1892), 361; Dickenson, 'The naturalist on the River Amazons and a wider world: reflections on the centenary of Henry Walter Bates', *Geographical Journal* 158(2) (July 1992), 211.

pp.325–26 *accomplished',* Sharp, 'Henry Walter Bates, F.R.S.', *The Entomologist* 25 (April 1892), 79.

p.326 *followed',* Distant, Obituary, *Proceedings of the Royal Geographical Society* 14 (1892), 252. The beetle work was for Godman & Salvin, *Biologia Centrali-Americana* (1884–90).

p.326 *forests',* Baron de Santa Anna Néry to the foreign secretary of the Royal Geographical Society, Paris, 8 March 1893, in Clodd, 'Memoir', xxix.

p.327 *Amazons',* Allen, 'Bates of the Amazons', *Fortnightly Review* 58 (July–Dec. 1892), 799.

p.327 *heard him',* Allen, *idem*, 802–3.

p.327 *my house',* Markham, 'Richard Spruce', *The Geographical Journal* 3 (1894), 247.

p.327 *friendship',* comment by Wallace, in Spruce, *Notes* (vol.2), 343. When Daniel Hanbury died in 1875, his brother Sir Thomas Hanbury, Bt., lent Wallace almost a thousand letters that Spruce had sent him; and these were then lodged with the Royal Pharmaceutical Society.

p.328 *pretention',* Spruce to Hanbury, 1864, Royal Pharmaceutical Society, Daniel Hanbury papers P320, in Pearson, *Richard Spruce*, 67.

p.328 *proportions',* Wallace, 'Biographical Introduction', Spruce, *Notes* (vol.1), xlii.

p.328 *hours',* Spruce to Daniel Hanbury, Welburn, 1868, Spruce, *idem*, xxxiv.

p.328 *intestines',* Spruce to George Stabler, Welburn, 1867, Spruce, *idem*, xxxv.

p.328 *couch',* Wallace, 'Biographical Introduction', Spruce, *idem*, xxxiv.

pp.328–29 *companions',* Wallace, 'Biographical Introduction', Spruce, *idem*, xlii. The housekeeper's little girl later rented Spruce Cottage; Professor Schultes met her there in mid-twentieth century.

p.329 *moment',* Wallace, Obituary notice, *Nature*, 1 Feb. 1894; also 'Biographical Introduction', Spruce, *idem*, xliii.

p.329 *aid',* Spruce to Matthew Slater, Coneysthorpe, 25 March 1892, Edwards, 'Spruce in Manchester', in Seaward & FitzGerald, *Richard Spruce (1817–1893) Botanist and Explorer* (1996), 273.

p.329 *sympathy with him',* Wallace, 'Biographical Introduction', Spruce, *Notes* (vol.1), xl.

p.329 *Government',* Wallace, 'Introduction', Spruce, *idem*, xlii. Markham mentioned 'years of importunity' to get the second pension, but did not say that it was by him. He did, however, say that the Howard family gave 'unvarying kindness' to the invalid botanist: Markham, 'Richard Spruce', *The Geographical Journal* 3 (1894), 247.

p.329 *himself',* Spruce remark to William Weston, in Seaward, 'Introduction' to Seaward & FitzGerald, *Richard Spruce*, 8.

p.330 *rank',* Henderson, 'Richard Spruce and the palms of the Amazon and Andes', in Seaward & FitzGerald, *idem*, 189.

pp.330–31 *so good',* Gradstein, 'Spruce's *Hepaticae Amazonicae et Andinae* and South American floristics', in Seaward & FitzGerald, *idem*, 142.

p.331 *every one', subgenera', reference',* Stotler, 'Richard Spruce: his fascination with liverworts and its consequences', in Seaward & FitzGerald, *idem*, 124, 136.

pp.331–32 *foliage', functions',* Markham, *Peruvian Bark. A Popular Account of the Introduction of Chinchona Cultivation into British India, 1860–1880*, (1880); Honigsbaum, *The Fever Trail* (2001), 174, 179; Rocco, *The Miraculous Fever-Tree* (2003), 247.

p.333 *conscious',* Davis, *One River* (1996), 373–74.

p.333 *all times',* Schultes, 'Some impacts of Spruce's Amazon explorations…', *Rhodora* 70(783) (1968), 313.

p.333 *Amazon itself',* Prance, 'Preface', Seaward & FitzGerald, *Richard Spruce*, i.

p.333 *existence',* Spruce to Daniel Hanbury, Welburn, 10 Feb. 1873, in Wallace, 'Biographical Introduction', Spruce, *Notes* (vol.1), xxxix.

BIBLIOGRAPHY

Adalbert, Prince of Prussia. 1847. *Aus meinem Reisetagebuch 1842–43*. Berlin; trans. Sir Robert Schomburgk & John Edward Taylor, *Travels of His Royal Highness Prince Adalbert of Prussia, in the South of Europe and in Brazil, with a voyage up the Amazon and Xingu, now first explored* (2 vols), London, 1849.

Allen, David Elliston. 1994. *The Naturalist in Britain: A Social History*. New Haven: Princeton University Press.

Allen, Grant. 1892. 'Bates of the Amazons'. *Fortnightly Review* 58, 798–809.

Angel, R. 1978. 'Richard Spruce, Botanist and Traveller, 1817–1893'. *Aliquando* 3, 49–53.

Anon. 1892. 'Memoir of the late H. W. Bates, F.R.S.'. *Zoologist* 16, 184–88.

Anon. 1894. 'Obituary of Richard Spruce'. *Proceedings of the Linnean Society*, 35–37.

Balick, Michael J. 1980. 'Wallace, Spruce and *Palm Trees of the Amazon*: an historical perspective'. *Botanical Museum Leaflets* 28(3), 263–69.

Bates, Frederick. 1892. Contribution to 'Obituary. Henry Walter Bates F.R.S.'. *Proceedings of the Royal Geographical Society* 14, 245–47.

Bates, Henry Walter. 1850–1858. 'Extracts from the correspondence of Mr. H. W. Bates, now forming entomological collections in South America', and 'Proceedings of natural-history collectors in foreign countries'. *Zoologist* 8 (1850), 2663–8, 2715–9, 2789–93, 2836–41, 2940–4, 2965–6; 9 (1851), 3142–4, 3230–2; 10 (1852), 3321–4, 3352–3, 3449–50; 11 (1853), 3726–9, 3801–4, 3841–3, 3897–3900, 4111–7; 12 (1854), 4200–2, 4313–21, 4397–8; 13 (1855), 4549–53; 14 (1856), 5012–6, 5016–9; 15 (1857), 5557–9, 5651–62, 5725–37; 16 (1858), 6160–9.

——1852. 'Some account of the country of the River Solimoens, or Upper Amazons'. *Zoologist* 10, 3590–9.

——1859–1861. 'Contributions to an Insect Fauna of the Amazon Valley, Part I: Diurnal Lepidoptera'. *Transactions of the Entomological Society of London* 5 (1859), 223–28, and (1861), 335–61.

——1861. 'Contributions to an Insect Fauna of the Amazon Valley, Lepidoptera-Papilionidae'. *Journal of Entomology* 1, 218–45.

——1861–1866. 'Contributions to an Insect Fauna of the Amazon Valley, Coleoptera: Longicornes'. *The Annals and Magazine of Natural History* 8–9, 12–17.

——1862. 'Contributions to an Insect Fauna of the Amazon Valley. LEPIDOPTERA: HELICONIDAE'. *Transactions of the Linnaean Society of London* 23, 495–566.

——1862. 'Contributions to an Insect Fauna of the Amazon Valley, Lepidoptera-Heliconiinae'. *Journal of the Proceedings of the Linnean Society of London* 6, 73–77.

——1863. *The Naturalist on the River Amazons: A Record of Adventures, Habits of Animals, Sketches of Brazilian and Indian Life, and Aspects of Nature under the Equator, during Eleven Years of Travel* (2 vols). London: John Murray; second, abridged ed. (1 vol.), London: John Murray, 1864; commemorative ed., with introductory Memoir by Edward Clodd, London: John Murray, 1892.

——1864. 'Contributions to an Insect Fauna of the Amazon Valley, Lepidoptera-Nymphalidae', *Journal of Entomology* 2, 175–213, 311–46.

——1867. 'New Genera of Longicorn Coleoptera from the River Amazons'. *The Entomologist's Monthly Magazine* 4.

——1867. 'On a collection of butterflies formed by Thomas Belt Esq. in the interior of the province of Maranham, Brazil'. *Transactions of the Entomological Society* 5, 535–46.

——1869. 'Contributions to an Insect Fauna of the Amazon Valley (Coleoptera, Prionides)'. *Transactions of the Entomological Society of London for the Year 1869*.

——1869. *Illustrated Travels: A Record of Discovery, Geography, and Adventure* (6 vols). London: Cassell, Petter and Galpin.

——1870. 'Contributions to an Insect Fauna of the Amazon Valley (Coleoptera, Cerambycidae)'. *Transactions of the Entomological Society of London for the Year 1870*.

——1871. 'Hints on the collection of objects of natural history, in "Hints to Travellers"'. *Proceedings of the Royal Geographical Society* 16, 67–78.

——1873. 'Notes on the Longicorne Coleoptera of Tropical America'. *The Annals and Magazine of Natural History* 11.

——1878. *Central America, the West Indies and South America, with Ethnological Notes by A. H. Keane.* London: Stanford (revised 1882).

Beddall, Barbara G. 1968. 'Wallace, Darwin, and the theory of natural selection'. *Journal of the History of Biology* 1, 261–323.

——ed. 1969. *Wallace and Bates in the Tropics: an Introduction to the Theory of Natural Selection.* London and New York: Macmillan.

——1972. 'Wallace, Darwin, and Edward Blyth: further notes on the development of evolutionary theory'. *Journal of the History of Biology* 5, 153–58.

Bentham, George. 1859–1876. Chapters in Carl Friedrich Philip von Martius, *Flora Brasiliensis*, based on Spruce's collections: '*Leguminosae* 1, *Papilionaceae*' 15(1) (1859–1862), 1–350; '*Swartzieae, Caesalpineae, Mimoseae*' 15(2) (1870–1876), 1–527.

——1875. 'VII Revision of the Suborder *Mimoseae*'. *Transactions of the Linnean Society of London* 30, 335–664.

Benton, Ted. 2013. *Alfred Russel Wallace: Explorer, Evolutionist and Public Intellectual – A Thinker for Our Own Time?* London: Siri Scientific.

Berry, Andrew, ed. 2002. *Infinite Tropics, an Alfred Russel Wallace Anthology.* London and New York: Verso.

Brackman, Arnold C. 1980. *A Delicate Arrangement: The Strange Case of Charles Darwin and Alfred Russel Wallace.* New York: Times Books.

Brooks, John Langdon. 1984. *Just Before the Origin: Alfred Russel Wallace's Theory of Evolution.* New York: Columbia University Press.

Browne, Janet. 1983. *The Secular Ark: Studies in the History of Biogeography.* New Haven, CT: Yale University Press.

——2003. *Charles Darwin: The Power of Place.* Princeton, NJ: Princeton University Press.

Burkhardt, Frederick, Sydney **Smith**, et al., eds. 1985–. *The Correspondence of Charles Darwin* (vols.1–11). Cambridge: Cambridge University Press.

Camerini, Jane R. 1993. 'Evolution, biogeography and maps: an early history of Wallace's Line'. *Isis* 84, 44–65.

——1996. 'Wallace in the field'. *Osiris* (2 ser.) 11, 44–65.

——ed. 2002. *The Alfred Russel Wallace Reader: A Selection of Writings from the Field.* Baltimore and London: Johns Hopkins University Press.

Cameron, Ian. 1980. *To the Farthest Ends of the Earth.* London: Macdonald.

Caro, Tim, Sami **Merilaita** & Martin **Stevens**. 2008. 'The colours of animals: from Wallace to the present day. I. Cryptic coloration'. In C. Smith & Beccaloni, 125–43.

——, Geoffrey **Hill**, Leena **Lindström** & Michael **Speed**. 2008. 'The colours of animals: from Wallace to the present day. II. Conspicuous coloration'. In C. Smith & Beccaloni, 144–65.

Clements, Harry. 1983. *Alfred Russel Wallace: Biologist and Social Reformer.* London: Hutchinson and Co.

Clodd, Edward. 1892. Contribution to 'Obituary. Henry Walter Bates, F.R.S.'. *Proceedings of the Royal Geographical Society* 14, 253–54.

——1892. 'Memoir' in H. W. Bates, *The Naturalist on the River Amazons* (commemorative ed.), xvii–lxxxix.

——1916. *Memories.* London: Chapman & Hall.

Colp, Ralph, Jr. 1992. '"I will gladly do my best." How Charles Darwin obtained a Civil List pension for Alfred Russel Wallace'. *Isis* 83, 3–26.

Darwin, Charles. 1839. *Journal of Researches into the Geology and Natural History of the Various Countries Visited by H.M.S. 'Beagle'.* London: Henry Colburn.

——1859. *On the Origin of Species by Means of Natural Selection, or the Preservation of Favoured Races in the Struggle for Life.* London: John Murray.

——1863. 'An Appreciation' [of H W Bates's *The Naturalist…*]. *Natural History Review* 3, vii–xiii.

——1871. *The Descent of Man, and Selection in Relation to Sex* (2 vols). London: John Murray.

——& Alfred Russel **Wallace**. 1858. 'On the Tendency of Species to Form Varieties, and on the Perpetuation of Varieties and Species by Means of Natural Selection'. *Journal of the Proceedings of the Linnean Society (Zoology)* 3, 53–62.

Darwin, Francis, ed. 1887–1892. *The Life and Letters of Charles Darwin* (3 vols). London: John Murray.

Davis, Wade. 1996. *One River: Explorations and Discoveries in the Amazon Rainforest.* New York: Simon & Schuster.

Dawkins, Richard. 2002. 'The Reading of the Darwin-Wallace papers commemorated – in the Royal Academy of Arts'. *The Linnean* 18(4), 17–24.

Dickenson, John. 1990. 'Henry Walter Bates and the study of Latin America in the late nineteenth century; a bibliographic essay'. *Revista Interamericana de Bibliografia* 40, 570–80.

——1992. 'Henry Walter Bates – the Naturalist of the River Amazons'. *Archives of Natural History* 19, 209–18.

——1992. 'The Naturalist on the River Amazons and a wider world: reflections on the centenary of Henry Walter Bates'. *Geographical Journal* 158, 207–14.

——1996. '"Getting on in his rambles in South America": the published correspondence of H. W. Bates'. *Archives of Natural History* 19(2), 201–8.

——1996. 'Bates, Wallace and economic botany in mid-nineteenth century Amazonia'. In Seaward & FitzGerald, 65–80.

Distant, William L. 1892. Contribution to 'Obituary: Henry Walter Bates, F.R.S.'. *Proceedings of the Royal Geographical Society* 14, 250–3.

Dover, Gabriel. 2000. *Dear Mr. Darwin. Letters on the Evolution of Life.* London: Weidenfeld & Nicolson.

Drew, William B. 1996. '*Cinchona* work in Ecuador by Richard Spruce, and by United States botanists in the 1940s'. In Seaward & FitzGerald, 157–61.

Eaton, George. 1986. *Alfred Russel Wallace, 1823–1913, Biologist and Social Reformer: A Portrait of his Life and Work, and a History of Neath Mechanics Institute and Museum.* Neath: W. Whittington Ltd.

Edwards, William H. 1847. *A Voyage up the River Amazon, Including a Residence at Pará.* London: John Murray.

Endersby, Jim. 2008. *Imperial Nature: Joseph Hooker and the Practices of Victorian Science.* Chicago: Chicago University Press.

Ewan, Joseph. 1992. 'Through the jungles of Amazon travel narratives of naturalists'. *Archives of Natural History* 19(2), 185–207.

——1996. 'Tracking Richard Spruce's legacy from George Bentham to Edward Whymper'. In Seaward & FitzGerald, 41–49.

Fagan, Melinda Bonnie. 2008. 'Theory and practice in the field: Wallace's work in natural history (1844–1858)'. In C. Smith & Beccaloni, 66–90.

Ferreira, Ricardo. 1990. *Bates, Wallace, Darwin e a Teoria da Evolução.* Brasília: Universidade de Brasília.

Fichman, Martin. 1981. *Alfred Russel Wallace.* Boston: Twayne Publishers.

——2004. *An Elusive Victorian: The Evolution of Alfred Russel Wallace.* Chicago and London: University of Chicago Press.

Field, David V. 1996. 'Richard Spruce's economic botany collections at Kew'. In Seaward & FitzGerald, 245–64.

Freshfield, Douglas. 1886. 'The place of geography in education'. *Proceedings of the Royal Geographical Society* 8, 698–718.

——1892. 'Death of Mr H. W. Bates'. *Proceedings of the Royal Geographical Society* 14, 191–92.

Galton, Francis. 1892. Contribution to 'Obituary: Henry Walter Bates, F.R.S.'. *Proceedings of the Royal Geographical Society* 14, 255–56.

Gander, Richard, ed. 1998. *Alfred Russel Wallace at School, 1830–37.* Rustington, West Sussex.

Gardiner, Brian G. 1995. 'The joint essay of Darwin and Wallace'. *The Linnean* 11(1), 13–24.

George, Wilma. 1964. *Biologist Philosopher: A Study of the Life and Writings of Alfred Russel Wallace.* London, Toronto, New York: Abelard-Schuman.

——1979. 'Alfred Wallace, the gentle trader: collecting in Amazonia and the Malay Archipelago 1848–1862'. *Journal of the Society for the Bibliography of Natural History* 9(4), 503–14.

Gepp, Antony. 1894. 'In memory of Richard Spruce'. *Journal of Botany* 23 (new series), 50–53.

Godman, Frederick D. 1892. 'President's Address for 1892'. *Proceedings of the Entomological Society of London*, xlvi–lix (l–lv about Bates).

Goodger, Kim, & Phillip **Ackery**. 2002. 'Bates, and the Beauty of Butterflies'. *The Linnean* 18(1), 21–59.

Gould, Stephen Jay. 1980. 'Wallace's fatal flaw'. *Natural History* 89(1), 26–40.

——2002. 'A Biographical Sketch' of Alfred Russel Wallace. In Berry, 1–26.

Grant Duff, Mountstuart Elphinstone. 1892. 'Obituary. Henry Walter Bates, F.R.S.'. *Proceedings of the Royal Geographical Society* 14, 244–57.

Hagen, Victor W. von. 1944. 'The great mother forest: a record of Richard Spruce's days along the Amazon'. *Journal of the New York Botanical Garden* 45, 73–80.

——1949. 'Richard Spruce, Yorkshireman'. In *South America Called Them: Explorations of the Great Naturalists, Charles-Marie de La Condamine, Alexander von Humboldt, Charles Darwin, Richard Spruce*, London, 296–352.

——1951. *South America, The Green World of the Naturalists.* London: Eyre & Spottiswoode.

Hartman, H. 1990. 'The evolution of Natural Selection: Darwin versus Wallace'. *Perspectives in Biology and Medicine* 34, 78–88.

Hemming, John. 1995. *Amazon Frontier. The Defeat of the Brazilian Indians.* London: Papermac.

——2008. *Tree of Rivers. The Story of the Amazon.* London and New York: Thames & Hudson.

Henderson, Andrew. 1996. 'Richard Spruce and the palms of the Amazon and Andes'. In Seaward & FitzGerald, 187–96.

Hogben, Lancelot T. 1918. *Alfred Russel Wallace: the Story of a Great Discoverer.* London: Society for Promoting Christian Knowledge.

Holmes, John Haynes. 1913. *Alfred Russel Wallace: Scientist and Prophet.* New York.

Honigsbaum, Mark. 2001. *The Fever Trail. The Hunt for the Cure for Malaria*. London: Macmillan.
Hooker, Joseph. 1881. 'Presidential address to the Geographical Section, British Association for the Advancement of Science'. *Proceedings of the Royal Geographical Society* 3 (new series), 545–608.
——1883. 'Speech at 1883 Anniversary Meeting'. *Proceedings of the Royal Geographical Society* 5 (new series), 419–20.
Hooker, William J. 1854. 'Notices of Books – review of Wallace's *Palm Trees of the Amazon*'. *Hooker's Journal of Botany* 6, 61–62.
Humboldt, Alexander von. 1816–1831. *Voyage aux régions équinoxiales du Nouveau Continent, fait en 1799, 1800, 1801, 1803 et 1804* (13 vols). Paris: Librairée Grecque-Latine-Allemande; trans. Helen Maria Williams, *Personal Narrative of Travels to the Equinoctial Regions of the New Continent during the Years 1799–1804* (7 vols), London: Longman, John Murray and H. Colburn, 1814–1829; also trans. Thomasina Ross (3 vols), London: George Bell & Sons, 1907.
——1849. *Ansichten der Natur, mit wissenschaftlichen Erläuterungen*. Stuttgart and Tübingen; trans. E. C. Otté & H. G. Bohn, *Views of Nature, or Contemplations of the Sublime Phenomena of Creation*, London: Bell & Daldy, 1850.
Jackson, B. Daydon. 1906. *George Bentham*. London: J. M. Dent.
Jones, Greta. 2002. 'Alfred Russel Wallace, Robert Owen and the Theory of Natural Selection', *British Journal for the History of Science* 35, 73–96.
Knapp, Sandra. 1999. *Footsteps in the Forest: Alfred Russel Wallace in the Amazon*. London: The Natural History Museum.
——2008. 'Wallace, conservation, and sustainable development'. In C. Smith & Beccaloni, 201–22.
Koch-Grünberg, Theodor. 1909–1910. *Zwei Jahre unter den Indianern. Reise in Nordwest-Brasilien 1903/1905* (2 vols). Berlin: Ernst Wasmuth Verlag.
——1916–1928. *Vom Roroima zum Orinoco. Ergebnisse einer Reise in Nordbrasilien in den Jahren 1911–1913* (5 vols). Berlin and Stuttgart: Verlag Strecker und Schröder.
Kohn, David, ed. 1985. *The Darwinian Heritage*. Princeton: Princeton University Press.
Kottler, Malcolm J. 1974. 'Alfred Russel Wallace, the origins of man, and spiritualism'. *Isis* 65, 145–92.
——1985. 'Charles Darwin and Alfred Russel Wallace: two decades of debate over natural selection'. In Kohn, 367–431.
Lambert, Aylmer Bourke. 1821. *An Illustration of the Genus Cinchona... including Baron de Humboldt's account of the Cinchona forests of South America*. London: J. Searle.
Lee, Monica. 1985. *300 Year Journey: Leicester Naturalist Henry Walter Bates, F.R.S., and his Family, 1665–1985*. Leicester: Castle Printers.
Linsley, E. Gorton. 1978. *The Principal Contributions of Henry Walter Bates to a Knowledge of the Butterflies and Longicorn Beetles of the Amazon Valley*. New York: Arno Press.
Madriñan, Santiago. 1996. 'Richard Spruce's pioneering work on tree architecture'. In Seaward & FitzGerald, 215–26.
Mallet, James. 2001. 'The speciation revolution'. *Journal of Evolutionary Biology* 14, 887–88.
Marchant, James, ed. 1916. *Alfred Russel Wallace: Letters and Reminiscences* (2 vols). London and New York: Cassell and Company (reprinted 1975).
Markham, Clements Robert. 1862. *Travels in Peru and India while Superintending the Collection of Chinchona Plants and Seeds in South America and their Introduction into India*. London: John Murray.
——1874. *A Memoir of Lady Ana de Osorio, Countess of Chinchon and Vice-Queen of Peru (AD 1629–39), with a Plea for the Correct Spelling of the Chinchona genus*. London: Trübner and Co.
——1874. *Peruvian Bark. A Popular Account of the Introduction of Chinchona Cultivation into British India, 1860–1880*. London: Trübner & Co. (and London: John Murray, 1880).
——1892. Contribution to 'Obituary. Henry Walter Bates, F.R.S.'. *Proceedings of the Royal Geographical Society* 14, 256–57.
——1894. 'Richard Spruce'. *Geographical Journal* 3, 245–47.
McKinney, H. Lewis. 1970–1980. 'Bates, Henry Walter'. In Charles C. Gillespie, ed., *Dictionary of Scientific Biography* (16 vols), New York: Charles Scribner's Sons, vol. 1, 500–4.
——1971. 'Introduction' to A. R. Wallace, *Palm Trees of the Amazon and their Uses*. Austin, TX: Coronado Press (reprint).
——1972. *Wallace and Natural Selection*. New Haven and London: Yale University Press.
Michaux, Bernard. 2008. 'Alfred Russel Wallace, Biogeographer'. In C. Smith & Beccaloni, 166–85.
Milliken, William, & Bruce **Albert**. 1999. *Yanomami, A Forest People*. Kew: Royal Botanic Gardens.
Moon, Henry P. 1976. *Henry Walter Bates F.R.S., 1825–1892: Explorer, Scientist and Darwinian*. Leicester: Leicestershire Museums.

Musgrave, Toby, & Will **Musgrave**. 2000. *An Empire of Plants. People and Plants that Changed the World.* London: Cassell & Co.

O'Hara, James E. 1995. 'Henry Walter Bates – His life and contributions to biology'. *Archives of Natural History* 22, 195–219.

Ohsaki, Naota. 2005. 'A common mechanism explaining the evolution of female-limited and both-sex Batesian mimicry in butterflies'. *Journal of Animal Ecology* (British Ecological Society) 74, 728–34.

Oosterzee, Penny van. 1997. *Where Worlds Collide: The Wallace Line.* Ithaca: Cornell University Press.

Owen, Denis. 1980. *Camouflage and Mimicry.* Oxford and Melbourne: Oxford University Press.

Papavero, Nelson. 1973. 'H. W. Bates'. In Papavero, *Essays on the History of Neotropical Dipterology* (2 vols), São Paulo, vol.2, 256–62.

Pearson, Michael B. 1990. 'Richard Spruce's "List of Botanical Excursions"'. *The Linnean* 6, 18–20.

——1993. 'A. R. Wallace's "Sketches of the Palms of the Amazon with an account of their uses and distribution"'. *The Linnean* 9, 22–23.

——1996. 'Richard Spruce: the development of a naturalist'. In Seaward & FitzGerald, 27–35.

——2004. *Richard Spruce, Naturalist and Explorer.* Settle, Yorkshire: Hudson History.

Porter, Duncan M. 1996. 'With Humboldt, Wallace and Spruce at San Carlos de Río Negro'. In Seaward & FitzGerald, 51–63.

Poulton, Edward B. 1923–1924. 'Alfred Russel Wallace, 1823–1913'. *Proceedings of the Royal Society of London* (ser. B) 95, i–xxxv.

Prance, Ghillean T. 1996. 'A contemporary botanist in the footsteps of Richard Spruce'. In Seaward & FitzGerald, 93–121.

——1999. 'Alfred Russel Wallace, 1823–1913'. *The Linnean* 15, 18–36.

——& Thomas E. **Lovejoy**. 1985. *Key Environments: Amazonia.* Oxford and New York: Pergamon Press.

Raby, Peter. 1996. *Bright Paradise: Victorian Scientific Travellers.* London: Chatto & Windus.

——2001. *Alfred Russel Wallace. A Life.* London: Chatto & Windus.

Raffles, Hugh. 2002. *In Amazonia: A Natural History.* Princeton: Princeton University Press.

Rice, Anthony L. 1999. *Voyages of Discovery. Three Centuries of Natural History Exploration.* London: Natural History Museum.

Rocco, Fiammetta. 2003. *The Miraculous Fever-Tree. Malaria, Medicine and the Cure that Changed the World.* London: HarperCollins Publishers.

Sandeman, C. 1949. 'Richard Spruce, portrait of a great Englishman'. *Journal of the Royal Horticultural Society* 74, 531–44.

Schomburgk, Robert H. 1840. 'Report of the Third Expedition into the Interior of Guiana, comprising the Journey to the Sources of the Essequibo, to the Carumá Mountains, and to Fort San Joaquim, on the Rio Branco, in 1837–8'. *The Journal of the Royal Geographical Society* 10(2), 159–267.

——1841. *Twelve Views in the Interior of Guiana* (drawings by Charles Bentley). London: Ackermann & Co.

——(ed. Peter Rivière). 2006. *The Guiana Travels of Robert Schomburgk, 1835–1844* (2 vols). London: The Hakluyt Society (3 ser.), 16–17.

Schultes, Richard Evans. 1951. 'De festo seculari Ricardi Sprucei, America Australi adventu, commemoratio atque de plantis principaliter vallis Amazonicis: diversae observationes' (Plantae Austro-Americanae VII). *Botanical Museum Leaflets, Harvard University* 15(2), 29–78.

——1953. 'Richard Spruce still lives'. *Northern Gardener* 7, 20–27, 55–61, 89–93, 121–25 (reprinted *Hortulus Aliquando* 3, 1978, 13–47).

——1968. 'Some impacts of Spruce's Amazon explorations on modern phytochemical research'. *Rhodora* 70, 313–39 (also in *Ciência e Cultura* 20, 1968, 37–49).

——1970. 'Preface' in reprint of Spruce, *Notes of a Botanist* (ed. A. R. Wallace). New York: Johnson Reprint Corporation.

——1978. 'An unpublished letter by Richard Spruce on the theory of evolution'. *Biological Journal of the Linnean Society* 10, 159–61.

——1978. 'Richard Spruce and the potential for European settlement of the Amazon: an unpublished letter'. *Botanical Journal of the Linnean Society* 77(2), 131–39.

——1979. 'Discovery of an ancient Guayusa plantation in Colombia'. *Botanical Museum Leaflets, Harvard University* 27(5–6), 143–53.

——1983. 'Richard Spruce: an early ethno-botanist and explorer of the northwest Amazon and northern Amazon and northern Andes'. *Journal of Ethnobiology* 3(2), 139–47.

——1985. 'Several unpublished ethnobotanical notes of Richard Spruce'. *Rhodora* 87, 439–41.

——1987. 'Still another unpublished letter from Richard Spruce on evolution'. *Rhodora* 89, 101–6.

——1990. 'Notes on the difficulties experienced by Spruce in his collecting'. *Rhodora* 92, 42–44.

——1996. 'Richard Spruce, the man'. In Seaward & FitzGerald, 15–25.
——& Robert F. **Raffauf**. 1990. *The Healing Forest – Medicinal and Toxic Plants of the Northwest Amazon*. Portland: Dioscorides Press.
——1992. *Vine of the Soul*. New York: Synergetic Press.
Schuster, R. M. 1982. 'Richard Spruce (1817–1893): a biographical sketch and appreciation'. *Nova Hedwigia* 36, 199–208.
Schwartz, Joel S. 1984. 'Darwin, Wallace, and the *Descent of Man*'. *Journal of the History of Biology* 17, 271–89.
——1990. 'Darwin, Wallace, and Huxley, and *Vestiges of the Natural History of Creation*'. *Journal of the History of Biology* 23, 127–53.
Seaward, Mark R. D. 1995. 'Spruce's Diary'. *The Linnean* 11, 17–19.
——1996. 'Bibliography of Richard Spruce', in Seaward & FitzGerald, 303–14.
——& Sylvia M. D. **Fitzgerald**, eds. 1996. *Richard Spruce (1817–1893): Botanist and Explorer*. London: Royal Botanic Gardens, Kew.
Sharp, David. 1892. 'Henry Walter Bates, F.R.S.'. *Entomologist* 25(847), 76–80.
Shermer, Michael. 2002. *In Darwin's Shadow: The Life and Science of Alfred Russel Wallace*. New York: Oxford University Press.
Slater, Matthew R. 1906. 'The mosses and hepaticae of North Yorkshire'. In J. G. Baker, ed., *North Yorkshire: Studies of its Botany, Geology, Climate and Physical Geography*, York: Yorkshire Naturalist Union.
Sledge, W. Arthur. 1971. 'Richard Spruce'. *Naturalist* 96, 129–31.
——& Richard Evans **Schultes**. 1988. 'Richard Spruce: a multi-talented botanist'. *Journal of Ethnobiology* 8(1), 7–12.
Slotten, Ross A. 2004. *The Heretic in Darwin's Court. The Life of Alfred Russel Wallace*. New York: Columbia University Press.
Smith, Anthony. 1986. *Explorers of the Amazon*. London: Viking.
Smith, Charles H., ed. 1991. *Alfred Russel Wallace: An Anthology of his Shorter Writings*. Oxford: Oxford University Press.
——& George **Beccaloni**, eds. 2008. *Natural Selection and Beyond. The Intellectual Legacy of Alfred Russel Wallace*. Oxford: Oxford University Press.
Smith, Nigel J. H. 1996. 'Relevance of Spruce's work to conservation and management of natural resources in Amazonia'. In Seaward & FitzGerald, 228–37.
——1999. *The Amazon River Forest. A Natural History of Plants, Animals, and People*. New York and Oxford: Oxford University Press.
——2002. *Amazon, Sweet Sea*. Austin: University of Texas Press.
——, Rodolfo **Vásquez** & Walter H. **Wust**. 2007. *Amazon River Fruits. Flavors for Conservation*. Saint Louis: Missouri Botanical Garden Press.
Spruce, Richard. 1850. 'Journal of an excursion from Santarém, on the Amazon River, to Obidos and the Rio Trombetas'. *Hooker's Journal of Botany* 2, 193–208, 225–32, 266–76, 298–302.
——1855. 'Note on the *India-rubber* of the Amazon'. *Hooker's Journal of Botany* 7, 193–96.
——1860. 'Notes of a visit to the Cinchona forests on the western slope of the Quitonian Andes'. *Journal of Proceedings of the Linnean Society. Botany* 4, 176–92; also in Spruce, *Notes of a Botanist*, vol.2, 228–50.
——1861 [i.e. 1862]. *Report of the Expedition to Procure Seeds and Plants of the Cinchona Succirubra, or Red Bark Tree* (with a brief note about the map, by Clements R. Markham, India Office, 3 Jan. 1862). London: George E. Eyre and William Spottiswoode, for Her Majesty's Stationery Office.
——1864. 'On the River Purus, a tributary of the Amazon'. In Clements Markham, trans., *The Travels of Pedro de Cieza de León*, London: Hakluyt Society (1 ser.), 33, 339–51.
——1869. 'Palmae Amazonicae, sive Enumeratio Palmarum in Itinere suo per Regiones Americae Aequatoriales Lectarum'. *Journal of the Linnean Society. Botany* 11, 65–183.
——1874. 'On some remarkable narcotics of the Amazon Valley and Orinoco'. *Ocean Highways: Geographical Review* 1 (new series), 184–93; also in Spruce, *Notes of a Botanist*, vol.2, 413–55.
——1884–1885. *Hepaticae Amazonicae et Andinae*, a separate volume of *Transactions and Proceedings of the Botanical Society of Edinburgh* 15, 1–588; published as *The Hepaticae of the Amazon and the Andes of Peru and Ecuador*, London: Trübner & Co., 1885; reprinted New York: New York Botanical Garden, 1984.
——(ed. Alfred Russel Wallace). 1908. *Notes of a Botanist on the Amazon and Andes* (2 vols). London: Macmillan; reprinted with a Foreword by Richard Evans Schultes, 2 vols., New York: Johnson Reprint Corporation, 1970.

Stabler, G. 1894. 'Obituary notice of Richard Spruce Ph.D.'. *Transactions and Proceedings of the Botanical Society Edinburgh* 20, 106.

Stecher, Robert M. 1969. 'The Darwin-Bates Letters. Correspondence between two Nineteenth-century Travellers and Naturalists'. *Annals of Science* 25(1), 1–47; 25(2), 95–125.

Stotler, Raymond E. 1996. 'Richard Spruce: his fascination with liverworts and its consequences'. In Seaward & FitzGerald, 123–40.

Wallace, Alfred Russel. 1850. 'On the Umbrella Bird (Cephalopterus ornatus), "Ueramimbé"'. *Proceedings of the Zoological Society of London* 18, 206–7.

——1852. 'On the monkeys of the Amazon'. *Proceedings of the Zoological Society of London* 20, 107–10.

——1849–1850. Letters from the Amazon, 23 Oct. 1848, 12 Sept. 1849, 15 Nov. 1849 and 20 March 1850. *Annals and Magazine of Natural History* (2 ser.), 3 (1849) 74–75; 5 (1850) 156–57; 6 (1850) 494–96.

——1853. *A Narrative of Travels on the Amazon and Rio Negro, With an Account of the Native Tribes, and Observations on the Climate, Geology, and Natural History of the Amazon Valley.* London: Reeve & Co.; also London, Melbourne and Toronto: Ward Lock & Co., 1911.

——1853. *Palm Trees of the Amazon and their Uses.* London: John Van Voorst; reprinted with an introduction by H. L. McKinney, Austin: Coronado Press, 1971.

——1853. 'On the Rio Negro'. *Journal of the Royal Geographical Society* 23, 212–17.

——1853. 'On some fishes allied to Gymnotus'. *Proceedings of the Zoological Society of London* 21, 75–76.

——1855. 'On the law which has regulated the introduction of new species' (The 'Sarawak' paper). *Annals and Magazine of Natural History* 16, 184–96; also Berry, *Infinite Tropics*, 36–49.

——1858. 'On the Tendency of Varieties to Depart Indefinitely from the Original Type' (The 'Ternate' paper, Feb. 1858). *Journal of the Linnean Society, Zoology* 3 (20 Aug. 1858), 53–62; also in Wallace, *Natural Selection and Tropical Nature: Essays on Descriptive and Theoretical Biology*, London: Macmillan & Co., 1891, 20–30; also Berry, *Infinite Tropics*, 52–62.

——1869. *The Malay Archipelago: the Land of the Orang-utan and the Bird of Paradise: a Narrative of Travel with Studies of Man and Nature* (2 vols). London; reprinted General Books, 2009.

——1870. *Contributions to the Theory of Natural Selection*. London and New York: Macmillan & Co.

——1889. *Darwinism: An Exposition of the Theory of Natural Selection with some of its Applications.* London and New York: Macmillan & Co.

——1892. 'Obituary: H. W. Bates, the naturalist of the Amazons'. *Nature* 45, 398–99.

——1894. 'Richard Spruce, Ph.D., FRGS'. *Nature* 49, 317–19.

——1905. *My Life. A Record of Events and Opinions* (2 vols). London: Chapman & Hall; Cambridge: Cambridge University Press, 2011. Updated and abridged by Wallace, 1 vol., London: Chapman & Hall, 1908; reprinted Whitefish: Kessinger Publishing, 2004.

Williams, David. 1962. 'Clements Robert Markham and the introduction of the Chinchona tree into British India'. *Geographical Journal* 128, 431–42.

Williams-Ellis, Amabel. 1966. *Darwin's Moon: A Biography of Alfred Russel Wallace*. London: Blackie.

Wilson, John G. 2000. *The Forgotten Naturalist: In Search of Alfred Russel Wallace*. Melbourne: Australia Scholarly Publishing.

Wood, John George. 1874. *Insects Abroad.* London: Longman, Green and Co.

Woodcock, George. 1945. 'Henry Bates on the Amazons'. *Adelphi*, April–June, 115–21.

——1969. *Henry Walter Bates, Naturalist of the Amazons*. London: Faber & Faber.

ACKNOWLEDGMENTS

I want to thank the many Brazilians who have helped and befriended me during my travels in Amazonia, as well as the helpful staff of British archives, libraries and museums. For this book, I am deeply grateful for expert advice and comment from the entomologist William Overal, the tropical botanists Sir Ghillean Prance and William Milliken of the Royal Botanic Gardens, Kew, and Wade Davis, the ornithologist Mark Cocker, and Sir David Attenborough on the final chapter. With pictures, I am indebted to Mark Honigsbaum for letting me use his photograph of a cinchona tree and Dudu Tresca for his of tropical vegetation; to Wade Davis for arranging permission to use pictures by his mentor Richard Evans Schultes; and to Frank Meddens for the balsa raft. Very many thanks also to the book's constructive and careful editor Ben Plumridge and picture editor Maria Ranauro.

ILLUSTRATION CREDITS

Maps on pp.6–9: Ben Plumridge
© Thames & Hudson Ltd.

1 (frontispiece): Bates mobbed by toucans, engraving by J. W. Whymper after a drawing by Bates, *Naturalist* (1st ed., 1863) vol.1, frontispiece; 2, 3. Riverside vegetation, photo John Hemming; 4. Buriti palms, photo Dudu Tresca; 5, 6. Paxiúba, açaí palms, photo John Hemming; 7, 8. Black-water stream, Indian in canoe, photo John Hemming; 9. Cucui outcrop, photo John Hemming; 10. *Monstera* leaves, photo Dudu Tresca; 11. Cinchona tree (Ecuador, Pacific slopes of the Andes), photo © Mark Honigsbaum; 12. Wallace, photo by Thomas Sims. Wallace, *My Life* (1905), vol.1, 264; 13. Bates, photo Wellcome Library, London; 14. Spruce (profile in Glengarry). Gray Herbarium, Harvard University, Cambridge, MA; 15. Humboldt, photo Wellcome Library, London; 16. Markham, photo Wellcome Library, London; 17. Three Wise Men, painting by Victor Eustaphieff. Down House, Kent. English Heritage; 18. Von Martius and Mura. J. B. von Spix and C. F. P. von Martius, *Reise in Brasilien* (1831); 19. Bates, night encounter. Bates, *Naturalist*, 2nd ed. (1864), 370; 20. Bates, turtle lagoon. Bates, *Naturalist,* 2nd ed., 359; 21. 'A forest scene…', engraving by J. B. Zwecker after sketch by Wallace. Wallace, *Geographical Distribution of Animals* (1876) vol.2, 29; 22. Bates, *Naturalist*, 2nd ed., 156; 23. Zeonia batesii butterfly. J. G. Wood, *Insects Abroad* (1874), p.627, fig.364; 24. Moth and hummingbird. Bates, *Naturalist,* 2nd ed., 114; 25. Mason wasp. Bates, *Naturalist,* 2nd ed., 226; 26. Tiger beetle, J. G. Wood, *Insects Abroad,* pl.1, p.11; 27. Mygale spider. Bates, *Naturalist,* 2nd ed., 97; 28. Catfish by Wallace. Natural History Museum, London; 29, 30. Fish drawings. Wallace, *My Life*, vol.1, 285–87, Archive of The Linnean Society; 31. Nazaré. Wallace, *Narrative of Travels* (1853), frontispiece; 32. Belem photo, private collection; 33. Sta. Isabel. Spruce, *Notes of a Botanist* (1908), vol.1, 427. Archive of The Royal Society, London; 34. Santarém, painting by Hércules Florence (1827), Boris Komissarov, *Expedição Langsdorff ao Brasil* 1821–1829 (1988), Langsdorff papers, Russian Academy of Science Archives, St. Petersburg; 35. Tarapoto. Spruce, *Notes of a Botanist*, vol.2, 41; 36. Flooded forest. Franz Keller, *The Amazon and Madeira Rivers* (1874), 80; 37. Rapids. Theodor Koch Grünberg, *Zwei Jahre* (1904), vol.2, 71; 38. Riverside camp. Theodor Koch Grünberg, *Zwei Jahre*, vol.2, 252–53; 39. Steamship, photo © 2014 the Estate of Richard Evans Schultes. All rights reserved. Used with permission; 40. Shaman and boy, photo © 2014 the Estate of Richard Evans Schultes. All rights reserved. Used with permission; 41. Schultes taking snuff, photo © 2014 the Estate of Richard Evans Schultes. All rights reserved. Used with permission; 42. Barasana boy and engraving, photo © 2014 the Estate of Richard Evans Schultes. All rights reserved. Used with permission; 43. Boys on engraving by Tiquié river. Koch Grünberg, *Zwei Jahre*, vol.2, 70; 44. Uaupés maloca. Koch Grünberg, *Zwei Jahre*, vol.1, 108–9; 45. Maquiritare maloca. Koch Grünberg, *Vom Roroima zum Orinoco* (1923), vol.3, 320; 46. Callistro. Spruce, *Notes of a Botanist*, vol.1, 325; 47. Tussari. Spruce, *Notes of a Botanist*, vol.1, 412; 48. Maku girls. Spruce, *Notes of a Botanist*, vol.1, 345; 49. Desana with egret plume. Koch Grünberg, *Zwei Jahre*, vol.1; 50. Yekuana Maquiritare in regalia. Koch Grünberg, *Vom Roroima zum Orinoco* (1923), vol.3; 51. Cigar holder, Koch Grünberg. *Zwei Jahre* (1904), vol.1, 281; 52. Boys dancing, photo © 2014 the Estate of Richard Evans Schultes. All rights reserved. Used with permission; 53. Ticuna wedding, engraved by J. B. Zwecker after a sketch by J. W. Whymper after Bates, *Naturalist,* 2nd ed., 451; 54. Tipiti making. Koch Grünberg, *Vom Roroima zum Orinoco*, vol.3, pl.57, p.352; 55. Making manioc grater. Koch Grünberg, *Vom Roroima zum Orinoco*, vol.3, pl.61.3, p.360; 56. Woman roasting manioc, photo John Hemming; 57. Piaçaba bundles, photo © 2014 the Estate of Richard Evans Schultes. All rights reserved. Used with permission; 58. Shooting fish, Koch Grünberg, *Vom Roroima zum Orinoco*, vol.3, 344; 59. Cinchona bark-collectors' camp, engraving by C. Laplante, after Adrienne Fuguet (1867), photo Wellcome Library, London; 60. Chimborazo group. Anonymous photo, inserted by Wallace into Spruce, *Notes of a Botanist*, vol.2, 201; 61. Spruce botanical specimen. Herbarium, Royal Botanic Gardens, Kew; 62. Balsa raft. From Jorge Juan and Antonio de Ulloa, *Noticias Secretas de América* (Madrid, 1748; London, 1826); 63. Wallace by Thomas Simms, from *My Life*, vol.1, 385; 64. Bates, *Naturalist* (1892 ed.), frontispiece; 65. Spruce in old age. Spruce, *Notes of a Botanist*, vol.1, frontispiece; 66. Spruce Cottage, photo John Hemming; 67. Hyacinth macaw, photo © Michael Lane. 123RF Stock Photo; 68. Cock-of-the-rock, photo © Daniel Mortell. 123RF Stock Photo; 69. Toucan, photo John Hemming; 70. Manatee, photo © burdephotography; 71. Camouflaged grasshopper, photo William Milliken; 72. Cutia, photo © Maria Daloia. 123RF Stock Photo; 73. São Gabriel. Charles Bentley for Robert Schomburgk, *Twelve Views in the Interior of Guiana* (1841). Private collection/The Stapleton Collection/Bridgeman Images; 74. Spruce hammock, photo John Hemming; 75. Esmeralda. Charles Bentley for Robert Schomburgk, *Twelve Views in the Interior of Guiana*; 76. Batesian mimicry. *Transactions of The Linnean Society* 23 (1862), pt.3 (also introduction to 1892 edition of Bates, *Naturalist*).

INDEX

Illustrations are indicated in *italics*; only those plant and animal species that are illustrated are included in the index.

Adalbert, Prince of Prussia 25, 37, 109
Aellopos titan (hawk moth) *91*
Albert, Bruce 281
Allen, Grant 16, 326–27
Amazonas province, Brazil 113, 121, 147*n*, 193, 194, 223, 237, 326
Andoa (indigenous people) 274, 277
Annals and Magazine of Natural History 31
anteaters 63–64
ants 105, 129, 160, 243; fire-ant (*Solenopsis saevissima*) 217–18, 229; leaf-cutter (saúba) 77, 188; tocandira (*Paraponera clavata*) 201
Antonij, Henrique (Senhor Henrique) 114–15, 138, 142, 193, 299
Aracú, José 219, 235
Aranha, João Batista Tenreiro 194
Arawak language 161, 177
Arnaoud, Padre 229
Aveiro, Brazil 217–18
ayahuasca (hallucinogenic plant) *see caapi*
Ayres, Pedro 159

Baião, Brazil 48–49
Banisteriopsis caapi (plant specimen) *261*; *see also caapi*
Baniwa (indigenous people) 151–52, 157, 277
Baré (indigenous people) 277
Bates, Frederick 13–14, 15
Bates, Henry Walter *2*, *87*, *90*, *91*, *257*, *262*; background and family 13–16; meets Alfred Wallace 21–23; inspiration for Amazon exploration 23–27; arrival in the Amazon 10–12, 37–41; and entomology 13–14, 21, 41, 43–44, 62; parts with Wallace 55–56; visits Caripí 61–65; visits Cametá 72–74; travels up the Amazon 83–84, 99–102, 107–12, 117–28, 235–66; consignments sent to England 213–16, 222–23; collecting on the Tapajós river 216–23; collecting on the Amazon 239–43, 246–49; on indigenous peoples 254, 263–64; suffers malaria 264; leaves Brazil 266; returns to England 313, 321; collections in the Natural History Museum 313–15, 317; on mimicry in the natural world 318–20, *338*; and Charles Darwin 315–18, 321; Assistant Secretary of the Royal Geographical Society 323–26; honours and awards 326; legacy 321–27; *The Naturalist on the River Amazon* (1863) 245, 321, 334
Bates, Sarah Ann 313
Batesian Mimicry 320, *338*
bats 62–63, 203; vampire 58, 63, 77–78, 189
Beagle, HMS 24
Beccaloni, George 312
bees: *Melipona fasciculata* ('sweat bees') 241; mimicry in 318
beetles 43, 62, 121, 239–40, 246–47
Belém do Pará, Brazil 10, 25, 31–32, 38–40, 44, *167*, 223–24, 265–66
Bendelak, Abraham 134
Bentham, George 29–30, 75, 78, 115, 137, 282, 284
Berry, Andrew 312
birds 57–58; *see also individual species*
Bismarck, Otto von 25
black-fly (*Simuliidae*) 111, 118, 179, 201, 217, 249, 273
Bonaparte, Prince Charles-Lucien, *Conspectus Generum Avium* (1850) 303
Bonpland, Aimé 24, 154, 203
Borrer, William 29
Brandão, José Antonio 129–30, 141
Brazil: independence 32; as Portuguese colony 24, 31–32
British Association for the Advancement of Science 23
British Museum (Natural History) *see* Natural History Museum
Brocklehurst, George 265
Brooke, Sir James 303
Brown, Edwin 37, 44
Burton, Richard 325
butterflies 15–16, 74, 215–16; *Adelpha* 121; *Asterope batesii* 214; *Asterope leprieuri* 101; *Asterope sapphira* 215; *Callicore* 101; *Cethosia* 43; *Chorinea* (*Zeonia*) 248; *Gonepteryx* 43; *Heliconii* 11, 100–1, 319, 322; *Morpho* 43, 71, 100, 121, 158, 247; *Papilio* 43, 120–21, 316–17; *Parides sesostris* 74, 120–21; *Philaethria dido* 11, 43; *Phoebis* 11, 43, 71, 101; *Pieris monuste* 71; *Tineidae* 247; *Tortricidae* 247; Bates's collection of 314–15
butterfly mimicry 120, 318–20, 338

caapi (hallucinogenic plant) *173*, *261*, 277–80
Cabanagem rebellion, Brazil 32, 37, 73, 130
cacao trees 96–97, 108
cadana (hallucinogenic plant) *see caapi*
caiman 54, 60, *90*, *91*, 127
Calistro, Senhor 69
Callistro (Tariana chief) 164–65, 197, *256*
Cametá, Brazil 45–46, 72–73
Campana mountains 271
Campbell, Archibald 57–58, 77
Campbell, James 76
cannibalism 175
Capim river 69–70
Carapanã (indigenous people) 161
Cardozo, Antonio *90*, 119, 124–26, 127
Caripí (farm) 61–65, 77–78
Carlisle, Countess of 232, *336*
Carlisle, Earl of 329
Casiquiare Canal 154, 203–4
Castle Howard, Yorkshire 233, 328, *336*
cecropia (trees) 79, 124, 141, 165, 204, 268
cedro (*Cedrela odorata*) 44
Cerro Pelado (mountain) 271
Chagas (trader) 196
Chambers, Robert, *Vestiges of the Natural History of Creation* 22–23
Chelsea Physic Garden 283
Chenu, Jean-Charles 239
Chlorocoelus Tanana (wood cricket) *91*
cichlid (fish) *94*
cinchona (*Cinchona succirubra*) *36*, *260*, 268, 282–89, 290–93, 327, 331–32
Clarke, Thomas 293
Clodd, Edward 13, 14, 15, 214, 316, 319, 327
coca (*Erythroxylum coca*) 139–40, 140*n*

cocks-of-the-rock (*Rupicola rupicola*) 151–53, *335*
Coleoptera *see* beetles
Condamine, de La, Charles-Marie 284
Conway, Martin 325
Cordeiro, Jesuino 193–95
Correia, Angelo Custódio 73, 83
Correia de Araújo, Filisberto (Lieutenant) 159, 160, 179
Crawforth, Anthony 56
crickets (*Tbliboscelus hypericifolius*) 101
Cross, Robert 289, 291–92, 331
Cross, Sir Charles 293
Cubeo (indigenous people) 161, 175, 177
Cucui, Brazil *36*
Cuduiari river 192
Cunucunuma river 206–7
Cupari river 219–22
Custodio (founder of San Custodio) 208–9
cutia (*Dasiprocta aguti*) 66, *335*
Cuyucuyú (catfish) *94*

Darwin, Charles 22, *88*, 297, 304–8, 316–18, 320, 323, 326; *Journal and Remarks* 24; *On the Origin of Species* 190, 307, 310–11, 315
Davis, Wade 279, 280, 333
Desana (indigenous people) 161, 162, *256*
Dias, Antonio 155, 232
Distant, William 242–43, 316, 321–22
Doubleday, Edward 26

Ecuador, Richard Spruce collects *cinchona* trees 284–94
Edward VII, King 312
Edwards, William 37, 42, 61, 115; *A Voyage up the River Amazon* 25–26
Ega (*later* Tefé), Brazil 118, 128, 246–47
El Niño 294
Elias, Ney 325
Entomological Society 302, 316, 322, 325, 326
Erepecuru river 103–7
Esmerelda, Venezuela 203, 205, *337*
ethnobotany 332
Evelyn, John 283

Felder, Catejan and Rudolph 314
ferns 77, 97, 103, 206, 250, 270–71, 275–76; *Azolla* 132; *Salvinia* 132; *Schizaea* 139
Fichman, Martin 308, 312
Fitch, Walter 296
FitzGerald, Silvia 333
Flores, president of Ecuador 288
Ford, Henry 221*n*
Freshfield, Douglas 325
frogs 49

Galton, Francis 323, 324
Gibbon, Lardner 299
Gibson, Sam 29
Godman, Frederick 314, 326
Goodyear, Charles 53
Gould, John 321
Gould, Stephen Jay 311
Gradstein, Rob 330–31
Gray, John 322
Guahibo (indigenous people) 280
Guamá river 68
Guayaquil, Ecuador 273, 275, 285, 288, 289, 291, 293
Guia, Brazil 150, 151, 154, 159
Guiana Shield (rock formation) 97, 148, 316–17
Gutierrez trading house, Guayaquil 293–94
Guyana 25

hallucinogenic plants 277–81, 332; *see also caapi, paricá*
Hanbury, Daniel 327, 329, 333
Hartt, Charles 300*n*
Hasskerl, Justus 284
Hayward, George 325
Henderson, Andrew 330
hepatics *see* liverworts
Herndon, William Lewis 299
Hertford, England 17
Hewitson, William 121, 214, 314
Hislop, Captain John 83, 95
Hooker, Joseph *88*, 282, 306, 317, 330
Hooker, Sir William 26, 29–30, 55, 75, 139, 140, 166, 284, 297, 322
Hooker's Journal of Botany 239
hornets 161
horse-flies *see* mutuca
Howard, George 333
Huallaga river 268, 271, 273
Huambiza (indigenous people) 274
human trafficking 193–94, 196
Humboldt, Baron Alexander von 23–24, 83, *88*, 127, 154, 200, 203, 205, 280
hummingbirds 58, 64, *91*, 192, 239, 246
Huxley, Thomas 22*n*
hyacinth macaws (*Ara hyacinthus*) 54–55, 219, 221, 227, *335*

Iauaretê, Brazil 164, 191
Ibis (journal) 309
Içana river 151
India, *quinine* production 284–85, 292–93, 331–32
indigenous peoples 253–64, *257*; of the Uaupés river 161–77, 193
inga (trees) 77, 85, 110, 135, 204, 272
Ipanoré (*formerly* São Jerónimo), Brazil 164, 176, 191, 196, 198

Javita, Venezuela 156–58, 228–29
Jeffries 95, 131
Jesuits in Brazil 32, 55, 60, 126, 282–83
Jívaro (indigenous people) 274
Joaquim, Manoel 179
José dos Santos Inocentes, Friar 159–60
Journal of Botany 29

Kaixana (indigenous people) 254
kapok (*Ceiba pentranda*) 78, 79, 122, 133
Kew, Royal Botanic Gardens 26, 29, 166, 268, 280, 282, 291, 302
King, Robert 30, 75, 77–78, 103–5, 115, 137–38
Knapp, Sandra 312
Kocama (indigenous people) 263
Kropotkin, Prince Peter 325

Lacerda, Albuquerque 194
Landi, Antonio 32
Langsdorff, Georg 25
languages, of the indigenous peoples 177, 211
Leavens, Charles 42, 44, 50
Ledger, Charles 332*n*
Leicester, England 13–15
Lepidoptera *see* butterflies
Lima, João Antonio 146–50, 154, 159–60, 178, 179, 190
Lindley, John, *Elements of Botany* 20
Linnaea Entomologica (journal) 245
Linnaeus, Carl 20, 282
Linnean Journal 308, 309
Linnean Society 306, 312, 320, 326, 330, 332, 333
liverworts (hepatics) 29, 77, 199, 200, 270–71, 275–76, 330–31; *Riccia* 132
Livingstone, David 324
London Journal of Botany 29
Lowe, Frederick 25
Luiz (freed slave) 67–68
Lyell, Charles 24, *88*, 304–7

maçaranduba (*Manilkara huberi*) 'cow tree' 42–43, 233
macaw 86, 192, 223, *335*; hyacinthine 54–55, 70, 219, 221, *335*
Maipurés, Colombia 229–30
Maku (Hupdu) indigenous people 197, *256*
malaria 38, 111, 132, 181, 183–84, 215, 230–32, 264; and *quinine* 268, 279, 282–84, 288, 293, 331–32; *see also* cinchona, quinine
maloca (indigenous dwelling place) 162–63, 175, 253, *255*
Malthus, Thomas 20; *Principles of Population* 305
Mamani, Manuel 332*n*
Manaquiri, Brazil 129–30, 140, 141
manatee (*Trichechus inunguis*) 181–82, *335*
Manau (Manoa) indigenous people 161
Manaus (Barra), Brazil 113–15, 138, 181, 223
manioc, being prepared for eating *258*
Maquiritare (indigenous people) 206, *255*, 277, *337*
Marajó island, Brazil 57
Markham, Clements *88*, 283–85, 288–89, 324, 326, 327, 329, 331–32
Marrieta, Don Victoriano 273
Martius, Carl von 24–25, 41, *89*, 141–42, 177, 299, 330
Maw, Henry Lister 25
Mechanics' Institutes 14–15, 18
Mexiana island, Brazil 58–60
Miller, Daniel 37, 145, 213
Milliken, William 281
Mitten, Annie 310
Mitten, William 310, 329
monkeys: *Aotus trivirgatus* 212–13; capuchins (*Sapajus*) 71, 102; coatás (spider monkeys) 71, 102, 219, 300–1; cuxiú marmosets (*Chiropotes satanas*) 73; howler 49, 102, 221; *Lagothrix humboldtii* 301; *Pithecia hirsuta* 250, 300; uacari (*Cacajao calvus*) 249
Montaigne, Michel de 166*n*
Monte Alegre, Brazil 96, 98
Morey, Don Ignacio 269, 273
mosquitoes 96, 111, 118, 132, 148, 181, 183, 201, 213n, 229, 249–50, 273, 283
mosses 77, 97, 103, 139, 199, 200–1, 270–71, 275, 329; Spruce's expertise on 28–29
Mucura, Brazil 175, 192
Müller, Fritz 320*n*
Munduruсu (indigenous people) 219–21
Mura (indigenous tribe) 110–11
Murchison, Sir Roderick 303
Murray, John 26, 323
mutuca (horse flies) 111, 118, 201, 217, 240

narcotics 277–81; see also coca, hallucinogenic plants, *caapi*, *paricá*
Nares, George 325
Natterer, Johann 25, 67, 151, 165, 181, 200
Natural History Museum 22, 23, 26, 245, 296, 302, 303, 311–12, 314, 322
natural selection 122, 132, 190, 304–8, 315–16, 320–21
Nazaré, Belém do Pará 39–40, 67, 70, *167*
Newman, Edward 302
Newton, Alfred 306
Niemann, Albert 140*n*
Nimuendajú, Curt 163
niopo (hallucinogen) see *paricá*

Oberthür, René 314
Óbidos, Brazil 99–100
Oliver, Daniel 282
Orinoco river 154, 155, 158, 199, 203, 229
O'Shaughnessy, Arthur 322–23
Overal, William 319*n*, 320*n*
Owen, Robert 18–19

Pacimoni river 207–11
Paine, Tom 21
palm trees 83; bacaba 209, 228; buçú (*Manicaria saccifera*) 45, 47; buriti (*Mauritia flexuosa*) *34*, 45, 209; açai (*Euterpe oleracea*) *34*, 43, 47, 59–60, 209; caiaué (*Elaeis oleifera*) 228; inajá (*Attalea maripa*) 43, 209; jacitara (*Desmoncus polyacanthos*) 82, 122; jauari (*Astrocaryum jauari*) 198; marajá (*Bactris brongniartii*) 129; mucajá (*Acrocomia aculeata*) 55; murumuru 104; patauá 228; paxiúba (*Socratea exorrhiza*) *34*, 43, 204; piaçaba (*Attalea funifera*) 155, *258*; pupunha (*Bactris gasipaes*) 164; tucuma (*Astrocaryum vulgare*) 221; oil from 228; Spruce's paper on 330; Wallace's book on 296–97, 330
Papuri river 196
paricá (hallucinogen) 110–11, 280–81
Pasé (indigenous people) 263
Pastaza river 274–77
pau mulato (*Calycophyllum spruceanum*) 267–68
Pavón, José 284
Pedra Pintada ('Painted Rock') 98
Pedro II, Emperor of Brazil 326
Peru: and *cinchona* bark 282–87; Richard Spruce in 269–73, 295
petroglyphs 97–98, 104, *174*; *see also* rock art
Petzell (German settler) 62, 65
Phytologist (journal) 27, 29
Pimichin, Venezuela 155–56
Pirara Incident 160*n*
Piura, Peru 295
Plowman, Timothy 279
pororoca (tidal bore) 68–69
Prance, Sir Ghillean 333
Prescott, William, *History of the Conquest of Peru* 283
Priestly, Joseph 52*n*
Pritchett, G.J. 284–85

quinine 282–84, 287, 288, 293, 327, 331–32

Raby, Peter 56, 312
Red Bark tree *see* cinchona
Revue bryologique (journal) 275
Rio Negro (river) *33*, 113, 117, 147–50, *336*
rock art 149, *174*, 196; *see also* petroglyphs
Roman, Padre Manuel 203
Roosevelt, Anne 98
Royal Geographical Society 25, 53, 227, 284, 299, 303, 323–25, 326
Royal Society 312
rubber trade 221, 238–39
rubber tree (*Hevea brasilensis*) 52–53, 79, 238
Ruiz, Hipólito 284

Salesian missionaries 163, 177
Salvin, Osbert 314
San Carlos de Río Negro, Venezuela 154, 199–200, 208, 228, 232–33
San Custodio, Venezuela 208–10
sand flies 158, 201, 203, 205
Santa Isabel, Brazil 148, 181, 186
Santa Isabel, Venezuela *168*, 208–9
Santander, Manuel 285
Santarém, Brazil 83, 86, 95, 130–31, *168*, 215, 235–37
São Gabriel da Cachoeira, Brazil 149–50, 155, 160, 180, 187, 188–89, *336*

São Jerónimo *see* Ipanoré (*formerly* São Jerónimo)
São Paulo de Olivença, Brazil 250–53, 268
sarsaparilla (*Smilax syphilitica*) 81, 81*n*, 193
Sceliphron fistularium (mason wasp) *92*
Schomburgk, Robert 25, 159, 160*n*, 200, 203, 206
Schultes, Richard Evans 78, *173*, 279, 280, 332–33
Screaton, Dr H. 13
Seaward, Mark 333
Seixas, José Antonio 48, 52
Sharp, David 321, 325–26
Shermer, Michael 302, 306, 311, 312
Silk, George 303
Sims, Thomas 22, 296
Slater, Matthew 141, 329
slavery 38, 59, 69–70
Slotten, Ross 312
Smith, Charles 312
Smith, John 186
Smyth, William 25
snakes 64–65, 252, 271–72
socialism 18–19
Solimões (river) 118, 129, 248
Sousa, Irineu de 237
Souza, Torquarto de 109
spiders 41, 241; *Mygale* (bird eating spider) 73–74, 93
Spix, Johann Baptist von 24–25, 41, 75, 141–42, 299
Spruce, Richard *87*, *262*; background and family 28–30; arrival in the Amazon 11–12, 75; on the Amazonian forest 76–83; travels up the Amazon 85, 99, 103–7, 130–37; friendship with Henrique Antonij 114–15; in Manaus 138–45; sends *botutos* (sacred trumpets) to Kew Gardens 166; on the Rio Negro 184, 185–90, 199; and evolution 190; on the Uaupés river 196–211; in San Carlos 228; drawings of Brazilian Indians *256*; tree species named for him 268; stays at Tarapoto, Peru 269; journey to Ecuador 273–77; observes uses of hallucinogenic plants 277–81; collecting habits 281–82; and *cinchona* bark 282–87, 290–93, 329; poor health 289–90, 294–95; loss of savings 293–94; return to England 295, 327–31; death and legacy 332–33; *Hepaticae Amazonicae et Andinae* 275, 276*n*, 330–31
Spruce Cottage, Coneysthorpe *262*, 328, 333
Stabler, George 329
Stanley, Henry 325
steamboat service *172*, 237–38, 246, 267–68
Stephens, James, *Manual of British Coleoptera* 16, 21
Stevens, Samuel 26–27, 31, 43, 53, 75, 117, 213–14, 296, 302, 313
Stotler, Raymond 331
Suites à Buffon (journal) 245
Swainson, William 20, 27
Swiss-cheese plant (*Monstera dubia*) *36*, 82

Tapajós river 86, 95, 216–19, 222
Tarapoto, Peru *169*, 269, 272
Tariana (indigenous people) 161, 164, 175, 177
Taylor, James 288–89
Taylor, Thomas 29
Teasdale, John 142, 143, 185, 205, 267, 285
Tefé (river) 118
termites 188–89, 201, 243–45
Tetracha punctata (tiger beetle) *92*
ticks 105, 112
Ticuna (indigenous people) 253–54, *257*
Tocantins river 44–53
Tomo, Venezuela 155
Torquato de Souza, Padre 109, 136
toucans *2*, 53, 58, *91*, 122–23, 130, 223, 228, 250, 301, *335*
Trotter, Henry 325
Tukano (indigenous people) 161, *171*, 177, *256*; Tukano language 161, 177
Tupinambarana island, Brazil 136
turtle-egg harvest 124–27
Tussarí, Ramon 206–7, *256*

Uarucapurí, Brazil 175
Uaupés river 160, 161, 164, 183, 193
umbrella bird (*Cephalopterus ornatus*) 91, 116–17, 129, 151, 178, 190, 192, 250, 302
Unitarianism 14
Urubuquara falls, Brazil 164, 191

vampire bats 58, 62–63, 77–78
Vila Nova (Parintins), Brazil 102, 109, 112, 136, 245–46

Wallace, Alfred Russel *87*, *262*; background and family 16–23; inspiration for Amazon exploration 23–27; arrival in the Amazon 10–11; parts with Bates 55–56; observations on birds 57–58; visits Marajó and Mexiana 57–60; travels to Santarém 85–86, 95–99; fish drawings *94*, 159, 311–12; travels up the Amazon 109, 129–30; in Manaus 115–17; on the Rio Negro 146–84; return to England and loss of personal collection 223–27, 296; maps the Rio Negro and Uaupés river 299–300; on indigenous peoples 300; Amazonian legacy 296–302; in South East Asia 302–10; and evolution by natural selection 190, 304–8; and Charles Darwin 304–8, 310–11; and spiritualism 311; edits Spruce's papers 138–39, 186, 275, 332; published works 311–12; awards and honours 312–13; *Darwinism* 308; *Geographical Distribution of Animals* 311; 'On the Law which has Regulated the Introduction of New Species' ('Sarawak' paper) 304; *The Malay Archipelago* (1869) 305, 310; *A Narrative of Travels on the Amazon and Rio Negro* 296, 297–98; *Palm Trees of the Amazon and Their Uses* 296–97, 330; 'On the Tendency of Varieties to Depart Indefinitely from the Original Type' ('Ternate' paper) 305–6
Wallace, Fanny 18, 20, 22, 296
Wallace, Herbert 30, 75, 85, 95, 98, 115, 146, 213
Wallace, Thomas 17
Wallace, William (Alfred's brother) 18, 19, 22
Wallace Line, The (biogeographical divide) 304, 309
Wanan (indigenous people) 161, 163
wasps 49, 105, 189, 203–4, 240–41, 319
Waterton, Charles 25
Watson, William 29
Wickham, Henry 221*n*

yagé (hallucinogenic plant) *see caapi*
Yanomami (indigenous people) 280–81
Yumana (indigenous people) 126
Yuri (indigenous people) 263

Zeonia batesii (butterfly) *91*
Zoological Society, London 117, 326
Zoologist (journal) 16, 27, 30, 75, 248